EXERCISING OUR WORLDVIEW

Brief Essays on Issues from Technology to Art
from One Christian's Perspective

CHARLES ADAMS

DORDT COLLEGE PRESS

Cover by Scott Vande Kraats
Layout by Carla Goslinga

Printed in the United States of America.

Dordt College Press   www.dordt.edu/DCPcatalog
498 Fourth Avenue NE
Sioux Center, Iowa 51250
ISBN: 978-1-940567-10-5

The Library of Congress Cataloging-in-Publication Data is on file with the Library of Congress, Washington, D.C.

Library of Congress Control Number: 2014957816

CONTENTS

FOREWORD

Some stories are made for the movies; others should have been left in ink. Some tales are made for the radio; others are best left around a campfire. Some lectures deserve publication; others crave real-time listeners. Some arguments can navigate the one-way editorial post; others beg for the refining presence of dissension and the eager hand raised in question amidst the engaged classroom. As a technological artifact, each medium of communication has its own unique tone that cannot be replicated by others. Professor Adams knew this well. Therefore, I must acknowledge up front, that any attempt at capturing the whole message of this compilation of reflections in print will be inadequate. Even the medium of radio, on which these essays were initially recorded and broadcast, never did justice to these words, which became what they were fully created to be once they entered the classroom. They are words that call out for an active audience willing to step out, guided by the light of Scripture, into the expansive arena of culture as a whole. These are words that beg for a competitor, not to defeat them, but to receive them as a rhetorical adversary,* work them over, and return them stronger than when they left.* So as you read these words, do not forget where these words belong – in the classroom of life. So interrupt, raise your hand often, and disagree along the way. But be sure to leave the words stronger. That is what they were created for.

Professor Adams was a great teacher. Some teachers are liked, others are respected, others simply know how to engage and challenge. Professor Adams was a bit of all of these. He could make you angry, or exasperated, or frustrated, but he wasn't going to let you rest in complacency or inaction until you tasted a vision for the renewed creation project that God calls us to work and play in. He was confident that while his words were finite, the Word of the Lord that he was called to proclaim would "not come back empty-handed" (Isaiah 55:8–11, *The Message*). Good teachers love what they teach. Great teachers love their students. These teachers know that a mind can only change if the heart** is owned by the creator,

---

* This is clearly reflected in the prefaces that he wrote to each edition. His personal thanks extend specifically to those who agreed with him, but also included students and colleagues who disagreed with him. They were co-owners of these words.

** Professor Adams would understand the human heart as "the concentration of the meaning of all terrestrial reality" as H. Dooyeweerd would say: "Just as all of the

redeemer, and sustainer of the universe. Professor Adams knew this to be true. More than anything, he wanted his students to see where their hearts were directed, both individually and culturally. The uncomfortable stripping away of old layers of thinking and old habits of doing was a small price to pay for a heart open to the Word – the age-old word of the Lord for our particular time and place. He challenged us to our office of prophet. He invited us to be delivered from the crowd.

On the surface, this compilation of classroom essays has no particular theme. That is exactly the point Professor Adams is trying to make. There are no disciplinary boundaries for the life lived *coram Deo*. Most writers leave the technicalities of life to be worked out by the masses of disciplinary technicians (from philosophy to engineering to literature to chemistry) seeking to write "as the crow flies," with high altitude doctrinal themes, theoretical frameworks, or inspirational platitudes, keeping a safe distance from the workings of everyday life. Professor Adams was not content to maintain altitude as a writer. Any pie-in-the-sky assertion always accompanied a reckless dive into the existential technicalities of everyday life. If the Christian life did not look different on the ground, there was no reason to assert it should look different from the sky.

To have Professor Adams as a teacher was to be team-taught by the unlikely quartet of Dooyeweerd, Shakespeare, Kierkegaard, and Dickens, all supervised by the teacher in Ecclesiastes. Neither technophile or technophobe, he was more than an engineer, he was an artisan of life.

Whether you saw him as engineering instructor or technological philosopher, a trademark of his classes was "never forget the basics." The basics might be Newton's 2nd law, which attempted to humanly express only one facet of physical law for the created order, or it might be the basics of our ontology, a remembering that there are only two types of being: creature (all things created) and Creator (God). In Professor Adam's classrooms, without knowledge of the basics, the problem will never be solved. This led me back to the final essay that Professor Adams wrote before an accident left him unable to join us in ongoing dialogue until the new creation. It is an essay on the basics. It is about love, the prerequisite lesson for all who walk the path of wisdom. The same love that spoke the world into being, the love that spoke through the word made flesh, and the love that continues to speak through his church today – if we don't

creation is centered in God as its unified, integral origin, so God has created within man a unitary center, which is the concentration point of his temporal existence. . . . This is the heart, in the religious sense of the word, the source from which radiate the streams of life, the soul or the spirit of our temporal, that is to say, of our bodily existence" (H. Dooyeweerd, *Secularization of Science*, Toronto: AACS, 1979, pp. 4-5).

get in the way. So, in contrast to conventional advice, do not be afraid to read the last chapter first. The simple theme of this last essay (*Mugby Junction*) makes a handy bookmark to use through the rest of the text. Welcome to class.

Ethan Brue
Sioux Center, Iowa
October 2014

Editor's Preface to the "Fifth" (2008) Edition

Updating this collection of essays at least one more time seems appropriate – primarily to serve the [Dordt College] course *Technology & Society* (EGR 390 and CMSC 390). Six essays have been culled from Part I, five from Part II, one from Part III, fifteen from Part IV, and one from Part V. Four essays have been added to Part I, five to Part II, three to Part III, ten to Part IV, and one to Part V.

This preface gives me the opportunity to express my gratitude to Professor Nick Breems for team-teaching EGR 390 / CMSC 390 with me last year. Professor Breems sat in on (i.e., audited, so to speak) the course back in 2006, team taught it with me for the first time in 2007, and will do so again in this spring semester of 2008. In doing all of that he has not only read through much of the material in this book, but he has also, through interaction and discussion with the author D. Livid Vander Krowd (yes, the author does still haunt the hallways of the Dordt College Science and Technology Center), inspired a number of the new essays contained herein, especially in Part IV.

– C. Adams
8 January 2008

Editor's Preface to the Fourth Edition

Academic courses need to be refreshed regularly, kept up to date so that students have the best and most relevant information as they develop the insights and abilities they need to serve as Christ's agents in that time and place to which he has called them. So too with the materials used in those courses. Lecture notes are revised each year and textbooks go through new editions every few years. Thus it is time to do a little culling and a little adding to the present book. Because this book serves primarily as a supplemental text in EGR 390 Technology & Society / CMSC 390 Computer Technology & Society, the section on technology has been expanded the most. But additions and subtractions have been made to the other sections as well. As the new year (2005) begins, questions at Dordt College revolve around the topic of biotechnology. Thus the reader will find at least a few new essays addressing that topic.

Prefacing a new edition gives me the opportunity to say (on behalf of D. Livid Vander Krowd) "thanks" to numerous people who have inspired (and in part, "written") many of these essays. The students who served to motivate and inspire many of these essays are innumerable, but since the last edition, special thanks must go to Dave De Haan (1999), Adam Blankespoor (2000), Matt Roozeboom (2000), Ben Groenewold (2002), and Beth (Vander Ziel) Blankespoor (2002). In addition, a special word of thanks to my colleagues in the Engineering Department: Nolan Van Gaalen, Doug De Boer, Kevin Timmer, Ethan Brue, and Matt Dressler. None of these persons are responsible for the stumblings and fumblings found in these pages. But without their communal support, nothing of any value that might be found here would have appeared.

– C. Adams
January 2005

Editor's Preface to the Third Edition

It seems somehow appropriate, as we begin the year 2000, to revise this collection of essays. There has been a focused concern during the last two years about the role of technology, particularly as we move into the 21st Century. Only the least of those concerns involve the phenomenon called "Y2K." So a number of more current essays involving the computer, as well as other manifestations of our cultural preoccupation with technology, have been added to this volume.

The primary use of this text is for supplemental reading in the course EGR 390 Technology and Society. In that regard I would like to thank those students who have been enrolled in that course and who, through their journals, personal comments, and other forms of communication, have caught errors and typos and suggested improvements. In particular, I would like to thank Mr. Brian Glover (Class of 1998), who, during the summer after which he took the course, wrote to me with an extensive list of corrections and suggestions.

Perhaps at this point a little revelation is in order. For those who do not recognize him, the sketch on the cover is a representation of the Danish philosopher Søren Kierkegaard. More than that: there is a not-so-subtle connection between the form and content of Kierkegaard's writings and the pen-name of the author of the present text. If that connection is not immediately apparent, then you are invited to explore Kierkegaard's writings a bit more.

– C. Adams
January 2000

Editor's Preface to the Second Edition

The collection of these essays into a second edition is occasioned by the passing of time, the availability of more essays (originally five-minute radio addresses) and the need, in a class that I teach to Dordt College Engineering and Computer Science seniors, for perspectivally qualified writings that are neither arrogantly and aggravatingly academic nor condescendingly and annoyingly Sunday-schoolish. (That still leaves room, of course, for the mindlessly mediocre. But you, dear reader, will have to be the judge as to whether such disapprobation applies.) The mentioned course, ENGR 390 Technology and Society, is the reason for the new section (PART IV) on technology. There are new chapters added to the other sections as well.

As the author notes in the Preface to the First Edition, due consideration ought to be given to the date associated with each of the essays (which range from December 1980 to September 1997). The date of a given essay's writing explains, for example, the odd numbers used in economic illustrations, and perhaps the intensity of concern for a particular issue. A second point made by the author in the Preface to the First Edition that deserves continued emphasis is that these essays are the, at times idiosyncratic, musings of a somewhat eccentric teacher. Vander Krowd was (*is*, since the halls of Dordt College are, on occasion, haunted by his material as well as ghostly countenance) an engineer who turned to teaching early in his career. To many of the latter profession, that is enough to explain the hyperbolic nature of many of his ideas. Decades of daily encounters with students have a tendency to bend curiously even the most resilient of

> *...boughs which shake against the cold,*
> *Bare ruin'd choirs, where late the sweet birds sang.*

– C. Adams
January 1998

Preface to the First Edition

The essays in this little book were originally written as five-minute radio addresses ["Plumblines," on Dordt College's radio station KDCR], most of them during the decade in history known as "The Eighties." They make no claim to doctrinal purity, absolute truth, inerrancy, infallibility, or orthodoxy. They represent, rather, a humble attempt to wrestle with some of the problems faced by Christians as they try to live in faithfulness to the Word of God. It is the reader's responsibility to decide whether these essays are faithful to that Word.

The reader is cautioned to give due consideration to the date associated with each of these essays, since many of them reflect on *current events* – "current," that is, at the time they were written and broadcast. They have been selected and arranged here to serve as a supplemental textbook for the course GEN 300: Calling, Task, and Culture.

The author is grateful for the opportunity to express his views – both on radio and in this written form – in the atmosphere of freedom, collegiality, and trust that characterizes Dordt College. Moreover, this booklet is especially offered to the following persons in gratitude for "sharing the vision," and as an "offering of repentance" for the author's unforgivable inability to sustain the traditional forms of written communication.

Ray Commeret	Don Coray
Harry Schat	Bob Vander Vennen
Roger Veenstra	Steve Stegink

– DLVK
January 1991

The Biblical interpretation of mediocrity goes on interpreting and interpreting Christ's words until it gets out of them its own spiritless (trivial) meaning – and then, after having removed all difficulties, it is tranquilized, and appeals confidently to Christ's words!

It quite escapes the attention of mediocrity that hereby it generates a new difficulty, surely the most comical difficulty it is possible to imagine, that God should let himself be *born*, that the Truth should have come into the world . . . in order to make trivial remarks. And likewise the new difficulty as to how one is to explain that Christ could be crucified. For it is not usual in this world of triviality to apply the penalty of death for making trivial remarks, so that the crucifixion of Christ becomes both inexplicable and comical, since it is comical to be crucified because one has made trivial remarks.

– Kierkegaard

PART I
DEVELOPING PERSPECTIVE

"You hypocrites!" says Christ, "You strain gnats out of your wine but swallow the American, suburban way of life whole, like a camel!"
                                        – Calvin Seerveld[*]

---

[*] "A Modest Proposal for Reforming the Christian Reformed Church in North America," in *Biblical Studies and Wisdom for Living*, John H. Kok, editor (Sioux Center: Dordt College Press, 2014), 239; first published in *Out of Concern for the Church*, Olthuis, et al. (Toronto: Wedge, 1970), 52.

CHAPTER I

SEEKING THE CHRISTIAN MIND

MAY 4, 1987

Harry Blamires introduces his book *The Christian Mind* (London: SPCK, 1963, vii) with the following bleak assertion:

> We speak of "the modern mind" and of "the scientific mind", using the word *mind* [for] a collectively accepted set of notions and attitudes. On the pattern of such usage I have posited a Christian mind, chiefly for the purpose of showing that it does not exist. That is to say, in contradistinction to the secular mind, no vital Christian mind plays fruitfully, as a coherent and recognizable influence, upon our social, political, or cultural life.

Blamires wrote this in 1963. I would argue that it might be as true today as it ever was, in spite of the fact that, as a Christian community, we have become more conscious of the need for what Blamires called a "Christian mind," and what we sometimes today call a "Christian worldview." But consciousness of a need does not necessarily imply the meeting of that need. And this is certainly the case with the need for a Christian mind among God's people. We can worship together on Sunday because we share a common Christian faith. We can even form some Christian organizations together, like Christian schools, Christian social agencies, and even a Christian political organization. But we do these things because our common Christian faith, with its understanding of God's call to live distinctive and obedient lives, and its appreciation of the antithesis, impels us at a gut level to do more than just worship together. Unfortunately, to do more than worship together on Sunday requires of us more than just a gut level motivation, it requires a rationally expressible Christian mind.

Let me attempt an explanation. On a daily basis each one of us is involved in various activities. We drive our cars here and there, shop for food and other material things, meet and talk with friends and with strangers, work, play, read, watch TV, eat, and sleep. And while we are doing these and innumerable other things, we engage in thinking of one sort or another. What is it that allows each one of us to engage in this myriad of activities as one whole person and have the comforting knowledge that whether we are driving our car or reading our Bible we are the same person, and that these two activities have some meaningful relationship to each other? It is our worldview – the overall picture of reality that exists in our minds in such a way that driving a car at one point during the day and reading the Bible at another make sense.

If I told you, e.g., that I had an appointment this afternoon between 3 and 4 PM on the planet Pluto and I was going to meet with a tuna fish who I was hiring to design a castle on a Plutonian sea shore so I could get away from Sioux Center on weekends, you probably wouldn't believe me. The reason you wouldn't believe me is that the ideas of instantaneous interplanetary travel and tuna fish architects are incompatible with your worldview.

Now let me try another example. Imagine that you go to see a film in a movie theater in Sioux Falls. The movie is about a young man who is in the process of "making it" on Wall Street. The overall theme of the movie might be to show that frantically seeking your first million on the stock market before you reach the age of 30 years can be harmful to relationships you might be trying to build with other human beings. It seems to me that there have been quite a few movies lately with a theme similar to this.

As we watch the movie, our worldview naturally filters and interprets what we see and hear. Perhaps our worldview causes us to react negatively to the self-centeredness of the main character in the early part of the film, and then react with pleasure as he discovers the error of his ways and in the end gives up the frantic pursuit of his first million so that he can have a more positive relationship with the girl he has fallen in love with. That's probably the way the producers of the film expected us to react.

But what about the various elements of the film that were left unstated? For example, it was simply assumed that everyone in the audience would understand why the young man would want a million dollars. The film questioned his methods and perhaps the intensity of his desire, but not the desire itself. Did your worldview do a double take and force you to question why anyone would want a million dollars the same way you might question the existence of tuna fish architects? If not, then the idea of a person wanting a million dollars must be compatible with your worldview.

The point I wish to make is this: all of the experiences we have in life either conflict with or are compatible with our worldview. Those that are compatible tend to reinforce it. Those that conflict cause us to reflect further on both the experience and our worldview, and that reflecting will result in either a reinforcement of or an adjustment to our worldview.

For this reason it is of ultimate significance that we understand our own Christian worldview. That means more than simply reciting the Apostle's Creed or even memorizing the catechism. If our worldview is to be distinctively Christian, and if it is to remain so, it needs to be consciously exercised.

There is no one Christian worldview. Rather, there are traditions of them. Some of them are strong, in the sense of being solidly biblical, and some of them are weak, in the sense that they conform to worldly patterns. I was brought up as a Christian, but I had no Christian worldview until sometime between the ages of 17 and 20 I was, one might say, spiritually adopted by a radical Dutch-Canadian Christian friend, twenty years my senior. He introduced me to a particular worldview tradition. I usually think of it as a "reformational" worldview. Rooted in the Scriptures, it traces its historical path through the Protestant Reformation, particularly the thinking of John Calvin, to a sort of reformation that occurred in the Netherlands in the nineteenth century, spearheaded by such men as Groen Van Prinsterer and Abraham Kuyper. Its history can be traced into the twentieth century where it was further developed at the Free University in Amsterdam by men like Bavinck, Dooyeweerd, and Vollenhoven. Today it is represented all over the globe, and I believe it is the worldview tradition out of which Dordt College was born and continues to grow.

But I'm concerned that, despite its strong historical roots and the important work that is being done today by many who seek to further develop it in submission to the Word of God, this reformational worldview will become a dead tradition unless all of God's people who call themselves reformed will consciously exercise it. So this essay is a call to get busy at the task of thinking through our reformational worldview, to sharpen it. One thing that means is that we all must be busy reading more. If you haven't read Calvin, Kuyper, Dooyeweerd, or other reformational giants, then do so. Then one day, perhaps the suggestion that a million dollars is something good to have will be as strange to you as the suggestion that when you need some architectural work done you should hire a tuna fish.

Chapter 2

Perspective I

September 4, 1987

The beginning of September is a unique and intense time for those of us in the teaching profession. Some people have the misperception that teachers approach the first weeks of September like most people approach the first hours of Monday morning, especially people who believe that for teachers summer is simply an extended weekend. But such is not the case. In fact, I would like to suggest that while astronomically and climatically spring occurs for us inhabitants of the Northern Hemisphere in March, for teachers spring occurs during the early days of September. That's the time when the air is fresh with the aroma of new beginnings. Whether the task is to teach in grade school, high school, or college, we have learned from the frustrations and mistakes of prior years and relegated them to the past. There is an eagerness to try new ideas and an anticipation of new faces.

On the first day of class the teacher feels a bit like an explorer; like Columbus setting foot on the shores of America, or Neil Armstrong taking the first step on the moon. And even before that, at the first faculty meeting of the new year, the atmosphere is like at a family reunion, only better, because the meeting is not just for the sake of meeting. Rather there is a purpose, a mission to be accomplished. To be part of that is to share in a common task that has significance not only for today, but for history. For teachers, the beginning of September is a time when they have a unique opportunity to broaden and clarify their perspective, to reinforce their deepest convictions concerning who they are, why they are, where they are going, and how the course of history will be changed because they were there.

But I can hear some of you saying to yourselves, "Well, that's mighty groovy if you're a teacher. But what about the rest of us?" And that gets me to the point of this essay. What about the rest of us? Do we have to be teachers to exercise our perspective, to view our place and task in history? Certainly not. The reformers made it clear centuries ago that knowledge of oneself is dependent not on who we are, but on our knowledge of

God. I would guess that almost everyone reading this essay is a committed Christian. So we all know who we are, why we are, and where we are going, right? Well, I wonder. I've had a couple of experiences this summer that led me to believe that American Christians, on the whole, have a rather poor perspective on their role in the history of the universe.

Consider this: when was the last time you gave any serious thought to the question of what things will be like in the year 2100, assuming, of course, the Lord doesn't return before then? That's only 113 years from now. I'm afraid that many American Christians, if asked that question, would answer the same way that your average, twentieth century, self-centered humanist would. They would say, "Why should I concern myself with the year 2100? I won't even be around then!" Such an answer betrays a lack of vision, a very narrow perspective. If we are truly God's servants, called into being to serve him at this period in history and in this general area on earth, then we are responsible for shaping the future. What will happen in the year 2100 is to some extent dependent on what we do now.

Let me give you a couple of examples. I have three sons. If by God's grace, my and my wife's feeble attempt to raise them as radical, Christian, world-reformers bears fruit, and if they in turn, with their wives, raise three radical, Christian world-reforming children each, and if that pattern repeats itself through the generations, then by the year 2100, there will be present on earth at least 350 radical, Christian world-reformers that my wife and I will have to some extent been responsible for. The scary part is that if they don't turn out to be radical, Christian world-reformers, my wife and I will still be somewhat responsible for them.

As another example, consider the Dordt College Engineering Department. It started eight years ago with one professor and a handful of students. Today we have four professors, over 50 students, and we have graduated 40 Christian engineers since 1983. We would like to think that by the year 2100 the department will have grown way beyond the present size, perhaps enough to rival even the big universities. More importantly, by the year 2100 the Dordt College Engineering Department, by God's grace, will have graduated thousands of radical, Christian engineers who will be responsible for shaping and giving direction to the technology of the twenty-first and twenty-second centuries! To be part of that effort now is an awesome, humbling, but also exhilarating experience.

Last Friday morning I met my freshman engineering class for the first time. As I looked out at the eighteen faces in the classroom, I realized that this was the first time in the history of the universe that those eighteen world-shapers were sitting in an engineering class. I wondered how the world and history would be affected by their presence, and I hoped that what I had planned to say to them would prove to be a positive and God-glorifying influence.

Each one of us, when we are called into being, is given a chunk of history to shape and develop. How we shape it is multiplied by the decades and centuries that follow us. That's why we need to work at developing our historical perspective – not simply knowing what took place in the past, but also developing a vision for the future, and how the way we respond now to the mandate the Lord gives us will affect, for good or for bad, that future.

CHAPTER 3

PERSPECTIVE II

OCTOBER 2, 1987

So you say that you woke up this morning with a sore throat and a bad case of laryngitis, but you've got work to do and you're already way behind schedule, so you can't go back to bed. Then you read the morning newspaper. On the local scene you read about how many Christian schools are facing economic crisis because of new laws initiated by the Department of Education and passed by the legislature. In national news you read that the liberals in Congress have managed to hand the pro-life cause another defeat, and the ACLU has won a case that pretty much gives free reign to pornographers. In international news you learned that the conservatives in Congress have managed to thwart efforts made by the United States and the Soviet Union to reduce nuclear armaments. And with the tensions in the Persian Gulf heating up, the probability of some maniac shooting off a nuclear missile and starting World War III seems to increase daily.

It's mornings like this that cause many Christians to despair of their calling to be agents of reconciliation and restoration in the world. The task of bringing every area of life under the lordship of Christ and redirecting the world to obedient service seems so hopeless when the agents of evil appear so powerful. It's easy to fall prey to an otherworldly perspective, which ignores our calling to this world, focuses on the eternal destiny of our souls, and essentially writes the world off as belonging to the devil and bound for utter destruction. But such despair and hopelessness at the prospect of working to change the world for good runs counter to the message of the Scriptures.

In a recent essay, I suggested that "each one of us when we are called into being is given a chunk of history to shape and develop," and "how we shape it is multiplied by the decades and centuries that follow us." The implication is that in spite of our perception that we have virtually no impact on the course of history, in the long run it is really we, God's humble and seemingly ineffectual servants, who through the power of his sustaining and redeeming Word will lead the creation to its ultimate destiny: the reconciling renewal of all things in Christ on the day of his return.

I believe that this is the message of God's Word: not for Christ to condemn the world to utter destruction while preserving our souls for a life somewhere else, but as Colossians 1:20 so clearly puts it, "to reconcile to himself all things, whether things on earth or things in heaven, by making peace through his blood, shed on the cross."

The Scriptures tell us that the Lord has a very quiet but powerful way of doing that – he uses his humble servants, without a lot of pomp and circumstance and making a big noise in the world. Rather, over the course of many generations, the combined effect of his people is to bring the world to the point of readiness for its King to return.

That's why 1 Corinthians 1 and 2 can be so helpful on mornings such as I described earlier. Consider 1 Corinthians 1:27–29:

> But God chose the foolish things of the world to shame the wise; God chose the weak things of the world to shame the strong. God chose the lowly things of this world and the despised things – and the things that are not – to nullify the things that are, so that no one may boast before him.

And perhaps nowhere do we sense better the very time-and-space-specific calling of the Lord to his servants than when we read the words of Mordecai in Esther 4:13–14:

> "Do not think that because you are in the king's house you alone of all the Jews will escape. For if you remain silent at this time, relief and deliverance for the Jews will arise from another place, but you and your father's family will perish. And who knows but that you have come to your royal position for such a time as this?"

"For such a time as this" – these words apply as much to us in 1987 America as they did to Esther in Persia during the reign of King Xerxes. And notice that there is no puppet-like predestination involved here. God's people are free to obey him or to hide their heads in the sand. If you or I do the latter, he will simply find another servant who will obey him, who will serve as his representative and agent for bringing the Kingdom to completion.

So in spite of the liberals and conservatives, in spite of nuclear warheads or Iranian terrorists, the Lord is using us, his people, to work out the final destiny of the universe – although not without pain, persecution, and seeming like evil has the upper hand. But by faith we ought to be able to see through the veneer of evil and suffering to the redemption that lies beneath, and sense the role that we, his humble servants, play as agents of that redemption.

That kind of faith is what drove Martin Luther in 1529 to write those magnificent lines at the beginning of the third verse of the hymn "A Mighty Fortress is Our God."

> And though this world with devils filled
>> should threaten to undo us,
> We will not fear for God has willed
>> his truth to triumph through us.

CHAPTER 4

NOSTALGIA, MUSEUMS,
AND THE COMING KINGDOM

AUGUST 2, 1989

Have you ever wondered why most people enjoy visiting museums? A few weeks ago my wife and I took a short weekend trip to the little town of Minden, just south of Route 80, deep in the wide open and relatively isolated prairie land of Nebraska. We visited a place called "Pioneer Village," which we had learned about in *National Geographic* magazine. Pioneer Village is a kind of museum; partly a reconstruction of a nineteenth century Nebraska village, and partly a series of warehouses containing exhibits of Nebraska prairie life from the early nineteenth century to the present. The village is made up of a general store, a sod house, a church, a train station, and a number of similar structures. Each is an original, having been transported from wherever it was first constructed to its present site at Pioneer Village. And each building contains original artifacts, arranged just as they were in the eighteenth century. The warehouse museums contain exhibits of various elements of American cultural history. For example, there are three warehouses that display automobiles, or perhaps I should say family transportation vehicles, for the earliest exhibits are of stagecoaches, Conestoga wagons, and horse-drawn buggies. I have never seen such a large collection of cars, and was fascinated as I studied them – imagining what it was like in the early part of this century to drive one across the prairie on what were for the most part unpaved roads.

But this essay is not meant to be an advertisement for Pioneer Village. Rather, I want to explore why it is that I (and many other people) are so fascinated with exhibits of historical artifacts. Put simply, the question I'd like to answer is "Why do people enjoy visiting museums?"

Historians will tell you that those who ignore the past are condemned to repeat the mistakes of the past. That's one good reason to study history and to be interested in people, events, and artifacts that belong to an era that predates our own. But by itself that reason is too cold and pragmatic to explain the attraction most people have for museums. When I spent a half hour pondering over the Conestoga wagon, or ten minutes reacquainting myself with a genuine 1958 Edsel, it was not to learn the mistakes of the past. I'm certainly in no danger of repeating Ford's error in judgment, and I can't say that I found anything particularly unworthy in the Conestoga wagon. In fact I was very impressed that such a vehicle could be designed and built given the technology of the times.

No, it wasn't a pragmatic desire to learn from the mistakes of the past that held my interest in front of that Conestoga wagon, that Edsel, or the Victrola wind-up phonograph that reminded me of one that my grandmother handed down to me when I was a young child. The lure was something far less tangible, something almost aesthetic in nature: like the lure of a Beethoven symphony or a drive through the Catskill Mountains in New York State on a clear autumn afternoon. Perhaps the lure is something akin to nostalgia.

I've never considered myself a particularly nostalgic person. I have some pleasant memories of the past, but I enjoy the present far too much to ever spend a great deal of time doting on things that have been. On the other hand, I know of quite a few people who get very nostalgic over such things as their childhood house or school.

Is the reason we like to visit museums primarily that of nostalgia? Again I think not – at least not nostalgia in the usual sense of that word. The thesaurus in my computer tells me that nostalgia has to do with "recollection," "remembrance," and "reminiscence." The dictionary on my desk defines nostalgia as "severe or morbid home-sickness." That can hardly be the reason for my fascination with the Conestoga wagon. (Unless by some psychological gymnastics one suggests that such fascination is really tied to a period in my childhood when I spent a lot of time watching *Wagon Train* on the television.)

I think there's a much deeper reason. For anyone who has ever built a house, painted a painting, cleaned and polished a car, or raised a child, there exists a certain sense of gratification when you step back and contemplate the fruit of your labor – even if it wasn't all that enjoyable while you were doing it, and even if things didn't quite turn out exactly as you planned them. Now I know that everything that exists – Conestoga wagons, Edsels, and people – is part of God's creation. He designed and brought into being all of these things. And the Scriptures tell us that in the beginning – before sin entered the picture – God was very pleased with this creation. In Genesis 1:31 for example we read that after the work of the sixth day "God saw all he had made, and it was very good."

Other passages in Scripture like Proverbs 8, Job 38, and Isaiah 40 convey the same idea.

It seems to me, therefore, that the pleasure we get by considering the works of our hands is simply a reflection of the pleasure God gets when he considers his creation. I know there's a danger of anthropomorphism here – fashioning God in our own image – but after all, this is the way he reveals himself to us in his Word. And the opposite danger is far worse – the danger of thinking that God is so transcendent, so beyond our comprehension, that we can never know him as a person. I'm not saying that God is like us. I'm saying that we are like God – made in his image, to use the language of Genesis; or "a little lower than the angels," to use the language of Psalm 8.

So why do we like to visit museums? Why do we take pleasure in contemplating the cultural artifacts of the past? Because each of these things represents a brushstroke in the great painting that is the creation. And being God's image-bearers, his representatives in this creation, we can't help but reflect, every once in a while, on the beauty of it all. Just as it does in the morning sunrise, in spite of sin, the beauty and goodness of God's creation shines through in the cultural artifacts that have been left behind by those image-bearers who went before us.

I believe that when the Lord returns to usher in his Kingdom, part of that Kingdom will include something like museums. Not places where artifacts from the dead and distant past are kept; but rather places where history will live and enable us to praise our Creator throughout all eternity.

So if you get a chance to visit a place like Pioneer Village any time soon, keep in mind that the collection of historical artifacts on display are just part of the picture album belonging to the Creator of this whole universe. And don't be surprised at how easy it is to worship in what may seem like the strangest of places.

Chapter 5

Starkeeping Too!

December 4, 1989

One of the most important tasks that any of us living in twentieth century America have is that of setting priorities. Imagine it's Saturday morning. The lawn needs to be mowed, the car ought to have its oil changed, you still have to prepare that Sunday School lesson for tomorrow, and you owe your mother in Wisconsin a phone call. How do you decide what to do, or in what order to do them? Or, suppose it's the first of the month. You've just been paid and you're working on your monthly budget. You find you have $50 in your "extra donations" category. You've recently received donation requests from Bread for the World, Bethany Christian Services, Friends of Iowa Public Television, and the World Home Bible League. Who should you give the $50 to?

The way we answer questions such as these – priorities questions – depends very much on our worldview, and the hierarchy of values that characterize that worldview. A couple of weeks ago I was involved in a discussion with a group of friends regarding what it means to live in a consistently Christian way. One friend began to read aloud from some material that was part of the Sunday School lesson he had taught to a class of teenagers. The material stressed the importance of evangelism, of spreading the Gospel by word of mouth. At the same time, however, it made the claim that this kind of evangelism was so important that it made such things as concern for the state of the environment insignificant in comparison. In fact, it went so far as to say that those people who evidenced a great deal of environmental concern had their priorities mixed up.

Are environmental concerns of a lower priority than evangelism concerns? I don't think we can answer that question. In fact, I don't think we *should* answer that question. It's a question that assumes that evangelism and the environment are part of two separate areas of life – a spiritual area and a physical area. Now, if you are a disciple of the ancient Greek philosopher Plato, perhaps you would agree with that assumption. And then you would probably answer the question with a "yes." But, if like me, you are a disciple of the Lord Jesus Christ, the Creator-Sustainer-Redeemer of the whole universe (Colossians 1:15–20), the one about whom the Psalmist wrote when he said "The heavens declare the glory of God" (Psalm 19), then you won't want to separate evangelistic and environmental concerns quite so easily.

But clearly we must make judgments regarding priorities. If my brother's children are in danger of being denied the education they deserve because of insufficient funds, then I'm going to address those needs before I mail off a donation to the Audubon Society. If I'm busy working with the local board of APJ on an environmental or public justice issue and my church asks me to donate a significant amount of time to a newly created evangelism effort, I'll very likely turn down that church request. In making these decisions I have set priorities. But they are my priorities and it would be wrong of me to think that everyone else must set the same priorities. In fact, I can make these priorities decisions without feeling guilty precisely because I know that others are making opposite decisions.

Each of us is called to be a steward of his or her time and resources. But each of us is a unique image bearer of God, called to a particular time in history and to particular situations for which we are best equipped. Some of us are evangelists, some environmentalists, some teachers, some social workers, and a few are astrophysicists. We all need to be concerned with evangelism, and we all need to be concerned with earthkeeping. But we don't all need to have the same priority-structure for those concerns. That's why we can't knock our astrophysicist sister for what appears to us to be her inordinate concern for the stars. You see, being a good steward of God's creation may just broaden the definition of earth-keeping to include *starkeeping* too!

CHAPTER 6

WHERE ARE THE CHRISTIAN *OUTLAWS?*

MARCH 2, 1990

One of the most worthwhile and exciting programs ever to be broadcast on television aired its last episode this week on PBS. Entitled *Eyes on the Prize: Part II*, it was a 6-part documentary of the civil rights movement. It began with the days of the late fifties when black Americans, as a self-conscious body, asserted their God-given dignity and challenged the American system of apartheid. Watching the episodes unfold, I was amazed at how, in retrospect, the issues now seem so clear; the heroes and villains so easily identifiable. In 1964 things were not so transparent.

Today I think of Martin Luther King as one of the greatest American Christians who ever lived: a man committed to speaking and living the truth much as the Old Testament prophets had, and at the same time guided by the teachings of Christ – particularly those Sermon-on-the-Mount teachings dealing with nonviolence. But in 1964 King was suspect. The FBI had already begun a campaign to discredit him, and the media presented him as just another "Negro agitator."

Today I realize that the greatest villains were not only those who wore hooded bed sheets and went around lynching my black brothers. Equal in wickedness were national leaders like FBI Director J. Edgar Hoover, Birmingham Sheriff Bull Connor, Georgia Governor Lester Maddox, and those savage police who with dogs, fire hoses, clubs, and guns enforced the unjust laws. In 1964 however, my support for the civil rights movement was inhibited by my lack of clarity with regard to who were the "good guys" and who were the "bad guys."

Why is this? Why is it that Christians, particularly evangelical Christians, are so slow in responding to the call to exercise their prophetic office, to stand up and be counted on the side of justice and righteousness? I'd like to suggest that one of the main reasons for our lack of prophetic sensitivity and vigor is the distorted, almost idolatrous, concept we have of *obedience to law* and *obedience to governing authorities.*

Let me give you a very current example. In attempting to speak prophetically against the evil of "abortion on demand," many Christians have resorted to forms of civil disobedience. They picket abortion clinics even in the face of those occasional laws that declare such activity illegal. Now I do not wish to condone the unloving verbal and, sometimes, physical abuse that some engage in during such demonstrations. But I find it close to idolatrous that many Christians refrain from joining such acts of civil disobedience on the grounds that we should always obey the law or, misquoting Paul, that we should always obey the governing authorities.

In Romans 13 Paul tells us that we ought to be "subject to the governing authorities, for there is no authority except that which God has established. The authorities that exist have been established by God." But being subject to the governing authorities is not the same as always obeying the laws. In creating the world, the Lord called into being the potential for government institutions so that societies would be able to function properly. He also called into being the potential for schools, for art galleries, for churches, and for businesses. These institutions, when functioning according to God's norms, contribute to the health of society – just as normative government does. The leaders of these various institutions are "governing authorities" in much the same way that the leaders of the state are. Paul calls us to be subject to them just as much as to the lawmakers and police. In verse 4 of Romans 13 we learn why. They are ministers of God for our good. In other words, by their work they (ideally) enable society to be what the Lord intends for it to be. So just as you obey the lawmakers who tell you that you must be age 18 to register to vote, you also obey the engineers who tell you that you must change the oil in your car regularly, or the educators who tell you that your child must learn mathematics as well as music and P.E., or the church leaders who tell you that you must be a professing Christian to take part in the Lord's Supper.

None of us hold these other authorities in the same absolute esteem that we do the government. We think nothing of questioning the church order, the school curriculum, or even the maintenance manual that comes with our new car – even if we do so in relative ignorance. And, in a way, that's good! We ought never to accept uncritically the instructions or laws set down by others. We need to remember that there are always two levels of authority higher than any we might find given by any institution: the authority of our own conscience, and finally God's authority as expressed in his Word. It is better to go to jail for disobeying a particular law than to obey it and in so doing violate either of these.

None of this, however, ought to be construed as advocating anarchy or lawlessness. Anarchy and lawlessness are one extreme. The opposite is the kind of passive and unquestioning obedience that many German citizens gave to their government during the Second World War, that many American citizens gave to their government during the civil rights movement, and that, I fear, many Christians give to the government today. If we are going to be faithful to the Lord we will need to exercise our prophetic office by protesting injustice and pointing the way toward righteousness whenever possible. To let that prophetic task be inhibited by passive and unquestioning obedience to civil laws is sin.

CHAPTER 7

BOB DYLAN AND BEETHOVEN

AUGUST 2, 1991

Two weeks ago, while visiting in Lincoln, Nebraska, I stopped in at a record store to look for a particular set of compact discs. To my delight, not only did the store have what I was looking for, but it was on sale as well. So after spending a half-hour perusing the complete compact disc display, I walked out of the store $40 poorer, but rich in anticipation of the aesthetic pleasure I would have by listening on my home stereo system to the almost three hours of music that I had just purchased.

Now that may seem like a rather routine experience, hardly one to write an essay about. Yet there are two elements of this experience that I consider quite remarkable. And they both have to do with the idiosyncrasies of the chief cultural group with which I identify: middle age, reformed Christians.

The first element of the experience that I found a bit remarkable – if not disconcerting – was the consciousness I had, while examining the store's abundance of compact discs, that I was the only one in the store over 25 years old (20 years over, to be precise). I've reasoned that this sense of estrangement, which I often feel, is related to the days and hours during which I usually shop. Having a more flexible job than most people, I have the freedom to avoid shopping with large crowds on weekends or during holidays. That means I usually visit these kinds of stores on a weekday, most often in the summer or during Christmas or springbreak. These, of course, are the same times that teenagers find themselves free to frequent such stores. But lately I've been feeling that there's more to it than that. As I look around me in the record store I have the sense that I am in a place designed for young people, not for us older fellows.

The second element of my compact disc shopping experience that created in me a kind of unsettling feeling of generational loneliness had to do with the music I purchased. You see, that three disc set was not Handel's *Messiah*, or Bach's *Mass in B*, both of which I have and enjoy immensely. No, it wasn't classical music at all. It was Bob Dylan's *Bootleg Series – Volumes 1–3*; 58 songs that he recorded from 1962 through 1989 and had not been previously released. I already have a number of Dylan's LP's and CD's and so when I noticed an article in *Stereophile Magazine* reviewing this latest release, I read it carefully. It was so positive in its praise that I had to hear the music for myself. And, speaking for myself, it was worth every one of those 40 dollars I spent on it. But the point is this: there are very few people around with whom I can share my enthusiasm for this kind of music. In fact there are only two! One is a colleague of mine in the Engineering Department at Dordt who has yet to reach the tender age of 30 years. And the other is my wife, who remains the vibrant and hip nineteen-year-old that I married 24 years ago.

So what's the explanation? Why do music stores seem to be the exclusive domain of young people? And why is it so hard to find anyone who appreciates Bob Dylan's music among my middle aged, reformed acquaintances? (After all, Dylan hit the half-century mark himself, just this year!) You may be tempted to respond, "it's because the stuff young people are listening to today, the stuff record stores primarily sell, isn't even music, it's just noise. And as for Bob Dylan, well, he can't even carry a tune!" And then, alas, I would have to, for the most part, agree. Much of what is today called pop music is like most of American television: a lot of sound and fury, signifying nothing. As one critic has called television "potato chips for the eyes," so we might refer to most pop music as "potato chips for the ears." And it's true that Bob Dylan's voice has none of the sonorous qualities that we associate with the voice of Pavarotti. But I would argue that there is far more to it than that. We cannot simply write off all pop music as inferior aesthetic drivel any more than we can write off all television as irredeemable and worthless visual junk.

Here we have what I believe is the crux of the matter. The generation gap in music, the polarity between classical and pop music, is, I believe, related to another polarity that we as a Christian community have blindly bought into – the dualistic polarization of life into sacred and secular. Not only has this dualism led us to compartmentalize our faith so that it is active only for a few hours on Sunday, but it has poisoned and atrophied our appreciation of the rest of life. And nowhere is this more the case than in the area of music, where the field itself is shamelessly divided into sacred music and secular music.

Now I'm no musician. I don't have the authority or the ability to analyze a musical piece, classical or popular, and subject it to a reasoned aesthetic critique. But just as I can appreciate a sunset, the sleek lines of a well-designed car, or my grandson's smile as he recognizes me and grasps for my glasses, so too I can appreciate what my intuition tells me are elements of good music. Brahms' *German Requiem* stirs in me a powerful awareness of the pain of death, even as it subtly reinforces my faith that death has been conquered. Dvorak's Ninth Symphony conveys fleeting images of a pristine creation, of great expanses of fertile land and flowing rivers of pure water, which suggest not only the unspoiled wilderness that once was North America, but also the eternal expanse in which the New Jerusalem will one day soon appear. And from *The Bootleg Series,* Bob Dylan singing "No More Auction Block" creates in my mind parallel images of nineteenth century American slaves who have escaped to freedom, and today's slaves of oppressive ideologies, beliefs, and customs, who come to their senses, casting off the shackles of cultural conformity. My point is that you don't have to be an expert to appreciate good elements in music. You may not as easily be able to say what are bad elements in music, but with patience and listening experience, even those of us with musically deprived childhoods can begin to know the difference between good and bad elements in music.

And here's my main point. I've used two adjectives, "good" and "bad" to describe "elements" of music. My claim is that just as there are good and bad elements in engineering designs, in political theories, in worship services, and in economic systems, so likewise there are good and bad elements in musical pieces. By good and bad I mean obedient or disobedient, normative or antinormative, in accordance with the way the Lord intends for his creation to work, or contrary to those intentions. To categorize music as "sacred" and "secular," on the other hand, is to create sweeping and absolute categories that automatically accept or reject whatever you put in them. Doing so blinds us to the disobedient elements in some of the hymns we sing and leads us to reject many of the obedient elements in what we label as "secular" music, whether that be classical or pop. As reformed Christians we need to approach music with the same redemptive openness that we say we have for all other areas of life. We need to encourage musicians and music theorists who are Christians to become Christian musicians and Christian theorists by searching for and articulating God's norms for music. Then they will be able to help the rest of us discern more clearly what is good and what is bad in not only the creations of Beethoven, but also those of the Beatles and Bob Dylan.

CHAPTER 8

THE EMPEROR'S NEW CLOTHES SYNDROME
DECEMBER 4, 1986

> Many years ago there was an Emperor so exceedingly fond of new clothes that he spent all his money on being well dressed. He cared nothing about reviewing his soldiers, going to the theatre, or going for a ride in his carriage, except to show off his new clothes. He had a coat for every hour of the day. . . .

So begins one of the best known fairy tales by Hans Christian Andersen, *The Emperor's New Clothes*. We all remember how the emperor, his worthy old minister, his very bright officer, and eventually all the adults in the great city were duped by two rogues pretending to be weavers. Convinced that the cloth produced by the self-proclaimed weavers was "the most magnificent fabrics imaginable. Not only were their colors and patterns uncommonly fine, but clothes made of this cloth had a wonderful way of becoming invisible to anyone who was unfit for his office, or who was unusually stupid." The people in the city, including the emperor, pretended that they saw the clothes, not wanting to admit their presumed stupidity. In fact, the emperor decides that he must wear these new clothes in a great procession through the city.

I'm sure you remember the end of the story, but read again as Anderson teaches us all a lesson in self-deception:

> So off went the Emperor in procession under his splendid canopy. Everyone in the streets and the windows said, "Oh, how fine are the Emperor's new clothes! Don't they fit him to perfection! And see his long train!" Nobody would confess that he couldn't see anything, for that would prove him either unfit for his position, or a fool. No costume the Emperor had worn before was ever such a complete success.
>
> "But he hasn't got anything on,' a little child said.
>
> "Did you ever hear such innocent prattle?" said its father. And one person whispered to another what the child had said, "He hasn't anything on. A child says he hasn't anything on."
>
> "But he hasn't got anything on!" the whole town cried out at last.
>
> The Emperor shivered, for he suspected they were right. But he thought, "This procession has got to go on." So he walked more proudly

than ever, as his noblemen held high the train that wasn't there at all.*

I'm fascinated with this fairy tale because I think it describes, as well as anything, the condition in which we Christians often find ourselves as we seek to live obediently in the modern world – delusion brought on by self-deception. In the future I hope to develop this idea further in essays dealing with specific instances of what we might call "The Emperor's New Clothes Syndrome." In the next few paragraphs, however, I want to relate an incident which forced me to think of Andersen's tale and, in fact, made me feel a little bit like one of the townspeople during the Emperor's procession.

It occurred a few weeks ago when I was away on a weekend business trip. After a long day of meetings we had Saturday evening free. A colleague of mine and I were staying overnight at the same home and we decided that it might be nice to take in a movie. Three others involved in the meetings joined us, and together we looked for a theater to go to. One of the members of our group suggested we see a film that had recently been released and was playing at one of the largest theaters in the city. It was entitled *The Name of the Rose*, and the colleague with whom I was staying had read the book, so he was quite eager to see it. He had also heard a good report on the film from a friend who had seen it.

Well, almost a good report. Apparently there was one short scene in it that was thrown in to satisfy modern society's lust for vicarious sex, but otherwise he was told that it was an excellent movie. So the five of us went to the 7 o'clock show.

I must admit that, for the most part, it was an intriguing movie. Not only was it a reasonably good mystery, but being set in the Middle Ages, it attempted to portray very graphically what life was like in those times. And in my estimation it succeeded. But there was the scene that we had been lightly warned about. I confess that I was embarrassed – to say the least.

But there was more to it than that. I had the very real sense that something was wrong and that I really didn't belong sitting there in the theater with the rest of the crowd, watching what was occurring on the screen. I think it is one thing to depict a monk being burned at the stake by the inquisition, in all its gory and violent detail. It's possible to do that in such a way that merely panders to the sadistic tendencies in modern man. But it's also possible to do it in a way that portrays evil in the world for what it is. And that is part of the task of obedient art.

In any case we know that it is only a depiction – no one is really be-

---

* http://www.andersen.sdu.dk/vaerk/hersholt/TheEmperorsNewClothes_e.html.

ing burned to the stake so that a movie can be made. But in the sex scenes that are ritually thrown in many of today's films, this is not the case. In order to depict illicit activity, the actors are forced to engage in a similar kind of illicit activity.

But I'm getting away from my main point. The question of what can legitimately be depicted in the film arts is a topic for another essay. My point is this: after the movie ended, my four Christian friends and I went to a restaurant for dinner. Over dinner we naturally discussed the film. But no one seemed to want to talk about the issue raised by the sex scene. We discussed the cinematography, the realism, the depiction of life in the Middle Ages, and the quality of the mystery aspect of the film. At one point I tried to get a response from the other four by suggesting that while I thought most of the film was very well done, I couldn't in good conscience recommend it. But no one picked up on that. No one wanted to talk about what perhaps needed to be talked about most. We sort of pretended that it either wasn't important or that it never occurred.

That's when I began thinking about Andersen's fairy tale. We Christians, in our desire not to appear stupid in the eyes of the world, play pretend. And we deceive ourselves into thinking that everything is fine. Perhaps we need the courage to be like little children and cry out, "The Emperor has no clothes on!" when the situation calls for it. The situation called for it the evening I saw that movie. I guess my friends and I fell prey to "The Emperor's New Clothes Syndrome." But there are many other far more important situations where we are guilty of doing the same thing day in and day out.

[**Postscript, 2005**: This essay was written almost 20 years ago (perhaps longer, depending on when you are reading this). It is a sad commentary on our Christian culture – the author included – that gratuitous sex scenes thrown into modern movies no longer embarrass us. Perhaps it is an indication of our more complete accommodation to *the Emperor's New Clothes Syndrome*. And we wonder why faithful Muslims in Iraq and Iran consider us to be an immoral culture!]

CHAPTER 9

NATIONALISM

SEPTEMBER 2, 1988

My 1956 Funk & Wagnalls dictionary defines *nationalism* as "devotion to the nation as a whole" and then gives as a synonym, "patriotism." Isn't it interesting how history subtly transforms the meaning of words? Despite the recent memory of World War II, Adolph Hitler, and the German National Socialist Party, "nationalism" in 1956 was a word that had fairly positive connotations.

Nationalism. Today the word is not usually used as a synonym for patriotism, but is instead used to express an extreme kind of devotion to country that is exclusive of everything else. As Christians we might say that the word "nationalism" expresses a form of idolatry whereby the state is exalted and served in place of God. Perhaps the best current expression of it is in Iran, where the people are so brainwashed and whipped into such a religious fervor that they are willing to sacrifice their lives, and even the lives of their children, for the good of the state.

Nationalism. It is a form of heart disease. And by heart I don't mean that organ in your chest that pumps blood to the rest of your body. By "heart" I mean the central core of your selfhood, that part of you that directs the rest of your total self to act either in obedience or disobedience to the will of the Lord, the part that is either under the influence of the Holy Spirit or the Evil One. Nationalism is when the powers of darkness turn your heart away from its true origin and fixate it on the state, the nation, the country of which you are a citizen.

Nationalism. It expresses itself in such phrases as "My country, right or wrong!" and "Love it or leave it." Its fires are fanned by parades of marching and uniformed citizens, the waving and saluting of flags, the reciting of oaths of allegiance, and the singing of national anthems. It feeds itself on conflict with other nations. Sometimes that conflict finds expression in relatively harmless forms such as sport. International soccer matches and the Olympics are two examples. Another expression that is becoming more common today is in the economic sector: restrictions on trade, such as import tariffs and quotas, are an example.

But for all of recorded history, the one expression of conflict that has dominated all others, the one on which nationalism has gorged itself, and the one that has caused more suffering in the world than all the earthquakes, tornadoes, hurricanes, and other natural disasters combined, is war. War – the senseless destruction of human beings, image bearers of God, by other image bearers of God. War – usually fought for such abstractions as "the defense of freedom" or worse, "for the fatherland," or worst of all, "for God and country."

World War I was a particularly horrible example of the kind of violent conflict that is both spawned by and nourishes nationalism. In 1914 fellow Christians in Europe slaughtered each other with nationalistic fervor for futile nationalistic ends. The English poet Wilfred Owen captured the horror and futility of that nationalism-fed conflict in his poem, *Dulce et Decorum Est*. The poem tells the story of a platoon of young World War I soldiers who are suddenly attacked with poison gas. One member of the platoon was unable to get his gas mask on in time. The poem describes his suffering and dying. The last lines go like this:

> If in some smothering dreams you too could pace
> Behind the wagon that we flung him in,
> And watch the white eyes writhing in his face,
> His hanging face, like a devil's sick with sin;
> If you could hear, at every jolt, the blood
> Come gargling from the froth-corrupted lungs,
> Obscene as cancer, bitter as the cud
> Of vile, incurable sores on innocent tongues, –
> My friend, you would not tell with such high zest
> To children ardent for some desperate glory,
> The old Lie: Dulce et decorum est
> Pro patria mori.*

The last line is an old Latin saying which means, "It is sweet and proper to die for one's country."

Nationalism and patriotism are two things that ought to be very clearly differentiated today. It's true that Samuel Johnson once said, "patriotism is the last refuge of a scoundrel." But I think he had in mind nationalism. Patriotism, as I wish to define it, is a virtue. It is our obedient response to the Lord calling us to responsible citizenship. Nationalism is a vice. It is the sin of communal pride and selfishness.

Nationalism and patriotism need especially to be distinguished

---

* Owen, Wilfred, "Dulce et Decorum Est," in Sanders, et al., *Chief Modern Poets of England and America*, Volume I, 4th ed. (New York: Macmillan, 1957), 293.

today because they are being increasingly confused in the insignificant sound and fury of the current presidential campaign. The Democrats and the press have recently attacked vice-presidential candidate Dan Quayle because he managed to avoid getting drafted and sent to Vietnam in the late 1960s. Do you hear the old lie once more being told: "Dulce et decorum est"? Back in the fifties and sixties a bunch of fearful and nationalistic old men got us into a futile war. Thousands of young men, some "ardent for desperate glory," most simply doing what they were told to do, went off to the jungles of Vietnam and either died or wound up killing their Asian contemporaries because they didn't question those fearful and nationalistic old men. A few young people avoided the conflict – some out of cowardice but many out of a conscientious decision to not be the puppets of nationalism. If Senator Quayle opposed the Vietnam War, found a way to avoid being drafted, and worked to end the war, then he ought to be honored, not castigated.

Another confusion of nationalism and patriotism is the Pledge of Allegiance issue. The Republicans are attacking Governor Dukakis because he supports the Supreme Court decision that the State cannot force teachers to salute the flag. The Republicans want you to think that anyone who will not salute the flag, or anyone who will even support a person's right not to salute the flag, is somehow unpatriotic. That is a horrible distortion that exalts mindless conformity and threatens the free exercise of conscience. There are many patriotic Christians who see the flag as a symbol of the God of American nationalism, and, like contemporary Daniels, refuse to bow down to the state's idol.

It is our duty as Christian citizens to serve the nation in which the Lord has called us to live. This includes paying taxes, praying for our leaders, calling them to repentance when necessary, perhaps working in government ourselves, and of course, participating in the election process. But as we do so we need to clear the political air of the smoke screen of nationalism used by the Democratic and Republican parties to cloud over the real issues. Those real issues are well-known to all of us even as they were known to the prophets of Israel: justice, mercy, and righteousness. And they stand in radical opposition to the sin of nationalism.

Chapter 10

Progress: A Different Language

July 11, 1983

There is a word in our contemporary vocabulary that describes a cherished value or norm of our society, and is therefore very useful in characterizing who we are as a people. The word is "progress." One problem with the word, however, is that it needs a frame of reference for it to make any sense, and that frame of reference may vary considerably from one group of people to another.

For example, people who place their hope in science and technology for the solution to current world problems like energy shortages, pollution, hunger, and war, define progress in terms of acquiring more scientific knowledge and developing new technological products and procedures. Some see progress as getting closer to the taming of nuclear fusion so that we may have unlimited free energy. Others see it in terms of discovering a cure for cancer and other diseases.

Another group of people, among whom we find most business people and economists, see progress in terms of the steady increase of profits from year to year. On a national scale, progress is defined by this group as a steadily increasing gross national product.

On an individual scale many young people see progress as the steady achievement of certain goals – for example, to get married, get a college education, get a high paying and interesting job, get a new car, own a home, etc. Many workers consider larger salaries and being put in charge of more people as progress. Many parents count it progress when their children, one after another, finish college, find jobs, marry, and begin to raise their own families.

Clearly, progress often includes the unencumbered, steady moving through life from the point where you are to a point where you hope to be at some time in the future, along a pathway that is marked by certain expected events.

It's the same for a Christian. Progress is the steady movement along a path. But here is where the Christian's definition of progress diverges radically from that of one's average American neighbor. In the Sermon on the Mount, Jesus uses language that pictures for us a kind of path when he said, "But seek first his kingdom and his righteousness" (Matthew 6:33a). The path Jesus is speaking of is the one leading from where we find ourselves now to the coming Kingdom of our Lord. The landmarks, or events along the path, are characterized as righteousness.

The contrast between progress along the path toward the Kingdom and progress in a worldly sense is made very clear in numerous places throughout the Scriptures. For example, in 2 Chronicles 1:11–12 we read of God's reply to Solomon for having sought the true Kingdom rather than a worldly one:

> God said to Solomon, "Since this is your heart's desire and you have not asked for wealth, possessions or honor, nor for the death of your enemies, and since you have not asked for a long life but for wisdom and knowledge to govern my people over whom I have made you king, therefore wisdom and knowledge will be given you. And I will also give you wealth, possessions and honor, such as no king who was before you ever had and none after you will have."

In James 4:1–2 we read:

> What causes fights and quarrels among you? Don't they come from your desires that battle within you? You desire but do not have, so you kill. You covet but you cannot get what you want, so you quarrel and fight. You do not have because you do not ask God.

And the psalmist tells us in Psalms 37:16: "Better the little that the righteous have than the wealth of many wicked."

We Christians use the word progress easily and often without seeing the radical distinction between the context that we must give to it and the context given to it by the self-centered, materialistic culture we live in. For the Christian, progress is the obedient, steady movement toward God's Kingdom. When the humanist uses the word progress, he's speaking a different language.

CHAPTER 11

AN INDIVIDUALIST CONTRADICTION
OCTOBER 8, 1986

One of the spirits of our time that we in the Christian community ought to constantly be on our guard to overcome is individualism. Individualists believe that each person is made completely separate from every other person. They also have a tendency to view any kind of dependency relationship as a sign of weakness. Children are dependent on their parents because they are weaker and are unable to support themselves. Women, at least in their traditional roles, have been viewed as the weaker sex, explaining why it is supposedly natural for them to be dependent on their husbands. And whenever a person cannot independently obtain what he or she needs in this life, whether it be food, education, or simple pleasures, that person is generally looked down upon by the individualist as being inferior.

Individualism manifests itself in many ways. One of its more tragic manifestations in our modern culture is the extraordinary rate of divorce resulting in so many broken families. When a husband and a wife fail to see themselves as the Lord intended for them to exist, that is, as a unity – one flesh, one mind, one common set of goals, one faith, one common set of tasks, and so on – it is usually because they have succumbed to the individualist spirit. Each is looking for fulfillment in his or her own individual way. In today's society when so many women are joining men in the workforce, it very often means that the career goals of husband and wife may clash. The one may feel that he must work as an artist in Greenwich Village, while his spouse feels she must accept an appointment as head of a research team at Stanford University. Each thinks individualistically in terms of his or her individual goals.

In the eyes of God it's as absurd as my legs saying they must be involved in tropical agriculture to be fulfilled and my head saying that it must be involved fulltime in playing the trumpet, with the result that these two parts of my body go their separate ways. If that really happened, I as a body would die. And indeed, that is what is happening in society as so many marriages die. It should remind us, as Romans 6:23

does, that the wages of sin are indeed death. And individualism is very much a sin.

What provoked me to write this essay, however, was not the problem of divorce, but another manifestation of individualism to which I'm afraid we as a Christian community have succumbed much more readily. One of our local Christian high schools is trying to develop a scholarship program so that students from families for which tuition is a serious burden might be helped. The reason that such a program is necessary is twofold. First, we have generally accepted the Western notion that individuals, or individual families, ought to seek their own financial livelihood independent of other individuals or families. Thus we find ourselves in the situation where there exists a wide range of family income among our Christian school supporters.

The second reason is that we have bought into the egalitarian notion that individuals or individual families ought to bear an equal share of the cost of educating their children. Therefore, the tuition is set at the same rate for everyone. Now for some, paying $310 per month to send two children to Christian high school is a real burden, maybe even an impossibility. But for others that $310 represents only a small portion of their monthly income. Doesn't that seem to be a contradiction? When it comes to earning money, we insist that each individual have complete freedom to earn whatever they can. But when it comes to paying money for tuition, we insist that everyone be charged exactly the same amount. That may be consistent with the perspective of individualism, but it is an inconsistent way of dealing with people.

Thus, we have a community that is economically torn apart by the fruits of individualism. The board of the Christian high school I'm speaking of is faced with a dilemma. As expenses rise the board members can increase the tuition. But in doing so they make it even harder on families with limited incomes. The alternative – cutting back on expenses – results in a school forced to limp by on less than adequate materials, overworked teachers, and the constant danger of deficit financing. That's why the school is attempting to develop a scholarship program. In a small way, it is hoped it will remove the stress of increased tuition on families who would be hurt the most by a tuition increase. Naturally, only those with genuine need would qualify for a scholarship.

But why should we have to face such a problem in the first place? It's precisely because we have yielded to the spirit of individualism. We fail to see ourselves as God's covenant people, a true community. We are too busy looking out for ourselves as individuals or individual families, and worrying that the next guy is going to get a better break than we get. I find that very troublesome. If the more subtle manifestations of a community's capitulation to individualism are already among us, then the more overt manifestations can only be around the corner. Time is short. Only true repentance and very concrete acts of obedience will ward off the angel of communal death that hovers over our families, churches, and schools.

CHAPTER 12

MASTERS OF WAR

MARCH 4, 1991

The war in the gulf is all but over now. The *Nightly News* will have to find new stories to entertain us with. The headlines in the morning newspaper will soon be directing our attention to the 1992 presidential race. And those of you who regularly read newspaper editorials will read less and less on the topic of war, the Middle East, or the politics of oil. Even this essay isn't really *about* the war, although I guess you can say that it was provoked by the war.

Shortly after the troop buildup began last fall, I began digging out some of my older records to listen to. Songs by Bob Dylan, Phil Ochs, and Joan Baez seemed suddenly appropriate again. One particular verse from an old Bob Dylan song seemed especially fitting in light of the greed displayed both by Saddam Hussein and those high tech German and American companies that had sold him his armaments. It goes like this:

> Let me ask you one question: Is your money that good? Will it buy you forgiveness? Do you think that it could? I think you will find, when your death takes its toll, all the money you made will never buy back your soul.[*]

That verse was marvelously appropriate back in 1964 when it was written, as it was this past year in light of the destruction that resulted from the weaponry designed and sold during the last decade. The verse, and the song of which it is a part, decry those who become wealthy by selling weapons of destruction. But that inhuman pursuit is only the symptom of a deeper sickness, the worship and chase after wealth. In fact, the end of the gulf war has only served to spread the disease. Even before the fighting died down there was a mad scramble on the part of numerous multinational corporations to get a piece of the "reconstruction" pie. Kuwait will spend billions to rebuild itself. Some of those same companies that profited by the destruction of Kuwait will now turn around and

---

[*] Bob Dylan, "Masters of War," *The Freewheelin' Bob Dylan*, Columbia, Compact Disc #CK8786.

attempt to profit by its rebuilding.

It seems to me that it is this lusting after wealth that creates most of the world's problems. That was the reason Iraq invaded Kuwait in the first place. And despite all the rhetoric about halting aggression, the reason the United States and its allies were so united in opposing Iraq was because they were united in their fear that they would lose some of the wealth that the oil fields of the Middle East provide at this time in history.

But who are we to point the finger. We Christians bought into the pursuit of wealth long ago. By pretending that material wealth is a blessing from the Lord, we have turned our children into little pagans, worshipping at the altar of the almighty buck. Why do we send them to college? Because we want them to get a "good" job, right? And what is a "good" job? One in which they make "good" money, of course. And what is "good" money? More money – more and more and more and more. We rarely ask how much is enough. And we *never* ask how much is *too much*.

The Scriptures speak very clearly regarding this disease of ours. We don't read it much. In fact, I can't say I've ever heard a sermon on it. But Paul's first letter to Timothy is as clear cut as you can get when he says,

People who want to get rich fall into temptation and a trap and into many foolish and harmful desires that plunge men into ruin and destruction. For the love of money is a root of all kinds of evil.

And it's not only from the Word of God that we can glean such insight. It ought to be clear to us simply by watching how rich people live that there is something very dangerous about wealth. The American playwright Tennessee Williams never impressed anyone with his Christian commitment. Yet he perceived the fatal consequences of living for money. Shortly before his death he wrote,

One does not easily escape from the seduction of an effete way of life. . . . With conflict removed, man is a sword cutting daisies. It is not privation but luxury that is the real wolf at the door . . . and the fangs of that wolf are all the little vanities and conceits and laxities that success is heir to. . . . Security is a kind of death. It can come to you in a storm of royalty checks beside a kidney-shaped [swimming] pool."

I'd like to suggest that American industry volunteer to help rebuild Kuwait, and that they do it at cost, that is, without making any profit on

---

** Tennessee Williams, "The Catastrophe of Success," *Harper's Bazaar,* January 1984, p. 132, referenced in Peter C. Moore, *Disarming the Secular Gods* (Downers Grove: InterVarsity, 1989), 180.

it. And to pay that cost why don't we, the American taxpayers, offer to cough up the necessary money. Let's see, if every American agreed to set aside $100 during the next year that would provide . . . $25 Billion! Oh, that's much more than is necessary. See how easy it would be. Easy that is, if we were truly interested in correcting injustice. Easy if our compassion was truly blind to the character of our wounded neighbor. Easy if we didn't first think of ourselves and of how striving after wealth would be somewhat thwarted by the loss of $100.

I'd like to end this essay by quoting a Psalm – Psalm 49. These words were written long ago, but they seem strangely appropriate now – at the end of the Gulf war.

> Hear this, all you peoples; listen all who live in this world, both low and high, rich and poor alike:
>
> My mouth will speak words of wisdom; the meditation of my heart will give you understanding.
>
> I will turn my ear to a proverb; with the harp I will expound my riddle:
>
> Why should I fear when evil days come, when wicked deceivers surround me – those who trust in their wealth and boast of their great riches?
>
> No one can redeem another or give to God a ransom for them – the ransom for a life is costly, no payment is ever enough – so that they should live on forever and not see decay.
>
> For all can see that the wise die, that the foolish and the senseless also perish, leaving their wealth to others.
>
> Their tombs will remain their houses forever, their dwellings for endless generations, , though they had named lands after themselves.
>
> **People, despite their wealth, do not endure; they are like the beasts that perish.**
>
> This is the fate of those who trust in themselves, and of their followers, who approve their sayings.
>
> They are like sheep and are destined to die; death will be their shepherd (but the upright will prevail over them in the morning). Their forms will decay in the grave, far from their princely mansions.
>
> But God will redeem me from the realm of the dead; he will surely take me to himself.
>
> Do not be overawed when others grow rich, when the splendor of their houses increases; for they will take nothing with them when they die, their splendor will not descend with them.
>
> Though while they live they count themselves blessed – and people praise you when you prosper – they will join those who have gone before them, who will never again see the light of life.
>
> **People who have wealth but lack understanding are like the beasts that perish.**

CHAPTER 13

VALUES AND THINGS

AUGUST 1, 1988

Do you own a VCR? Do you own a television, a new Cadillac, a hand gun, or a bottle of expensive Scotch whiskey? If you do own any one of these things, do you ever feel guilty about it?

One question raised by Christians over the course of time is whether things contain values intrinsically or whether things are most often always neutral. For example, is a switchblade knife evil in itself, or is it simply a neutral tool that is very often put to evil use by its owner?

The broader question has been raised recently in a number of places. In the July 11 issue of *The Banner*, for example, Dr. Lee Hardy, a professor of philosophy at Calvin College, seems to suggest that the products of modern technology, like the automobile and computer, are not neutral, but have values that are subtly and very often unconsciously designed into them. On the other hand, in the July 25 issue of *The Banner*, in an article entitled "Polaroid People," by Rev. Mark Tidd, we read just the opposite. He says, and I quote:

> Nothing that God created or that those who mimic his creative power have produced by bringing form out of substance is intrinsically evil. Things are things. The thing has potential for good or evil, depending on the handler's use or abuse of it.

I believe that this is the most widely held view of the matter today, at least among reformed Christians. And there are good arguments that can be given to defend it. One of the most common is an appeal to Scripture, particularly 1 Timothy 4:4–5, where we read:

> For everything God created is good, and nothing is to be rejected if it is received with thanksgiving, because it is consecrated by the word of God and prayer.

The context of this quote is Paul's warning Timothy concerning those who teach that eating certain foods is wrong because those foods are evil in themselves, having, for example, been used as offerings to idols. As reformed Christians who emphasize the sovereignty of God, we often see these words of Paul as fitting within a broader context; one that recognizes God as the Creator and Sustainer of all things, sees his creation as fundamentally good, and views evil as not a force or substance, but rather as a heart direction in which one turns away from God.

But things are not so neat and tidy. (If they were, the question wouldn't keep popping up the way it does.) Timothy may be able to eat meat that has been offered to idols because he knows that it is not intrinsically evil, there are no "weaker brethren" around whom he might offend, and he eats it in thankfulness that God provides this form of nourishment for his body. But would we be able to say the same thing about a marijuana cigarette? Would you be able to receive one with thanksgiving, thereby consecrating it by the word of God and prayer? I have my doubts. There may be an extremely rare case where the marijuana is actually prescribed by a physician as treatment for a rare disease. But there are also extremely rare cases where Bibles can be used to harm people; think of those who feel compelled to handle poisonous snakes or drink poisons because they take literally and out of context the prophecy in Mark 16:18.

I think that we need to re-evaluate this notion that "things are things" and have no pre-determined values intrinsic to them. Most "things" we come into contact with today are products of modern technology. That means their origin is rooted in the process of our giving form to creation. This process, like any other human activity, is one that is done either in obedience or disobedience to God. What I want to suggest is this: the products of technology intrinsically retain a portion of the values operative in the process of technology. Very often these are broad values that characterize a culture as well as its technology. For example, the automobile embodies the western value of individualism, and the Cadillac in particular embodies the American value of status-consciousness.

In the excellent book *Responsible Technology* (Grand Rapids: Eerdmans, 1986), edited by Stephen Monsma, we read the following:

> At times the argument is made that technology and its tools and products are neutral, and that only the uses to which they are put involve valuing. But there can be no such neat division. Of course valuing affects how technological tools and products are used, but valuing begins long before the use stage. Technology itself is value-laden. Thus responsibility in a technological society necessarily means that we must have insights into the valuing that is reflected in the technology with which we are surrounded. Only then can we make responsible choices. (3)

Does this mean that Christians ought not bring products of modern technology, like VCR's, computers, and over-the-counter pain killers, into their homes? Certainly not. But we need to be aware that these products of modern technology are not neutral. They embody values that may very strongly influence the consumer. So don't look down on your neighbor who decides not to own a television, a gun, or a Cadillac. And do be very careful to consider the value implications of any product of our technological culture that you consider purchasing.

CHAPTER 14

TELEVISION: FOR ADULTS ONLY

OCTOBER 3, 1989

Last week Dordt College held a lecture series on the topic of television. Dr. Daryl Vander Kooi, of Dordt's Communications Department, presented "the case against television," summarizing the views of three well-known scholars. Dr. Clifford Christians, from the University of Illinois, then presented a case for television. One of the most unique characteristics of this particular discussion was the attempt to deal with more than just the *content* of TV, more than just the quality of the programming. Instead, both speakers concerned themselves with how the *structure* of television determines our television watching habits, as well as the kind of programs that are produced.

In this essay I want to consider two technical characteristics of television and show how they can, if not recognized and combated, lead to serious distortions in our lives as we attempt to be faithful image bearers of God.

The first characteristic, one identified by Dr. Christians in his lecture, is intimacy. Because of its relatively small size, the television is usually viewed, as the phrase goes, "in the privacy of one's own home." Most people watch TV either alone or in a small group. This tends to give it the quality of being almost an extension of the person. Perhaps you have seen small children totally absorbed in watching TV. They are not distracted in the least by events that are going on in the room around them. And, as adults, many of us have experienced the uncomfortable feeling of annoyance at being interrupted while watching TV, coupled with a sense of responsibility for dealing with a child's or spouse's question.

Now this preoccupation, or absorption, that often characterizes TV watching is not necessarily bad. In fact I believe it is the kind of thing we were created to do occasionally; the kind of engagement that we call, in different contexts, concentrated reflection or daydreaming. But there are limits to concentrated reflection and daydreaming. In fact, when a person's life is nothing but daydreaming we say that he is mentally ill. I suggest that due to its characteristic of intimacy, the TV fosters a kind of artificial daydreaming. The reason that this tendency occurs in otherwise normal people is due to the fact that most TV viewing requires only passive concentration. Healthy reflection and healthy daydreaming require active concentration. But television does our imagining for us. Even so, this would not be as serious if it were not coupled with the property of being addictive. TV's quality of intimacy along with its ability to do our imagining for us, gives it a narcotic character that entices us to come back for more. Like cocaine, television is addictive, and, if misused, it reduces our capacity to think and to engage in reflection − two rather critical human activities.

The second property of television that we need to consider is its *transient visual immediacy*. By that I mean to say that TV gives us the ability to be visually present virtually anywhere on earth, but does it only for the moment, after which we are whisked away to somewhere else, having no time to reflect on where we have been. For example, suppose we watch an historical drama dealing with the Civil War. We are given the visual sense of being present in, let us say, Gettysburg during the 1860's. We listen to conversation that might have taken place at that far off place and time. But after the program ends, and our minds are free once more to do their own thinking, we find ourselves in an apartment in downtown Chicago, or following a pack of wolves somewhere in the north woods, or on the Starship Enterprise about to do battle with Klingons. My point is that we are swept away from the Gettysburg of the 1860's without time for reflection or discussion on what it all meant.

A few years back one of the major networks tried to do something different. After showing a film entitled *The Day After*, which attempted to depict the aftermath of nuclear war, the network aired an hour long panel discussion. It was a feeble attempt to overcome the problem I've just described. But it didn't work. The panel discussion was trite and artificial and most viewers were bored. Perhaps it was doomed to failure; for instead of allowing viewers to engage in reflection, it merely provided for viewing the supposed reflection of others. Unless viewers are able to engage in their own genuine reflection on the events portrayed by a dramatization, the whole event becomes trivial − something that can be turned off and forgotten with the flip of a switch.

The problem with television's *transient visual immediacy* is that it trivializes everything it attempts to communicate. If I read a book I can easily stop and ponder the meaning of a paragraph, or I can stop and turn back to an earlier chapter to find relationships between the different thoughts of the author. But television forces me to mentally move along at its pace. There is no turning back and there is no stopping to ponder a point.

The remarkable quality of visual immediacy, irrespective of its transient nature, also easily leads us down the road to trivialization. Television allows us to be "live," in Washington DC during an important presidential address and to see the tragedy of war in real time. (I'm convinced that the TV played an important role in hastening the end to the Vietnam War in the '70's.) But television also makes it possible for us to be there for every airplane crash, for every erratic trend in the stock market, and for Zsa Zsa Gabor when she makes a fool of herself for the sake of publicity. If I'm watching "The Nightly News" I have little choice but to be trivialized by the reporting of such irrelevant events.

In summary then, I am saying that television, by its very nature, has properties that can easily be used to narcotize and trivialize us. As citizens of the Kingdom of God, what ought we to do about it? Throw out our television? No, television is a legitimate technological product, just like sharp knives and drugs. But just like sharp knives and drugs we ought to take some precautions. We ought not simply hand it to children. We ought not use it uncritically ourselves.

To overcome the problems that I've described in this essay we can do at least two things – and the first may surprise you. My first suggestion is to buy a VCR if you don't already have one, and to use it to tape the television programs you want to see. That way you can fast-forward through the commercials, you can replay segments that you want to analyze more carefully, and you can pause, stopping the program so that you can reflect on and discuss what you are viewing. My second suggestion is simply to never watch television spontaneously. Always plan in advance what you will watch and watch only programs that you can honestly say are not wasting that one precious commodity the Lord gives to all of us, time. If you do that, your TV will probably be on less than 1 to two hours per day. And perhaps one day you will hear the Lord say, regarding your stewardship of time, "Well done good and faithful servant; you have avoided being either narcotized or trivialized."

CHAPTER 15

BARBARITY REVISITED

DECEMBER 7, 1990

Forty-nine years ago today, on December 7, 1941 – a day that Franklin D. Roosevelt said would "live in infamy" – the nation of Japan, with a barbarity that would have shocked Genghis Kahn, attacked Pearl Harbor, killing and maiming hundreds of United States sailors and sinking or crippling a large number of ships. Three years and eight months later, the United States returned the favor, outdoing the Japanese in barbarity by vaporizing, broiling, and poisoning thousands of men, women and children – citizens of Hiroshima and Nagasaki. Both events demonstrate that even the more developed nations of the human race have not out-grown their penchant for brutality and inhumanity. If Attila the Hun could have foreseen what took place during those years of the Second World War, he would have swollen with pride, knowing he was part of a tradition that would so dominate history. From the day Cain slew his brother Abel, humankind, driven by the spirits of pride, greed and hate, has either enacted its ambitions, or defended itself against its neighbor's ambitions, by twisting the elements of nature into a tool to inflict bodily harm. Whether that tool be a hand-sized rock used to crush a single human skull, or a 50 Megaton thermonuclear missile, capable of incin-erating a metropolis of people in less than a millisecond, history demon-strates that humankind has consistently resorted to violence as the final arbiter of most conflicts.

Today the United States stands on the edge of a precipice. A nudge in the wrong direction will send thousands of God's image-bearers, most-ly Iraqi and American soldiers, to a violent death in the bottomless gul-let of that age-old beast called war. But as horrible as that prospect may seem, especially to the relatives of those human sacrifices, more shocking and more ironic still is, first, the fact that it will occur during a period in history when such brutal carnage is technologically unnecessary; and second, that it will have the enthusiastic support of millions of people in much the same way that the University of Iowa football team will be cheered on when they play in the Rose Bowl on New Year's Day.

Consider the fact that we live in a technologically accelerating age. Over twenty years ago it was demonstrated that we could send men to walk on the moon. In medicine, surgical operations undreamed of only thirty years ago are regularly performed. Chemistry and biochemistry have been developed to the point where yesterday's exotic materials and drugs are today so commonplace and so plentiful that agencies like the Food and Drug Administration have difficulty keeping up with them. We no longer go to the river for our water but to the kitchen faucet. We no longer read by the light of a torch but with the flick of a switch are able to turn night into day. And if you are like me you hardly use a pen or pencil anymore – the computer is much neater, faster, and capable of catching our spelling errors better than any proof-reader. Yet in spite of this incredible technological progress, we still wage war the same way that barbarians did millennia ago. Oh, the weapons have changed in form – but not in purpose. We still believe that defending against an aggressor like Saddam Hussein means violently killing people. I don't understand that. Why can't we develop safe defensive weapons? By safe, I mean that they would cause no more death and destruction than the automobile (hopefully less!). In an advanced technological age such as ours, there is no reason why, for example, we could not develop a system that immobilizes people without doing them bodily injury. Aggressors might then be stopped, their weapons confiscated, and the leaders removed to some prison supervised by the United Nations. This may sound like science fiction, but, I assure you, it is not technologically impossible. The chief difficulty is that we have a bias toward violent means of defense. As a race we have not even considered the gentle alternatives.

But perhaps the fruits of our advanced technological age are partially to blame for that bias as well. Remember that my second point of horrific irony is the claim that when the marines go to war in the Persian Gulf, they will be cheered on with the same enthusiasm given the Iowa Hawkeyes in the Rose Bowl. The marvel of instantaneous visual communication – particularly as it is provided by the television news media – will isolate us from the actual pain and suffering, the physical brutality of the war. The images of soldiers fighting and dying in the Iraqi desert will all too easily recall to our minds images of Sylvester Stallone, Chuck Norris, and Arnold Schwarzenegger. Right now the majority of American citizens – if not consciously then certainly subconsciously – are rooting for war. That's because the majority of American citizens spend much of their time being entertained by the television. And war makes good television. The television provides us with a vicarious experience of adventure. And as long as that experience is vicarious, the adventure can be horribly violent without causing us any real pain. In fact the experience deadens our sense of moral pain so that increasingly we crave more intense and gory forms of violence to satisfy what has become a sadistic national appetite for death and destruction.

In summary, what I am saying is that as the twentieth century comes to a close, the human race, in dealing with conflict, is using its advanced technological know-how more in a return to the barbaric past than in efforts to seek peace and justice. Instead of using our knowledge of creation – our cultural mandate – to develop humanitarian instruments for the securing of peace, we develop increasingly more sophisticated and barbaric means of violent destruction. And then we compound our sin by using our advanced understanding of communications technology to turn the reality of human suffering into a form of entertainment.

At times such as these we need to repent. And if we repent, we shall see that all is not darkness and despair. There is a light shining in the darkness of even the Persian Gulf conflict. Consider the words of the prophet Isaiah, so appropriate at this time of year, words that can bring shalom and redirection to even the most bloodthirsty of tyrants:

> The people walking in darkness have seen a great light; on those living in the land of deep darkness a light has dawned. . . . Every warrior's boot used in battle and every garment rolled in blood will be destined for burning, will be fuel for the fire. For to us a child is born, to us a son is given, and the government will be upon his shoulders. And he will be called Wonderful Counselor, Mighty God, Everlasting Father, Prince of Peace. Of the greatness of his government and peace there will be no end. He will reign on David's throne and over his kingdom, establishing and upholding it with justice and righteousness from that time on and forever. The zeal of the Lord Almighty will accomplish this. (Isaiah 9:2, 5–7)

CHAPTER 16

VIOLENCE AND "RESPECTING AUTHORITY"

MAY 8, 1992

For a while it seemed like 1967 all over again. Violence in the streets of a major city. Cars overturned, buildings set ablaze, looters running through the streets carrying merchandise that they'd stolen from inner city stores. The stench of racism filled the air as the specter of injustice cast its shadow over the whole nation. Even amidst the peace and stability of Sioux Center one couldn't help but shudder at the realization that beneath the affluent veneer of American society are crawling those demons of poverty, hopelessness, violence, racial hatred, greed, power-abuse, and injustice. What causes them to break through and fester on the surface of our inner cities every twenty years or so? There are the obvious answers. We live in a society where the rich get richer and the poor get poorer; where quality education is reserved for those with white skin, good English, and relatively wealthy parents; where justice, if it is blind, seems to be blinded by racism and greed, and where violence is considered entertainment.

But I don't want to deal with the obvious answers in this essay. I want to consider a much more subtle explanation for both the Rodney King verdict and the unrest that followed it. It's an explanation that those of us living comfortably in the rural Midwest, particularly those of us who are Christians, need to reflect on very carefully.

I suggest that one of the more insidious perversities at the root of such atrocities as the Rodney King verdict and the subsequent inner city violence is our general and uncritical acceptance of a distorted definition of authority. What is authority? In the Dordt College Statement of Purpose authority is carefully defined as the relationship that exists between offices, in the context of specific tasks, which determines the direction of responsibility and power. Now that's a highly abstract definition that needs some explaining. But before I try to explain it, let me tell you some other things the Dordt Statement of Purpose says about authority. First of all it says that authority is always "authority to serve." Those two words, "authority" and "service," must go hand in hand. The point is, as Jesus said to his disciples, we must never try to "lord it over" one another. That would be a misuse of authority. Secondly, there may be no such thing as absolute authority among us. Only God has absolute authority. Human authority is always given by God for the purpose of service and is always limited to a particular task. Even its character is limited by the particular task for which it exists. Let me give a couple of examples. A father has authority over his young children. That means that he is responsible for his children. He has the task of serving them by enabling them to grow up as the Lord would have them grow up. He also has the power to serve in this way. He provides their food, their shelter, the security of a loving and disciplined home environment, and the direction so that they may live their lives in obedience before God. But this authority that the father has over his children is limited. He does not have that authority over his wife. If he works outside the home, he does not have that authority over those workers for whom he is responsible. If he is an elder in his local church, his authority is different again. His different kinds of authority exist to serve different kinds of tasks. And none of those authority relationships gives him the right to lord it over another person, not even his own children.

Now let's look at that definition of authority again. Saying that authority is the relationship between offices in the context of specific tasks carefully limits the range of authority that any one person may exercise at any particular moment. The office of father is related to the office of child in a family. That relationship is characterized by authority in the sense that the father is responsible for his children and the children are responsible to their father. That father does not have authority over any other children. On the streets of a city the office of policeman is related to the office of citizen in the context of the task of maintaining civic order. The policeman is responsible for the safety of the citizens and the citizens are responsible to the police for their lawful behavior. But even in that relationship, the authority the police have is very different from that the father has in his family. The authority is limited and defined by the task area.

Authority determines the direction of power. The father is empowered to make certain limited decisions that his children must respect. The police are empowered to make certain limited decisions that citizens under their jurisdiction must respect. And the elders in a church are empowered to make certain decisions that the congregation must respect. By "respect" I do not necessarily mean "agree with." Rather, "respect" means "subject themselves to." But the father, the policeman, and the elder are not empowered to make decisions outside their respective task areas.

Again, the main point here is that no one has authority, no one is legitimately empowered to make decisions that affect others, outside of his particular task area. In fact, authority ought never to attach directly to persons, but to only to offices.

So what does all this have to do with the Rodney King verdict and the riots in Los Angeles? I want to suggest that we have a distorted view of authority in this country. We associate authority with persons, we allow it to grow beyond its legitimate task area boundaries, and we have a tendency to see all authority as somehow analogous to the authority a father has over his young children. This means that we tend to see all authority as absolute and paternalistic.

Why did the jurors in the Rodney King case return a verdict of not-guilty when the evidence clearly indicated that the police were guilty of brutality? One reason is that the defense was able to work on the jurors' deep-seated and paternalistic notion that somehow the police have absolute and unquestionable authority. Daddy has a right to spank Johnny when Johnny does something bad. If Daddy gets angry when spanking Johnny, well, that's not ideal, but after all, who's in charge here?

Why did the citizens of Los Angeles riot, running wild through the streets, burning and looting? One reason is that they needed to exercise their manhood and womanhood by saying no to a system of paternalistic injustice. Now the method they chose may be both sinful and stupid. But how many young adults, growing up oppressed by authoritarian parents, have the composure and maturity to say, after being beaten, "I'm sorry Dad, I'm twenty years old now and responsible for my own actions. I can no longer accept your abusive punishment." Most young adults under those circumstances will one day explode, taking their anger out physically on their immediate surroundings and verbally on their oppressive parent. They will then storm out of the house, slamming the door behind them. Sound familiar?

In this country we have treated the poor and minorities as children who we take care of so long as they "obey the rules of the house." We've accepted it when the police, government officials, teachers, church leaders, and who ever else has authority, abuse that authority, treating the rest of us paternalistically. And we've invented an eleventh commandment that says: "Thou shalt obey those in authority without question, for all authority is the same as that which your parents had when you were a child."

We need to hear again the words of Jesus in Matthew 20, when he said,

> "You know that the rulers of the Gentiles lord it over them, and their high officials exercise authority over them. Not so with you. Instead, whoever wants to become great among you must be your servant, and whoever wants to be first must be your slave – just as the Son of Man did not come to be served, but to serve, and to give his life as a ransom for many."

True authority is always authority to serve. Until we learn that, we may need to develop in ourselves and in our children a healthy skepticism regarding anyone who claims to be an authority. On my bulletin board in my home office I have a bumper sticker that I've saved from those turbulent days in the sixties. On it are written two simple words: QUESTION AUTHORITY. Perhaps it's time to start displaying it more conspicuously.

CHAPTER 17

HUMBLE CONFIDENCE

FEBRUARY 4, 1993

The troubles that seem to be plaguing the Clinton administration in these its first few weeks, make clear that even the best of intentions can be thwarted by foolishness and sin when the actions resulting from those intentions are not normed by the Word of God. Such has been the case with the nomination of Zoe Baird, the issue of homosexuals in the military, and the recent question of cuts in social security benefits. And these are only the issues that have risen to the surface of that cesspool of controversy we call the American political scene. Lurking below, occasionally bubbling up into view, are those momentarily less weighty issues such as the Israeli-Palestinian impasse, the problem of developing an energy policy, Iraq, health care, and so on and so on. I had hopes that the Clinton administration, with its relative youth and idealism, would begin to make some real progress on many of these issues. But wisdom is not a commodity that can be voted into office, bought with the dollars of billionaires, liberated by military power, or even coaxed into being by youth and idealism. True wisdom is rooted in the Word of the Lord – that Word by which the universe was created, in which all things find their meaning, and by which all things will be redeemed.

But if that Word is at the very center of reality, why does there seem to be such a dearth of wisdom? After all, as the writer of Proverbs asks, "Does not wisdom call out? Does not understanding raise her voice?" Well if so, it seems that many of our leaders are hard of hearing.

But let's not be too quick to judge these leaders. Especially here in the United States where we elect them ourselves. For in a very real sense our leaders are simply the product of the society for which all of us are responsible. And perhaps it is those of us who call ourselves Christian and who claim to seek, if not to know, the will of the Lord for our lives, who must bear the greatest blame. For if wisdom is to call out, it is through us that it must speak. Jesus, you will remember, did call us to be the light of the world. That means we must wrestle with the issues of the day, debate with each other, and in the light of God's written Word begin to indicate in which direction that Word points us. For our leaders to lead effectively, they need us to articulate the direction. Yes, that's right, little old us – little, old, *humble,* us. The trouble is we don't believe it. Paul tells us that God uses the weak things of this world but we look up to the strong and consider ourselves too weak to make a difference. He tells us that God chooses to work through the lowly of this world, and yet we remain enthralled by the rich and the powerful, the high and the mighty, and ashamed of our relatively low positions. God comes into this world as a poverty stricken, lower class Jew – and 2000 years ago there wasn't much worse in the civilized world. Yet we aspire to wealth, power, fame, and the admiration, if not adulation, of our society. Jesus associated with the drug addicts, pimps and prostitutes, homosexuals, and AIDS victims of his day. But we teach our children to associate with only the proper people and we hire police and build prisons for our outcasts.

In Chapter 8 of the book of Proverbs we read that Wisdom does indeed call out. She says,

> To you, O people, I call out; I raise my voice to all mankind.
> You who are simple, gain prudence; you who are foolish, set your hearts
> on it. . . .
> Choose my instruction instead of silver, knowledge rather than choice
> gold, for wisdom is more precious than rubies, and nothing you desire
> can compare with her.
> I, wisdom, dwell together with prudence; I possess knowledge and
> discretion.
> To fear the Lord is to hate evil; I hate pride and arrogance, evil behavior
> and perverse speech.
> Council and sound judgment are mine; I have insight, I have power.
> By me kings reign and rulers issue decrees that are just; by me princes
> govern, and all nobles who rule on earth.

But these words are too easily distorted by our obsession with power, prestige, and success. We need to read them in the context of Paul's words to the Corinthians where he says,

> Brothers and sisters, think of what you were when you were called. Not many of you were wise by human standards; not many were influential; not many were of noble birth. But God chose the foolish things of the world to shame the wise; God chose the weak things of the world to shame the strong. God chose the lowly things of this world and the despised things – and the things that are not – to nullify the things that are, so that no one may boast before him. It is because of him that you are in Christ Jesus, who has become for us wisdom from God. . . .

Instead of, on the one hand, timidly shrinking away from the public square, and, on the other hand, being smitten with those who wield power in this world, we Christians need to raise our voices, individually and collectively. We need to write to the newspapers – the *Des Moines Register* as well as the *Sioux Center News*. We need to write, phone, FAX, or e-mail our concerns to those in political leadership roles, starting with Lee Plasier and Wilmer Rensink, but also including Fred Grandy, Chuck Grassley, and Bill Clinton himself. We need to do so in humility and with confidence: humility that befits a servant, but confidence appropriate to a servant of the Lord.

"Does not wisdom call out? Does not understanding raise her voice?" Indeed she does. But only when God's people have the humble confidence to supply the words. Without that, our national leaders are condemned to stumble around in a silent world thwarted by foolishness and sin.

CHAPTER 18

SERVING MOLEK

APRIL 6, 1993

The book of Leviticus is probably not high on the list of favorite Bible readings for family devotions. With its detailed treatment of ancient Israel's ceremonial law, it is more fit for scholarly inquiry than for edifying discourse. But Leviticus is still God's Word; and there are times when it serves better than any other part of that Word to reveal to us the passionate concern of the Lord for his people or for a particular kind of activity. Consider Leviticus 20, verses 1–5:

> The LORD said to Moses, "Say to the Israelites: 'Any Israelite or any foreigner residing in Israel who sacrifices any of his children to Molek is to be put to death. The members of the community are to stone him. I myself will set my face against him and will cut him off from his people; for by sacrificing his children to Molek, he has defiled my sanctuary and profaned my holy name. If the members of the community close their eyes when that man sacrifices one of his children to Molek and if they fail to put him to death, I myself will set my face against him and his family and will cut them off from their people together with all who follow him in prostituting themselves to Molek.

Child sacrifice was a practice in Old Testament times associated with the Ammonite worship of the god Molek. It was a practice that the Lord found detestable (and says so in numerous Old Testament passages). Yet there are many parallels to this detestable behavior today, even among Christians. The most obvious, of course, is the slaughter of the unborn by the practice of abortion. The gods requiring this form of sacrifice, counting even many Christians among their servants, are those of hedonism (in the form of unbridled sexual indulgence), individual autonomy (the individual's supposed "right to choose"), and absolutized freedom (the assumed absolute right to privacy).

Another form of child sacrifice, one that we too easily forget or refuse even to recognize, is the slaughter of our nation's young men on the battlefields of war. Twenty-five years ago, for example, the leaders of this nation, to palliate the gods of nationalism, fear, and the American way of life, offered up thousands of my contemporaries to be slaughtered in a primitive and far off land called Vietnam. And I'm sure that few citizens recalled, this past March 16 – the twenty-fifth anniversary of My Lai – that there are ways to sacrifice children far worse than death, namely, inciting them to become butchers of unarmed men, women, and children. We may recoil in horror today at the atrocities committed in Bosnia, South Africa, and India – and rightly so. But before that horror turns to self-righteousness we need to remember that it has all happened before in a village called My Lai, and the perpetrators were American youth who were in the process of being sacrificed to those gods of nationalism, fear, and the American way of life. The Lord has called that *detestable*.

But there are still other, more subtle forms of child sacrifice practiced today. For example, in seeking the immediate gratification favors of the god of economic prosperity, we in the West have been willing to ignore, and thereby sacrifice our children's future. Our patterns of deficit spending, both as a nation and as individual families, mean that future generations will live under the yoke of debt – a debt incurred to finance their parents' materialistic pursuit of an ever higher standard of living.

Closely tied to this fiscal sacrifice of our children's future is the greedy and voracious appetite that those of us in the West have for consuming nonrenewable resources. Twenty years ago we first became aware of a fact that should have been obvious far earlier – our petroleum resources are finite. But two national episodes of energy shortage, with their gasoline lines and plant shut-downs, have long since receded from our memory. Today we guzzle gasoline, freely burn carbon based fuels, and toss most plastics into land-fills, ignoring the fact that in doing so we deprive our children of the raw materials they will need in the future to produce pharmaceuticals, fertilizers, pesticides, and herbicides. The Old Testament Ammonites beat drums loudly to drown out the screams of their children as they placed them on the fiery alters of Molek. We don't even do that. Our faith in the gods of economic and technological progress have deafened us to the prophetic voices that warn of our children's future suffering due to their lack of nonrenewable resources.

The service of false gods requires sacrifice. Whether those gods be individual autonomy, hedonism, nationalism, technological progress, or an ever escalating standard of living, there is a price to be paid. Most detestable of all is when, like the ancient servants of Molek, we make our children pay the price for us.

CHAPTER 19

POSTMODERNISM: ACADEMIC FIDDLING
WHILE THE WORLD BURNS

FEBRUARY 7, 1997

When I was a young child I had a recurrent nightmare in which I observed a person slowly playing a violin in the front room of a house while violence was being done to another person in a back room. The violinist seemed blithely unaware of the trauma occurring only two rooms away. I call it my dissonance dream, and, although I had the role of a mere observer, it caused me greater distress than even the more typical "monster" nightmares, where one is being chased by a large animal while seemingly wearing boots of lead. I would wake from that dissonance dream trembling, wet with sweat, and unable to get back to sleep. Thankfully, my mother seemed to recognize my state of high anxiety and was ready with the cure: a glass of warm milk and a few comforting moments sipping it, while seated on the couch in our living room.

I've recently had occasion to reflect on that dissonance dream. Although it was confined to my childhood, it is one of the few nightmares that accurately, if metaphorically, reflects a paradox of life in the twentieth century: the incongruous coincidence of tranquility and mayhem. At times I think that, in some mystical way, the dream is related to the circumstances of my birth. I've recently come to realize that in the month or so before my conception, my parents were enjoying the peaceful bliss of newlyweds while, on the other side of the world, more than 300,000 people were dead or dying from the atomic bombs that were dropped on Hiroshima and Nagasaki. Strange, isn't it – this confluence of joy and suffering that so often characterizes our world.

What brought all of this most recently to my – not usually so melancholic – mind was a Dordt College faculty discussion about postmodernism. Actually the discussion was about a book about postmodernism, a book entitled *Truth is Stranger Than it Used to Be* by Richard Middleton and Brian Walsh (Downers Grove: InterVarsity, 1995). In the book the authors do an excellent job of analyzing postmodernism from a Christian perspective and try to give some hopeful suggestions for how we, disciples of Christ, might live in a postmodern age.

For those of you who are wondering just what in the world I am talking about, postmodernism might be described as a worldview. Its central tenet is a rejection of the Enlightenment faith in reason, science, and authority. In some respects it's not new. The Romantic poets like William Blake ("The New Jerusalem") and William Wordsworth ("The Tables Turned") knew something was wrong with those "dark Satanic Mills" of the early industrial revolution and with a "meddling intellect [that] Mis-shapes the beauteous forms of things: – We murder to dissect." More recently the existentialist thought of Kierkegaard, Sartre, and Camus denounced the modernist worldview that glorified reason at the expense of individual freedom and responsibility. Even the flower children of the late sixties, with their rejection of structure and authority, might be seen as harbingers of postmodernism. In some sense the New Age movement is a kind of superficial expression of postmodernism.

The most positive thing that can be said about postmodernism is that it takes seriously the claim that all of one's life experiences are colored by one's worldview or perspective. There is no neutral area of life that we can retreat to and where we can all agree. Even Joe Friday's *facts* are shaped by a perspective. Even our simplest mathematical calculations are the product of a particular worldview. Of course, reformational Christians have been saying that, at least, since Abraham Kuyper. And that has made it a little easier for us to communicate with our secular academic contemporaries.

But there is a negative side to postmodernism as well. For many postmodernists, not only can we not know the reality around us except through the system of our worldview, but it is often believed that *there **is no reality** except* that which we construct from out of our worldview. Thus, for example, everyone's point of view must be as valid as everyone else's. And anathema upon anyone who tries to impose his perspective on anyone else, or, worse yet, suggests that his perspective might be the truth.

Still, the academic climate conditioned by postmodernism is exciting. It's fun to sit and discuss these issues with fellow Christians, and sometimes even with non-Christians – for at least now they are willing to listen. Do you hear the violin playing?

The trouble with all this is that postmodernism, while certainly driving much of our culture, is not driving all of it. Its most obvious influences are in the humanities and social science divisions of major universities, among the artists that produce such cultural expressions as films, books, music, and television, and certainly among those members of the media who daily tell us about ourselves: the journalists and writers for major newspapers, magazines, and TV and radio news broadcasting. Perhaps you may ask, "Well, who is left?" Well, I will tell you who is left. The majority of scientists, business owners, engineers, financiers, and military leaders know nothing about postmodernism. And the few that do are largely unaffected by it. These are the power brokers, the people in our culture for whom there *are* absolutes — absolutes such as the scientific method, the law of supply and demand, the drive to rise to the top of one's field, the almighty buck, faith that modern technology can solve all our problems, and the belief that democracy must be preserved — even if that preservation requires the destruction of other human beings. Is the violin still playing?

My dissonance dream today is that the masses in our wealthy American culture will be influenced by postmodernism into *tolerating* these power brokers. Thus economism, scientism, hedonism, and militarism will not be unmasked and seen as the evil spirits of our age that they are. And the poor will get poorer, the hungry will not be fed, the widow and the orphan will suffer injustice. And the masses will be kept content in their postmodern world by the mind-manipulating products of high technology — advanced versions of Nintendo, and lots and lots of choice in TV programming. And the Word of the Lord that came to the prophet Hosea will come to our land . . .

> For the Lord has a case against the inhabitants of the land,
> Because there is no faithfulness or kindness
> Or knowledge of God in the land.
> *There is* swearing, deception, murder, stealing, and adultery.
> They employ violence, so that bloodshed follows bloodshed.
> Therefore the land mourns,
> And everyone who lives in it languishes
> Along with the beasts of the field and the birds of the sky;
> And also the fish of the sea disappear. (Hosea 4:1–3 [NASB])

All the while the postmodern violinist plays on . . . and I feel very much in need of a glass of hot milk.

Chapter 20

The Superficial Healing of Brokenness

July 5, 1997

A short time ago I was watching the evening news on television and learned that the famous actor Jimmy Stewart had died. It was a bit of a coincidence, for that morning I had just finished reading William Romanowski's book *Pop Culture Wars* (Downers Grove: InterVarsity, 1996) in which Stewart's name had, on numerous occasions, arisen. From childhood, Jimmy Stewart had always been one of my favorite actors. One of my fondest memories is the time when, as very young children, my father took my brother and me to see the Jimmy Stewart western *The Man from Laramie*. And I must confess that, less than a year ago, I could not resist the urge to purchase the video of a Jimmy Stewart western that impressed me during my early adolescence – *The Man Who Shot Liberty Valence*. Stewart's characters were almost always gentle but strong, flawed in their finitude, but noble in their integrity, honesty, and fidelity. For a young kid growing up in the fifties – with very little capacity for critical thinking – his characters resonated with the picture of human goodness that I was learning in Sunday School.

Romanowski, however, provides another and different insight into Stewart's movies and the characters he portrayed. He argues that films such as *It's a Wonderful Life* and *Mr. Smith Goes to Washington* embody the myth of the American civic faith and, as such, militate against a Christian worldview. It's true that those films embody certain values having to do with personal morality that are congruent with biblical values, but they also embody the unbiblical values of individualism, human autonomy, and materialism. And while correctly pointing out some individual sins, the films leave untouched the more serious communal sins that have come to characterize modern America. In that sense it may be said that while many Jimmy Stewart films attempt to bring healing to some of the brokenness of American society, they do so in a superficial manner, leaving the really serious sins – such as faith in the false gods of democratism and capitalism – uncovered at best, and more often than not lending them support. Thus the films represent a *superficial healing of brokenness*.

This brings me to the heart of this essay. You see, in addition to watching the nightly news and finishing Romanowski's book, I also happened, on that particular day, to be reading from the prophet Jeremiah. And I was struck by a point made by Jeremiah in Chapter 6 and again in Chapter 8. The Lord is angry with his people Israel because they are living apart from his Word even while they pay lip-service to it. The Lord is particularly angry with the leaders of the people – the prophets and priests – who preach a kind of superficial "niceness and peacefulness" morality even while they practice greed and infidelity. This is what we read in Jeremiah chapter 6:

> From the least to the greatest all are greedy for gain; prophets and priests alike, all practice deceit. They dress the wound of my people as though it were not serious. "Peace, peace," they say, when there is no peace.

It strikes me that it is not only in Jimmy Stewart films that we are guilty of this same kind of falseness today. Among Christians there seems to be an unwritten rule that we always speak positively, never criticize, continually smile, and avoid discord at all costs. Whether in church, school, or on the street, we have developed pleasant little customs – cheerful but trite words and actions – that coat our interactions with a veneer of goodness and harmony and protect against serious probings that may reveal fundamental flaws.

More serious examples of this superficial healing of brokenness can be found in our Christian institutions. Recently I attended a conference celebrating Christian business enterprise. I was astonished, while attending one workshop, to hear a Christian business owner describing the purpose of his business as making money in order to support missionaries. Now I am all for supporting mission programs. But this ordinarily noble activity was rendered superficial by virtue of the owner's never raising questions regarding what kinds of product or service the business was providing or how it was providing it. In fact, it became clear that the products and management of the business differed in no way from those of the surrounding, secular culture. And the owner wondered why his employees, many of whom were not even Christians, could not get enthused about his goal of supporting missionaries.

There is one other example of this modern expression of the superficial healing of brokenness that I cannot end without mentioning. It is our system of Christian schools. Standing in the tradition of Abraham Kuyper, I am an ardent supporter of the Christian school movement. But if our schools merely add a veneer of personal morality to a basically secular education, then I fear that the words of the prophet Jeremiah become all too appropriate. Without a radically biblical understanding of history, natural science, music, literature, mathematics, and business, our tuition buys us only a superficial sense of goodness and peace, when in fact there is no genuine goodness and peace.

The next time you watch *It's a Wonderful Life*, or any similar "feel-good" movie, consider the words of the prophet Jeremiah. For the only truly wonderful life is the one that is Christ-centered from beginning to end. The alternative is a merely superficial healing of brokenness.

CHAPTER 21

REPEALING THE 2ND AMENDMENT

JANUARY 6, 1992

The collapse of the Soviet Union and the general impotence of Communism as a world military threat have profound implications for the way in which the United States ought to look at itself and at the future. Even President Bush has recently admitted that a substantial decrease in the military budget is in order. But we can learn more from these historic events than how to spend our national income. The fact that the world is filled with nations having relatively young systems of government indicates that civic permanence is an elusive quality. Rather than resting secure in the decreased threat of external aggression, the United States ought to consider that its system of government is almost three times as old as the one established by Lenin in 1917, and might therefore be ripe for a similar demise resulting from inner decay.

What is it that guarantees national permanence? Ask that question to a politician or to an average citizen and you will get a number of answers. Some will say democracy. Others will tell you it is our economic system of free enterprise. Still others will point to the Constitution, in a way similar to the way some Christians point to the Bible, and see in it the foundation for civic permanence.

But the problem with putting your faith in political or economic systems is that you have history working against you. After all, Lenin truly believed in the economics and politics of communism. Czar Nicholas believed, Napoleon believed, Hitler believed, and the leaders of the Roman Empire believed. So what's to make belief in democracy, free enterprise, or the American Constitution any less capricious? I'm reminded of some lines from the poet William Habington. Written in the 1600's and contemplating the stars as they look down upon the political leaders of his day, it goes something like this:

> For as yourselves your empires fall,
> And every kingdom hath a grave.
> Thus those celestial fires,
> Though seeming mute,

The fallacy of our desires
And all the pride of life confute.
    For they have watch'd since first
      The World had birth;
And found sin in itself accurst,
And nothing permanent on earth.*

Democracy does not guarantee permanence. Neither does the Constitution. And certainly the free enterprise system does not. If these contribute to civic permanence it is because they, in some way, rest on the solid foundation of God's Word for his creation. When norms like justice, mercy, and faithfulness undergird the kind of democracy that characterizes our system of government, then it may be said that democracy contributes to civic permanence. But notice, it is that undergirding normative foundation resting on the bedrock of God's Word that provides the stability, not the concept of democracy in the abstract.

One of the most critical norms that we citizens of the Kingdom must recognize as vital to our national citizenship is that of ongoing reformation. Rooted in the cultural mandate, it requires that we not rest satisfied with the fruits of our labor – political or otherwise – but always seek to examine where we are, examine how we are living in light of God's Word, and then make changes that re-form our living so that it is more in conformity to God's Will.

In keeping with that norm, I want to suggest that one of the best ways we can respond to the changing world situation is to amend the Constitution by repealing Article II of the Bill of Rights. Article II, you may recall, is the one that the NRA, that group of ethical Neanderthals, holds near and dear to its heart. Article II of the Bill of Rights, sometimes referred to as "the second amendment to the Constitution," supposedly guarantees individual citizens the right to own guns. I say "supposedly" because I'm not convinced that it really does. You see, the founding fathers were much more concerned with protecting *the people* (that's *people* in a communal sense) from the potential despotism of a central government, like the one they witnessed in the form of King George of England. Do note the way the article is worded:

> A well regulated Militia, being necessary to the security of a free State, the right of the people to keep and bear Arms, shall not be infringed.**

---

*   William Habington, "Nox Nocti Indicat Scientiam," from *English Poetry, Vol. I*, The Harvard Classics (New York: Collier, 1909).

** Zechariah Chafee, *Documents on Fundamental Human Rights* (New York: Atheneum, 1963), 72.

The primary purpose of the article is to provide for the common defense against outside aggression, and I might add, in the context of the society of 1789. It, no doubt, also was intended to insure that individual states would have the right to organize state militias, probably with a mind to checking the power of the federal government.

Since 1789, however, our nation has moved away from a more communal appreciation of the phrase "the people" and has increasingly embraced a radical individualism. But it has never outgrown its affinity for using violent means to end disputes. The "Gunfight at the OK Corral" remains the model for settling grievances. And so every man must have his guns.

But this is 1992 and this is the United States of America. We learned a long time ago that we can change the system, even when we are oppressed by it, without resorting to violent revolution. In fact, that is one of the greatest strengths of the American socio-political system. Unlike the Russian revolution of 1917, the recent struggles in Latin America, or the constant political strife in Africa, we have been able to make changes in this country without guns. Women's suffrage, the civil rights movement, the cultural revolution of the sixties, and Watergate all demonstrate the superiority of moral fervor and passive resistance as over against violence. And you have to be awfully self-deluded to believe that carrying a gun is going to protect you from crime. The presence of guns in this society is one of the primary reasons why we are in last place among the world's civilized nations when it comes to violent crime.

So let's repeal the 2nd Amendment to the Constitution. Let's teach the hunters among us the pleasures of reading good novels. Let's put the gun manufacturers and gun salesmen out of business, or in jail. And let's put Doc Holliday and Wyatt Earp into that compartment of our fading memory where we store other national embarrassments like the Dred Scott case and the war in Vietnam. By doing so we will have reformed not only the Constitution, but our national lifestyle. And because we will have acted justly, we will have increased the probability that the United States will exist to witness its 300th anniversary.

Chapter 22

War

February 4, 1991

My father was in the army during the Second World War and spent a large part of his time in the desert area of North Africa. When I was a boy, growing up during the early fifties, I learned about war and soldiering much the same way every other American boy did: by watching John Wayne and Audie Murphy movies on television. My father rarely, if ever, talked about his experiences as a soldier. He had a collection of "war memorabilia" – some German medals, a knife with a swastika on the handle (which I imagined that he took from the body of a dead Nazi – although he never said so), and his own bayonet knife: a huge, dull thing, so heavy that I couldn't imagine how it could be used. He kept this stuff in the bottom draw of his workbench, in the basement of our small Cape Cod house. He also had a book, mostly a compilation of war cartoons, that could be found buried among the issues of *Readers Digest* on one of the living room end tables. I probably leafed through it half a dozen times during my boyhood, never reading anything more than the captions beneath the cartoons. Sometime between my boyhood and teenage years it disappeared from that end table. I never missed it, and have never even thought about it until just now.

My mother told me more about my father's experience as a soldier than he did. Like any good wife and mother she wanted her four boys to respect their father and so she didn't dissuade us from imagining that he had been a valiant soldier, a hero, maybe a quiet and gentle version of Audie Murphy. Of course we were told that war was not a pleasant thing, and were led to believe that just as women had to bear the pain of childbirth, men had to bear the pain of being a soldier. We knew that our father had been awarded at least one, maybe two purple hearts – indicating that he had been wounded in action. But we never learned the details. My father had a sizable scar on the side of his neck. My brothers and I used to imagine that it came from a Nazi bayonet that he had wrestled from the enemy just in time. When asking him about that, he would just smile. I believe he once confessed that it was nothing more than the remnants of his encounter with a less than skillful doctor, when, as a boy, he had a cyst removed. We preferred the Nazi bayonet story.

One real injury that he carried with him the rest of his life was deafness in one ear. Apparently – and we never learned the details – he was in the close vicinity of a bomb or grenade when it exploded; close enough to damage his eardrum.

And finally there was that other injury that, as kids, my brothers and I could never quite understand. During our childhood our father had twice been taken to the hospital – the veterans' hospital in East Orange, New Jersey – because of a bleeding ulcer. It was a fairly serious ailment and each time he was hospitalized for what my memory tells me was a week or more. My mother told us that he developed that ulcer in the army. But that just didn't make sense to us. We had seen soldiers on TV getting shot, getting blown up, and occasionally even getting gassed. But no soldier on TV ever got an ulcer.

So I spent my childhood like most American boys: playing with guns, and thinking that maybe someday, like my father, I would be in the army – even though I knew my parent's hope was that there would be no war when their four sons came of age.

Dennis Matzkanin was a classmate of mine in high school. He had a different set of experiences that taught him about war and soldiers. His father was younger than mine and apparently had served in the army during the Korean conflict. His father was unlike mine in another way. He openly talked about his war experiences. In fact, one day in 1962, Dennis brought a stack of photographs to school. They were pictures that his father had brought home from Korea. I'll never forget how I felt when, after three or four other classmates of mine viewed them, they passed them to me. We were in biology class. My seat was next to the windows. As I looked at those pictures I saw horror for the first time in my life. The photos showed the maimed and mutilated bodies of dead men and women, the mutilations making it obvious to even my innocent mind that these people had been tortured before they died. The realization hit me like a sledge hammer – that I was not looking at TV actors or comic book animation. These were photos of real people – people who had once been alive like me, but now were dead – people whose death must have come as a relief from unimaginable torment. My fifteen-year-old psyche was not up to dealing with that. I handed the pictures back to Dennis and tried to calm down my stomach, which was feeling sick in a way that I had never experienced before. But I made it through that class, through that day, and a week later you would never know that anything had happened. Yet something deep inside me had changed that day.

In 1964 I went away to college. Since I was the oldest son, and the first to leave home for more than one night, the three hour drive to Troy, New York turned into a family event. Arriving on campus I learned that freshmen had to make a choice right away as to whether we would enroll in physical education or ROTC [Reserve Officers' Training Corps]. It was required that we choose one or the other. In the hours remaining before the family left for home my father gave me his advice. He urged me to sign up for ROTC – *Army* ROTC naturally. He reasoned that regardless of whether or not there was a war, I would be drafted after college and have to serve two years anyway. So why not do it as a 2nd Lieutenant. His own experience as a private in the army told him that officers had it much easier. Well, I wasn't very excited about taking gym class in college, so I took my father's advice instead: I signed up for army ROTC.

The first semester wasn't bad at all. Twice a week, at 8:00 in the morning, we had class at the local armory. Class amounted to indoor target practice with 22 caliber rifles. It was fun – just like when my friend Alan Carlson and I used to play with his BB-gun. Later in the semester they let us shoot 22-calibre pistols. That was harder, but still fun. What I didn't like about ROTC was the marching drills every Tuesday afternoon. Dressed in heavy wool uniforms, we would have to stand in formation and then march around the parking lot for three hours. But I was picked to carry the American flag (someone must have liked my all-American-boy looks), so at least in the beginning I felt pretty good about it.

During the second semester things began to change. But then a lot of things changed for me during that first year in college. For one thing I discovered that I wasn't as smart or as well-educated as I thought I was. I struggled with my classes. I began to see myself and my classmates as so many puppets going through motions, doing what we were *supposed to do*. The words of Macbeth, which we read in freshman English, haunted me:

> Life's but a walking shadow, a poor player,
> That struts and frets his hour upon the stage,
> And then is heard no more. It is a tale
> Told by an idiot, full of sound and fury,
> Signifying nothing. (Act V, Scene V)

But I had been brought up as a Christian. And when Shakespeare could give me no solace, I turned for the first time in my life to seriously read the Scriptures. Late at night in my dormitory room the light in my heart began to go on as I read the Sermon on the Mount and the book of Ecclesiastes. Ecclesiastes taught me that Macbeth was on to something. Life "is a tale, told by an idiot, full of sound and fury, signifying nothing" if you anchor your hope in this world. "Meaningless, meaningless" said the writer of Ecclesiastes, "All of it is meaningless, a chasing after the wind." Turning to the Sermon on the Mount I heard the Lord say "Seek first the Kingdom of God and righteousness, and all these things will be given to you as well."

But the Lord said many other things as well in that Sermon on the Mount. Among them were these revolutionary thoughts:

> You have heard that it was said, "Eye for eye, and tooth for tooth." But I tell you, do not resist an evil person. If anyone slaps you on the right cheek, turn to them the other also. . . .
>
> You have heard that it was said, "Love your neighbor and hate your enemy." But I tell you, love your enemies and pray for those who persecute you, that you may be children of your Father in heaven. (Matthew 5:38–39, 43–45)

That began to raise questions in my mind regarding what I was doing in ROTC. Then two events occurred that helped me answer those questions. During one ROTC lecture, we were introduced to a high ranking officer. He was young, tough-looking, and had just come back from fighting in Vietnam. He lectured us for about 30 minutes. One thing he said has stuck with me to this day. Recalling his experiences in Vietnam, he told us that when in combat, you have no time to make judgments. You simply act. He told us that we had to be trained to survive. And survival means – and this is the part I distinctly remember – survival means that if a women or a child approaches you that you think may be the enemy, you blow them away with you M-16. No time for sentiment, no time for judgment; it's either you or them, kill or be killed. That seemed horribly incompatible with the teachings of Christ.

Then one Saturday we graduated from 22's to M-1's. The M-1 is the rifle used in the Second World War and the Korean War. We learned to take it apart, clean it, put it back together again, and finally, to shoot it. You can't have indoor target practice with an M-1. It's simply too powerful. So out in a large field we had our first experience at shooting it. With the first shot I thought my arm would fall off at the shoulder. I couldn't imagine how it could be very accurate since it kicked so hard when you fired it. But the worst thing about it was the deafening sound. It was the loudest thing I ever heard. That night when I went back to my dormitory, my ears continued to ring like they had never rang before. It was hours before the ringing subsided. And with that ring there arose, in the pit of my stomach, the sensation that I had not felt since Dennis Matzkanin had showed me those gruesome pictures back in high school. I was sick.

But it wasn't just a physical sickness.

The last thing I remember of ROTC was marching in the end-of-the-year parade. Some gung ho upperclassman marching behind me kept yelling at me that my rifle wasn't straight. I took a perverse delight in responding by making it all the more crooked, and then deliberately marching out of step, actually skipping, glorying in the knowledge that this was the last time I would have anything to do with the military.

Three years ago my father died after struggling for six months with cancer. When helping my mother go through his papers I discovered his army discharge. I was surprised to learn that he had been discharged on medical grounds. It was then I learned for sure that my father was not anything like even a mild-mannered John Wayne or Audie Murphy. No, instead he was a sensitive image bearer of God who did what he thought he should do in 1942, but then recoiled in horror at the barbarity of the situation in which he found himself. He grew up in a generation that looked down on that kind of sensitivity as a form of weakness; and that was probably one of the reasons why he never told my brothers and me much about his war experiences. But what others might have called weakness I see as a kind of strength – and indicator of a residual humanity that holds fast, even when our circumstances and neighbors turn to savagery. And I'm thankful to have inherited at least a small portion of it.

Earlier on that same day my three brothers and I were seated in the office of the local funeral director where we were making the arrangements for our father's memorial service. The funeral director was trying to tell us that since my father was a veteran he was eligible for certain "honors" in connection with his burial – something about a flag, and some other things. Neither my brothers nor I could imagine that an association with the military or even the government could be honoring to him at that point. For the past twenty years he had ardently insisted that his first and primary citizenship was in heaven. Now that had become his only citizenship.

Some people when they contemplate the current war think of all the young soldiers who are doing what our culture has brought them up to believe is the right thing to do, and who will suffer terribly for it. I think of my father; and of those words spoken by the prophet Isaiah which promise that one day we will see things more clearly:

> He will judge between the nations and will settle disputes for many peoples. They will beat their swords into plowshares and their spears into pruning hooks. Nation will not take up sword against nation, nor will they train for war anymore. (Isaiah 2:4)

CHAPTER 23

SUING GRANDMA

APRIL 1, 1999

We live in a litigious society. The relationships between members of our society are increasingly being characterized by the lawsuit. The promise made between two mutually respectful image bearers of God has been replaced by the contract signed by two adversaries. It's not uncommon to read in the newspaper about children suing their parents. And growing numbers of couples are flocking to lawyers to have pre-nuptial agreements drawn up before they are married.

What's behind all this litigation language? Is it affecting the Christian community as well as the culture at large? And what do we do about it? Those are the questions I will try to answer in this essay.

The roots of the problem, I believe, can be traced to two spirits that pervade North American culture. They are the spirit of individualism and the spirit of consumerism. Individualism has been around since the time of the Enlightenment. It expresses the belief that the individual person is the source of all meaning, and that the highest good toward which any society ought to strive is the absolute freedom of the individual citizen. Consumerism, on the other hand, is relatively new. But it is closely related to the spirit of social Darwinism that was formulated in the late nineteenth century. Social Darwinism built on Charles Darwin's theory of natural selection and declared that people and societies, like animals and plants, compete for survival and, by extension, success in life. It has been used to support various expressions of imperialism, racism, and capitalism. Consumerism is a form of social Darwinism that includes the belief that success in life is to be measured by the number of material possessions acquired by the individual.

When these spirits prevail, relationships between people and groups of people become adversarial. Actions are grounded in attempts to gain advantage over other individuals and groups so as to maximize material gain, usually in the form of money. Adversarial relationships are the polar opposite of trust relationships. For genuine trust to exist between two parties there must be a mutually perceived prior commitment to selfless service. This is part of what Jesus meant when he told his disciples "Anyone who wants to be first must be the very last, and the servant of all" (Mark 9:35).

"Being first" might here be construed as the achievement of right relationship with one's neighbor and with God. On the other hand, seeking advantage for oneself results in distrust, in broken relationships.

It should be a surprise to no one that adversarial relationships and litigation language have permeated our Christian communities. That's inevitable, of course, given our fallen world and our inclination to sin. But it is not just our selfishness and greed that breeds such evidences of corruption. Our proclivity to abdicate our calling as salt and light and to be influenced by those in positions of prominence is a slightly less malevolent but perhaps more insidious cause for our joining the litigation choir. As we Christians find ourselves in positions of increased wealth and power, we want to protect that wealth and power. In part that is defensible on the basis of stewardship norms. But when looking for instruments of defense we have turned to the rich and powerful in our culture and learned the language of litigation. So we hire lawyers.

Now there is a place for lawyers and litigation among Christians. After all, Paul appealed to Caesar on the basis of his Roman citizenship when confronted with the threat of injustice. And that is why we need more good Christian lawyers – to fight injustice and to promote righteousness with the proper instruments of the law. But when we Christians hire lawyers, too often it's because we have been corrupted by the spirits of individualism and consumerism. The result, therefore, is not justice and righteousness, but distrust.

Consider the following example. Almost thirty years ago, at a Christian high school in a city far, far away, I began my teaching career. One of the first tasks that occupied us communally as a faculty was that of writing a student handbook. The first draft of that handbook was influenced heavily by advice given by a lawyer who served on the Board of Trustees. As a result it had about it the aroma of litigation language. It spoke of rights. Do's and don'ts were listed clearly. Its style was tight and precise. Thankfully, a local pastor, the parent of a freshman student, sniffed out what was going on and objected. He volunteered to rewrite the handbook and to make a presentation at a faculty meeting. At that meeting he discussed with us the difference between covenant language and litigation language. He convinced us that while litigation language was sometimes required in our dealings with the state, the documents that we used for communication with each other ought to convey mutual trust and shalom. To do that, we needed to actively resist the status quo formalities of litigation language.

In closing, I want to encourage all of us who claim to be disciples of Christ to become more sensitive to the spirits of individualism and consumerism, and to the adversarial and litigation language that we too often adopt when communicating with each other. Whether Christian institutions or Christian individuals, we need to nurture trust – not suspicion – among ourselves. That may mean becoming more vulnerable. Consider the words of Paul to the Corinthians. Regarding Christians suing other Christians, he writes:

> . . . one brother takes another to court – and this in front of unbelievers! The very fact that you have lawsuits among you means you have been completely defeated already. Why not rather be wronged? Why not rather be cheated? (1 Corinthians 6:6–7)

Of course that way of thinking is totally foreign to those who see their neighbors only as adversaries with whom they must compete in climbing the ladder of success. But Christians ought to be able to say with the psalmist, "Some trust in chariots and some in horses, but we trust in the name of the LORD our God" (Psalm 20:7). Or by extension to today: "Some trust in courts, and some in litigation language, but we trust in the name of the LORD, and in his shalom, which evidences itself in selfless servanthood and mutual trust."

CHAPTER 24

THE COVENANT AND HISTORICAL CONTINUITY

JUNE 2, 1999

In a sermon on the book of Daniel that I heard recently, the preacher explained the reason for King Nebuchadnezzar giving Daniel and his friends new names. He made a parallel with the renaming of Russian cities by Communist revolutionaries in the early part of the twentieth century. It seems that by renaming persons, or things that are important in the lives of persons, you can more easily influence the worldview – even the identity – of those persons. Thus, by renaming Daniel, Hananiah, Mishael, and Azariah as Belteshazzar, Shadrach, Meshach, and Abednego, King Nebuchadnezzar was trying to change them from faithful Israelites to faithful Babylonians. The renaming would enable them to more easily forget their past and melt into the surrounding Babylonian culture. Likewise, by renaming Volgograd and St. Petersburg as Stalingrad and Leningrad, the Communists sought to cut off the Russian people from their Czarist past and enable them to more easily adopt the Communist worldview.

This point struck me because it seems to me that we can find evidence of it in the Christian community into which I have been adopted. I grew up in the 1950s as a kind of rootless, middle class American, in the suburbs outside of New York City. My mother was a nominal Protestant, my father a Catholic, and the identities of my grandparents, great-grandparents, or other ancestors were essentially unknown and irrelevant to me. My worldview was shaped more by the Brooklyn Dodgers, Davy Crockett, and Hopalong Cassidy than by the ethnic or cultic origins of my relatives. But by God's grace I came under the influence of Reformed preaching as a child. Then, at just the right time in my adolescent development, I was befriended by a radical, counter-cultural Dutchman, who made me aware of the Kuyperian, Dutch neo-Calvinist worldview and community. That was almost 35 years ago. Since that time I've come to realize that the Lord works his covenant faithfulness through generations of people in community with each other, people bound together by a common desire to live obediently before his face.

That's nothing new, of course. One cannot read the Scriptures without being impressed how God worked through his people Israel and how he works through the Church, the body of Jesus Christ in this world. In the Scriptures the identities of those covenant communities are characterized by particular cultural patterns and by persons who stand out as representatives. Abraham, Moses, Elijah, and Daniel in the old covenant, John the Baptist, Peter, and Paul in the new covenant, are not famous individuals in their own right, but represent God's people. As representatives they are a source of inspiration and education. That is, we recall them to our contemporary minds to marvel at God's covenant faithfulness and to learn how we ought to live today.

Now, I believe it is precisely this fact, that God works through generations of *covenant communities*, represented by cultural patterns and particular persons, that twenty-first century Dutch Calvinist Christians are in danger of ignoring. I want to give you three examples of how this is happening – two having to do with cultural patterns and one having to do with particular persons. But first, consider that this practice of influencing a people by cutting them off from their cultural roots is the consequence of a *zeitgeist*, that is, a spirit of an age that seeks absolute allegiance. With King Nebuchadnezzar it was the spirit of Babylon. With the Russian revolutionaries of 1914 it was the spirit of communism. Today there is a different spirit at work. Perhaps you will recognize it before I am done with my three examples.

First consider the nature of the family. The family is a cultural pattern. The way it is manifest in twenty-first century North America is certainly different from the form it took at the time of Abraham, Isaac, and Jacob. And there are even differences between the family of the 1950s and the family of the 1990s. But there is an underlying creational structure that governs and limits the variations that are possible. The mythical nuclear family of the 1950s is just one cultural pattern – with both strengths and weaknesses. Undergirding every normative expression of the family, however, is the norm of faithfulness – covenant faithfulness. And that norm is not one that can be realized by individuals. It takes a community, the minimum expression of which is a husband, wife, and child. Today, however, we find the family as a cultural pattern – whatever its form – is in disarray. We are a people who have lost our grip on the norm of faithfulness. Thus we find our culture more and more characterized by single parent households. And that is an aberration. But in our reluctance to be critical, in our desire to be inclusive, we are almost to the point of accepting the aberration as the norm. This point struck me

a little over a year ago when I attended a baptism service for three covenant children. One of the three was carried to the front of the church
by only his mother. Now it is true that we in the congregation took part
in the ceremony, promising to do all in our power to help raise that
child to be a faithful servant of the Lord. But still, I felt something was
wrong. A mother and her infant do not constitute a family and are thus
an inadequate expression of the covenant. I think it would have been far
better if the grandparents had joined the mother and infant around the
baptismal fount.

My second example of our ignoring how God works through covenant communities is on the lighter side – although it is creating great
tensions in many congregations. I'm thinking particularly about the style
in which we worship the Lord on Sunday. More narrowly, I'm thinking
about the music that we use in our worship services. The tension exists
between those who prefer what is called a "contemporary" style and those
who are more comfortable with the hymns that they learned as children.
The problem is that there are two norms operating here. One is that of
growth and development. Just as individuals are called to grow in wisdom, so too communities ought to develop new ways to express themselves. So it is good for a church community to find new ways of communally praising the Lord, including new musical forms. But there is also
the norm of historical continuity, recognizing how the Lord has faithfully
led a community in the past to bring it to the point where it is in the present. That norm requires us to preserve the richest cultural expressions of
our past so that future generations might learn from them. With respect
to music, this means that we ought to be learning new musical ways of
worshipping while at the same time retaining the best of the older ways.
But the tendency today is to be embarrassed about the hymns of the past
and to assume that they will turn off the younger generations. Well, there
probably are some old hymns that we ought to be embarrassed about.
But there are also those that we need to preserve. Think, for example, of
the hymn *A Mighty Fortress is Our God*. With words written by Martin
Luther and music composed by Johann Sebastian Bach, it is a cultural
expression that serves both as a vehicle for praise and for teaching.

My final example concerns the manner in which some Dutch
Calvinists may be losing contact with their unique Christian world-
view because of its association with what they refer to as dead, white
Dutchmen. When I graduated from college thirty-one years ago, not
only did I read the Bible cover-to-cover for the first time, I also read John
Calvin's *Institutes of the Christian Religion*, Abraham Kuyper's *Lectures on*

Calvinism, and Herman Dooyeweerd's *In the Twilight of Western Thought*. What I found there was an intellectual gold mine that shaped my worldview from one that was fragmented and dualistic to one that would enable me to carry out my life's work with some degree of integrity. As an engineering teacher, I find the intellectual tradition of Dutch neo-Calvinism to be the best possible tool for enabling my students to develop their engineering abilities as an expression of their biblical faith. But recently I've been told that "that's not where today's students are at" and that we are going to turn them off by our references to those who are pejoratively called "dead, white Dutchmen." I can respond positively to some of the motives behind this advice: a desire to be inclusive, to respect other cultures, and to recognize that increasingly more students are coming to us from outside the Dutch Calvinist tradition. But despite the good motives, I see the advice as wrong-headed. I see it as another example of historical discontinuity, the forsaking of our cultural and faith identity.

So what is the deeper motive – assuming there is just one – behind the drive to distance ourselves from historic expressions of the family, reformation hymns, and reformational leaders who are dead, white, and Dutch? I believe it's a form of postmodernism. It's an embracing of inclusiveness with such absoluteness that we become embarrassed about the unique qualities of our own cultural heritage. Behind it all, of course, is the spirit of conformity to the pagan gods of our age. Like Nebuchadnezzar renaming Daniel and his friends or like the Communists renaming cities in Russia, the pagan gods of twenty-first century North America cannot tolerate the quality of "called-out-people-ness" with which the Lord has blessed Dutch Calvinism. But to conform to inclusi*vism* by denying their cultural heritage is for Dutch Calvinists to lose the richness of a tradition blessed by God. Worse, it is to deny to people like myself – outsiders – the opportunity for escaping rootlessness and finding integrity and true meaning in their worldview.

CHAPTER 25

DISCIPLINE, TRAINING AND EXERCISE

DECEMBER 1, 1999

The end of the first semester is a rough time for freshman engineering students. Most of them, used to getting "A's" throughout high school without having to work very hard, are close to completing the most arduous academic period of their lives and are apprehensive over the prospect of not quite making a "B" average – a prospect that puts scholarship retention in jeopardy. I must confess that I have an uneasy sympathy for these students – partly because I am one of those who are "cracking the whip," so to speak. But also because I've been there myself. Oh, there are a few students who are clearly more interested in their newly experienced social freedoms than in their academic work. And I see them as getting just what they need and deserve: a swift kick in the pants. But the larger number are serious, conscientious, and bright, Christian students who, for the first time in their lives, are feeling intimidated, uncertain, somewhat wretched, and possibly even humiliated. I know. I've been there.

So why do we allow this torment to continue? Simple: it's a necessary part of life. Call it discipline. Call it training. Call it exercise. Whatever your call it, it's a process in which every image bearer of God has to be engaged if he or she is going to be an effective servant. You see, we come into this world and survive only because others, primarily our parents, serve us. We grow from being completely helpless and self-centered to being unique and effective servants of others only by discipline, by training, by exercise. Anyone who has been successful at a sport knows this. The first time you try running the 400 meters, for example, you lose, and you feel a bit humiliated. But if you've got some ability, some vision for being who the Lord has blessed you with being, and a coach who is hard-nosed enough to push you even when it's obvious that you're hurting, well, then you will be in training the next day, running 600 meters. Early in the track season your lungs will feel like they're on fire after each 600-meter run. Your legs will ache and you will hardly be able to walk straight the day after a good workout. But eventually you will get into shape, and, with the wind rushing past your face and the track smoothly passing beneath your flying feet, you will be able to join with that 1924 Olympic champion Eric Liddell, and say, "I feel the Lord's pleasure."

But this essay is not about sports or academics. It's about the faith that gives meaning to those and to every other human activity. And my point is simply that to be an authentic Christian you need to exercise your faith. Discipline, training, and exercise: these are as necessary to the Christian as they are to the athlete or the scholar. And thus, along with hard-nosed coaches and teachers, each one of us needs to have faith leaders – those who are in authority over us with respect to our role in the Kingdom of God – who are willing and able to subject us to discipline, to train us, and to motivate us to exercise. Paul was one of those faith leaders. Consider the following words of encouragement – hard-nosed words of encouragement – that Paul offers Timothy:

> Have nothing to do with godless myths and old wives' tales; rather, train yourself to be godly. For physical training is of some value, but godliness has value for all things, holding promise for both the present life and the life to come. This is a trustworthy saying that deserves full acceptance. That is why we labor and strive, because we have put our hope in the living God, who is the Savior of all people, and especially of those who believe. Command and teach these things. Don't let anyone look down on you because you are young, but set an example for the believers in speech, in conduct, in love, in faith and in purity. Until I come, devote yourself to the public reading of Scripture, to preaching and to teaching. Do not neglect your gift, which was given you through prophecy when the body of elders laid their hands on you. Be diligent in these matters; give yourself wholly to them, so that everyone may see your progress. Watch your life and doctrine closely. Persevere in them, because if you do, you will save both yourself and your hearers. (1 Timothy 4:7–16)

Paul was obviously concerned for discipline, for training, and for exercising the faith of those over whom he was placed in authority. But what about today? Where are the leaders in the Church who insure that the body of Christ is disciplined, trained, and is exercising its faith? My sense is that they are very few and far between. It seems that many church leaders are reticent or fearful of imposing anything that may stress, intimidate, or shame those for whom they are responsible. For example, it used to be that before young Christians would be allowed to publicly profess their faith they would be required to meet with the church elders – all the church elders – at a regular elders meeting, for the purpose of being examined. Today it is more common for two or three elders to meet with a young person in that young person's home. The rationale is that it is less intimidating. I'm sure it is. I'm also sure that the young people who are "examined" this way miss out on a critical experience in the Christian life. It's the experience of being confronted by those in authority, of being forced to demonstrate your own authenticity as a Christian. Sure it's a bit intimidating. But with that intimidation comes, in part, a motivation for succeeding. And when the experience is over, one has the lasting satisfaction of having been tested and shown worthy.

But today we have become so afraid of either intimidating or turning off our brothers and sisters in Christ – particularly our young brothers and sisters in Christ – that we have become spiritually wimpish. We thus promote wishy-washiness in the church.

If we are wishy-washy about exercising our bodies, we will grow physically flabby and die young. If we are wishy-washy about exercising our minds, we will fall prey to the exploitative, brainwashing techniques of those that manage our culture. We become puppets, our strings yanked here and there by every television commercial. If we are wishy-washy about our faith, we become "lukewarm," worthy of nothing better than to lie at the bottom of the Lord's spittoon. But if we are wimpy leaders in the Church who are responsible for encouraging wishy-washiness in others, then we become candidates for having millstones tied round our necks and for being cast into the depths of the sea.

CHAPTER 26

PURSUING HAPPINESS

DECEMBER 4, 1997

> We hold these truths to be self-evident – that all men are created equal;
> that they are endowed by their creator with certain unalienable rights; that
> among these are life, liberty, and the pursuit of happiness.

So reads the first sentence in the second paragraph of the American
Declaration of Independence. It was written by Thomas Jefferson with
some editorial help from Benjamin Franklin. Jefferson was a deist, and
his understanding of "unalienable rights" was shaped by his deistic worldview. That worldview considers the universe to have been created by a
god who, after creating it, left it running like one big machine made up
of innumerable individual parts. Those parts behave, or are supposed to
behave, according to natural laws. Human beings are among the more
intricate parts of that machine, and like all the other parts, their lives are
governed by natural laws that guarantee them certain rights and require
of them certain standards of behavior.

Benjamin Franklin wasn't even a deist. He was closer to being
an atheist. His heart's commitment was to a worldview shaped by the
Enlightenment; a worldview undergirded by a deep faith in reason.
It is Franklin who inserted into the Declaration of Independence the
term "self-evident," considering it to be the best adjective to describe
the "truths" that the founding fathers held in common. That is significant. For, as anyone who has studied mathematics or logic knows,
the term "self-evident" has been an important scientific concept that
finds itself already in the writings of Newton and Descartes, two of the
Enlightenment's founding fathers.

So if the Declaration of Independence was written by deists and
atheists and is filled with language that evidences its Enlightenment
worldview, how is it that Christians have embraced it so warmly? One
reason, of course, is that it has the ring of biblical truth about it. And, in
its historical context, it was a major step away from an oppressive and disobedient social system, toward a more normative social system – one that
seemed to resonate with the Word of the Lord for love and justice. But

another reason why Christians have warmly embraced the Declaration of Independence is that most have uncritically accepted at least a part of the Enlightenment worldview, and thus fail to see the problematic elements in many Enlightenment documents.

In this essay I want to focus on just one particular phrase that I find problematic. That phrase is "the pursuit of happiness." The question I wish to raise is whether the pursuit of one's own happiness is a legitimate activity. I raised this question once with a class of Dordt students, a group of particularly intelligent and thoughtful Dordt seniors, I might add. Their immediate response was, "Yes, of course. After all, God wants us to be happy." I must admit I was a bit taken aback by the unequivocal and confident character of their answer. It was almost as if they had learned the answer in catechism class and had it reinforced year after year since then. Allow me to raise a couple of concerns about the notion that "it's OK to pursue one's own happiness because God wants us to be happy."

First of all, we live in an individualist culture wherein the overwhelming thrust of media messages that I receive is the assertion that the primary purpose of everything outside me is ultimately to serve me, to make me happy. That's the basis of virtually all advertisements, of course. But it is also at the basis of the development of the human characters that populate television shows and movies. All food and health products exist to serve my physical needs. Government exists to serve my need for security. School exists to serve me with an education that will get me a good job. The good job exists to provide the money I need to live a happy life. The national parks exist to serve my recreational needs. Church exists to serve my need for fellowship, and comforting if and when I am not happy. Do you see where I am going? To say that "it's OK to pursue one's own happiness because God wants us to be happy" suggests to me a worldview in which God exists to meet my needs. Now none of my students would ever embrace such a heretical attitude. But when we are uncritical of the surrounding cultural forces shaping our worldview, such heresy is a nuanced extension of the individualistic self-centeredness that we *have* unthinkingly embraced.

A second concern that I have is with the notion that Christians must always be happy; that sorrow, sadness, and grieving are dangerously close to being sinful activities. This has given rise to the cult of warm-fuzziness that pervades many Christian churches today, and explains why you rarely hear hymns like "Ah Dearest Jesus" and "O Sacred Head" sung anymore, even around Easter. Now, certainly we want to avoid what is generally perceived as the opposite extreme: the caricature of the seventeenth century Puritan who is obsessed with evil, walks around constantly with a long face, and has no confidence in his own salvation. But *its* polar opposite is equally absurd, unbiblical, and injurious to any attempt to live an obedient lifestyle.

A third concern I have is with the notion that it is even possible for an individual person to achieve happiness by pursuing it. From a biblically informed perspective, I believe that such a notion is self-contradictory. I believe that pursuing one's own happiness can only bring unhappiness.

So let me now explain the basis of my concerns. First and foremost, to be human is to be created in the image of God, to be a servant called to a particular time and equipped with particular talents in order to accomplish particular tasks. The very meaning of our being is to respond to the Lord by serving him, our neighbors, and the nonhuman creation in which we are placed. When we do that, we are being faithful to ourselves, to being the creatures the Lord calls us to be. That is the only basis of true joy and happiness. When we focus first on ourselves, we are being unfaithful to the Lord, to our neighbors, to the world and time in which we are called to serve, and ultimately to our selfhood. That can only bring misery.

This is the basis of Jesus' words in Matthew 16:24 where he says, "Whoever wants to be my disciple must deny themselves and take up their cross and follow me. For whoever wants to save their life will lose it, but whoever loses their life for me will find it." That is a paradox to the Enlightenment or twentieth century American mind. But to the mind of faith, it is truth. Only when we forego seeking our own good, our own happiness, and seek instead our neighbor's good, our neighbor's happiness, can we ever find true happiness.

Finally consider the words of Paul in Romans 12:12. "Be joyful in hope, patient in affliction, faithful in prayer." Those three commands are very different from each other. I can be faithful in prayer by striving after faithfulness. I can be patient in affliction by working at developing patience. But the joy I have in hope flows from the quality of that hope. When my hope is pure – unfettered by fatalism, selfishness, or lack of faith – the result is pure joy. When "My hope is built on nothing less than Jesus' blood and righteousness," then I have true joy. Likewise, when my life is lived selflessly, as a servant to neighbor, my world, and my Creator, then my life is truly happy. My own happiness is not something I need or ought to pursue.

CHAPTER 27

SAFE INVESTMENTS AND LIMITED LIABILITIES

FEBRUARY 4, 1998

Prudence, circumspection, and practicality: these are qualities we ordinarily admire and try to cultivate, especially in young people. When one carefully considers or calculates the likely outcomes of one's possible actions, the results usually are safe investments and limited liabilities. So prudence and circumspection are probably qualities we would expect to find in science and technology students, right? And where we don't find it, we would want to encourage it – am I right again?

Well, not always. Strange as it may sound at the outset, I find myself on occasion trying to discourage the alleged qualities of prudence and circumspection in some students. You see, I teach a course entitled *Technology and Society* here at Dordt College. It's a course primarily intended for senior engineering and computer science students where we attempt to wrestle with the broad social and moral issues surrounding modern technology – issues that the students may very well have to face soon after graduation. Let me give you two examples. The first is the question of how one ought to run a modern, technical business that is obedient to the Word of the Lord and therefore serves one's neighbor, as well as the rest of creation. We discuss such issues as the place of profit, the sin of being motivated by profit, how to avoid the adversarial climate that results when you view fellow-servants as employers and employees, and the idea that workers ought to be paid in a more or less equitable fashion rather than with some getting outrageously high salaries and others getting the minimum wage. The enigma encountered by the students, as they see it, is that running a business according to the norms of love instead of the profit motive could easily result in the failure of the business. They see a Christian business being taken advantage of by the competition, by the customers, and by its own workers. Thus prudence dictates sticking with the dog-eat-dog methods of Wall Street and Madison Avenue, or at best, tempering those methods with some Christian concern.

The other example has to do with the military. Since the military, and companies that contract with the military, hire engineers, computer scientists, and other technically trained people, a serious question for the Christian engineer is whether or not she will work for an organization that produces weapons of destruction. One solution to the problem that gives rise to that question suggests that Christians take an approach to defense that is radically different from the conventional. Given the remarkable strides that modern technology has made in aerospace science, computer development, and bioengineering, it seems reasonable to suggest that Christian engineers blaze a path to the development of nonlethal, nonviolent weapons of defense. But, again, the response of some of my students is, "it won't work." "As soon as you develop some nonlethal, immobilizing agent, the forces of evil will develop a violent counteragent to render it useless. The only prudent defense is to violently destroy those forces of evil."

Now I realize that there are other attitudes besides prudence and circumspection operative here. But fear, ends-justifies-means morality, and survivalist selfishness are not usually considered to be virtues to be cultivated – prudence and circumspection often are. So you see, I sometimes find myself in the unique situation of actually cautioning my students against being prudent and circumspect. That causes me no little anxiety, for most of the time I am cautioning my students in the opposite direction.

But as so often happens, I recently discovered, in a book, an analogous moral dilemma, a situation where in order to be truly obedient to the Word of the Lord, one has to throw prudence and circumspection to the wind. I was reading *The Four Loves*, a beautiful book by C.S. Lewis that ponders the meaning of affection, friendship, Eros, and charity. In the chapter on charity – the love of 1 Corinthians 13 – Lewis relates how St. Augustine mourned over the death of a beloved friend and drew the conclusion that one ought never give one's heart to anything but God. "Do not let your happiness depend on something you may lose. If love is to be a blessing, not a misery, it must be for the only Beloved who will never pass away." Lewis suggests that initially this makes good sense. But then he goes on to describe why this kind of prudence is very bad advice. Listen to what he writes:

> Of course this is excellent sense. Don't put your goods in a leaky vessel. Don't spend too much on a house you may be turned out of. And there is no man alive who responds more naturally than I to such canny maxims. I am a safety-first creature. Of all arguments against love none makes so strong an appeal to my nature as "Careful! This might lead you to suffering."

To my nature, my temperament, yes. Not to my conscience. When I respond to that appeal I seem to myself to be a thousand miles away from Christ. If I am sure of anything I am sure that His teaching was never meant to confirm my congenital preference for safe investments and limited liabilities. I doubt whether there is anything in me that pleases Him less. And who could conceivably begin to love God on such a prudential ground – because the security (so to speak) is better? Who could even include it among the grounds for loving? Would you choose a wife or a Friend – if it comes to that, would you choose a dog – in this spirit? One must be outside the world of love, of all loves, before one thus calculates.*

Lewis then explains away Augustine's error as a "hangover from the high-minded Pagan philosophies in which he grew up." And then with words that invite generalization to the issues I described earlier, issues dealt with by my engineering students, Lewis writes:

There is no escape along the lines St. Augustine suggests. Nor along any other lines. There is no safe investment. To love at all is to be vulnerable. Love anything, and your heart will certainly be wrung and possibly be broken. If you want to make sure of keeping it intact, you must give your heart to no one, not even to an animal. Wrap it carefully round with hobbies and little luxuries; avoid all entanglements; lock it up safe in the casket or coffin of your selfishness. But in that casket – safe, dark, motionless, airless – it will change. It will not be broken; it will become unbreakable, impenetrable, irredeemable. The alternative to tragedy, or at least to the risk of tragedy, is damnation. The only place outside Heaven where you can be perfectly safe from all the perturbations of love is Hell. (121)

Likewise, I would argue, to seek to do the will of the Lord in business or in military defense is to become vulnerable. It is to expose oneself (and those who one is responsible for) to the risk of tragedy. But that, of course, is precisely what the Lord taught in his sermons and parables. To turn the other cheek, to sell all that you have and give to the poor, to seek the pearl of great price, is to throw prudence, circumspection, and practicality to the wind. And Lewis was right when he wrote that the alternative to the risk of tragedy – is damnation.

---

* Lewis, C.S., *The Four Loves* (New York: Harcourt, Brace, 1960), 120.

CHAPTER 28

BE THYSELF

MAY 5, 1998

Last month my fifth grandchild was born: a granddaughter. Her name is Kinsley, and while holding her in my arms less than twelve hours following her birth, I was struck by the realization that here indeed is a twenty-first century person. When the year 2000 rolls around, Kinsley will be a mere 20 months old and will not even be aware of the millennial shift. Throughout her life she will know that she was born in 1998, but she will remember nothing of life in the twentieth century.

I'm struck by this realization because, as a teacher, I often find myself trying to convince my students that they are unique people, called to a particular time in history, equipped with particular talents and abilities, and appointed to a particular historical task. They don't just pop upon the scene by chance. Rather, the Lord calls them into being, as his servants, "for such a time as this" (Esther 4:14). For example, I have a group of senior engineering students who are graduating in a few days, some of whom were born in 1975. That means that one-third of their lives – the third during which they grew up and had their formal education – was spent in the twentieth century. The two-thirds of their lives during which they will make their primary contributions to this world, however, will be spent in the twenty-first century. It's particularly important for these Christian engineering students to develop a sense of their historical identity. That's because they have been called to the crucial task of giving direction to technology in, what future historians will refer to as, "the information age."

It's not always easy for a young person to have an awareness of her own identity, of her being called by the Lord to occupy a place in history. I suppose that's because when you are young, you haven't yet experienced much history. It's easier for those of us who have been around a while to perceive the threads that hold the fabric of creation together and to sense the Lord's pleasure, pride, hope, and anticipation with regard to his younger servants. I certainly sensed that of Kinsley as I held her that evening following her birth. I remember sensing that with my other grandchildren as well. And, as a teacher, I've had similar experiences with a few of my students over the years. It's rare, but occasionally a student that you have only really known as a face with a name at a particular location in a classroom will suddenly be transformed before your eyes into a person of acute historical significance. Perhaps, as the student sits in your office, a few words or a facial expression will provide the clue. Then, with an undeniable and awful clarity, you hear the Lord say "this is my young servant, in whom I have great delight and very high expectations. Take great care in discharging your responsibilities as teacher! I am watching!" Wow! No wonder that it is written in the epistle of James (3:1), "Not many of you should become teachers, my fellow believers, because you know that we who teach will be judged more strictly."

The point of this essay, however, is that it is difficult for young people to come to a clear awareness of their own identity as unique servants of the Lord. Thus, as I think about my granddaughter and my graduating seniors, I'm concerned with how we as the Body of Christ might enable these young people, as they grow and mature, to *know themselves*, and consequently, to *be themselves*.

In this the end of the twentieth century and the beginning of the twenty-first century, there are powerful forces that work against the kind of self-knowledge that the Lord requires of his servants. From all sides our young people are besieged with images of what, in the eyes of our consumerist society, is the ideal form of personhood that one ought to embrace. But those images are fabricated by self-serving devils who seek to enslave. They pervade the media such as TV, film, and the Internet. They rule the crowd; sadly, even in Christian schools and the instituted church. And they are even passed on, unwittingly, by loving Christian parents when we have not exorcised them from our own lives.

As I think of Kinsley, I'm doubly troubled by the thought that, in our age, the lineage of Eve seems to be more squarely in the sights of these image-hawking devils than the lineage of Adam. There are great gender lies that our culture tells to its daughters. For example, that to be happy you have to have a particular kind of appearance, a particular set of material possessions, and a man before a particular birthday. And these are difficult to resist without the benefit of a half-century of hindsight. When she is five, or fifteen, or twenty-five years old, how can her parents and grandparents be assured that Kinsley will embrace my message that "you have been given just the talents and abilities you need to be yourself. Don't try to be less – and more is less when it prevents you from being your true self."

---

* E-mail message: [*Subject*: Inarticulate ramblings, *Date*: Thu, 23 Apr 1998 20:57:10].

Well, the only hope for that happening is if she is taught from her youth to delight in *being her true self,* in response to the delight that she learns that the Lord has in her *being her true self.* In learning that, however, she will also have to learn to stand out from "the crowd," to be something of a nonconformist, perhaps even a little eccentric. And that will take some eccentric, nonconformist role models to help achieve. In ending this essay, I'd like to suggest one such eccentric, nonconformist role model – one that may be unfamiliar to many of you who listen to this radio station – the nineteenth century Christian philosopher, Søren Kierkegaard. Kierkegaard always appreciated the Socratic maxim, "Know thyself." But he insisted that it be supplemented with the maxim "Be thyself." Listen hard to the following difficult but important quote from a book about Kierkegaard.

> The term "eternal" in the history of philosophy and in Kierkegaard's thought refers (minimally) to the unchanging. The temporal refers to what is *in time* and therefore changing. Our lives, surroundings, worlds, are *in time* and thus subject to constant flux. Who each of us is is constantly changing at one level. We grow, mature, marry, divorce, change occupation, have children or grandchildren. But throughout all this I am still I, and you are still you. And I am who I am even if I never became that person. The eternal element in me is both my continuity and my true self. It is the self not only that I can become but that I *should* become. The very worst thing that can happen to me as a person, my most basic form of failure, is that I do not become who I am. This is the general nature of all spiritual corruption, "for despair is precisely to have lost the eternal and oneself."
>
> And so Socrates' maxim "Know thyself" must be supplemented with the maxim "Be thyself.""

I would only want to add, that to know and be your true self, you first need to know whom you belong to. That is perhaps most simply yet elegantly articulated in the answer to the first question of the Heidelberg Catechism, where we read "That I am not my own, but belong – body and soul, in life and in death – to my faithful Savior Jesus Christ."

** Mullen, John Douglas, *Kierkegaard's Philosophy: Self-Deception and cowardice in the present age* (New York: New American Library, 1981), 50–51.

CHAPTER 29

THE AGE OF COMEDY

AUGUST 4, 1998

When I was a boy growing up in the 1950s, the world was a serious place. True, *my* world – a child's world – was a playful place. Whether playing outside in the snow or in our room with the contents of our toy box, or calling on our friends to go play baseball at the nearest field – it didn't matter. My younger brother and I, in any case, were playing.

But the world outside of our childhood was a serious one. That's not to say it was an unfriendly world or a world that lacked joy and compassion. It was just serious. My father was serious about the job he went to each morning before I awoke. He was serious about what he called "providing for the family." And he was certainly serious about instructing my three brothers and me so that we would one day be able to "provide for a family." My mother has a naturally gregarious personality. But when it came to directing life at home, she was serious. Whether it was balancing the budget, preparing meals, getting her four sons to do their homework, or insuring that we got to bed on time, she was serious. Not severe, overly stern, grim, or dour, but definitely serious. My teachers were all serious about their work. One or two may have been overly strict and harsh, but the majority were simply serious about learning. It wasn't until my freshman year in high school that I encountered a teacher who one might call a "joker." And it quickly became obvious that his bantering with students was related to the fact that his contract was not being renewed.

Sam and his wife – the older couple who operated the "candy store" where I often stopped after school for a five-cent chocolate soda or simply to peruse the latest comic books – were serious about running their store. And our family doctor – who made house calls when I was very young, and whose specimen-displaying office filled me with an uneasy fascination – well, he was always serious. Especially so, it seemed, when he was administering shots of penicillin.

Finally, I should mention the minister in the little Orthodox Presbyterian Church in which I grew up. He was a warm and intelligent fellow who could communicate equally well with the teenagers in the "young peoples' group," the older women in the "women's missionary society," the group of very down-to-earth, lower middle class men from whom the elders and deacons were chosen, as well as the one intellectual in the congregation. Our pastor had a fine sense of humor, which he employed at just the right times. Yet his sermons on Sunday morning – as well as the general quality of his pastoral efforts – one would have to describe as serious. Not grim or dour, not "fire and brimstone," but serious.

My purpose in painting this picture of seriousness – that I would argue was characteristic of the 1950s and early 1960s – is to contrast it with the present age. Today, it seems to me, we live in an age of comedy. Our typical interactions with one another are characterized by the joke, the one-liner. Think about it! How much of our daily life is dominated by the one-liner, the clever-comeback, or the stand-up comedy routine? Imagine that you are out for a Sunday stroll in Sioux Center and you pass some friends doing the same thing but walking in the opposite direction. It seems as if it is no longer sufficient simply to smile warmly and say "good morning." No, like two comedic gunfighters we reach into the holsters of our imagination and see who can be the first to draw out a clever one-liner.

Well, that's really not so bad, is it? The problem, however, is that this way of interacting seems today to be the expected norm for almost all situations. Certainly television is dominated by it. Not only late-night talk shows and prime-time situation comedies, but even the nightly news is infected with a frivolous attempt to be entertainingly funny. This expectation of comedy has become so pervasive that teachers are forced to become entertainers. Now I'm not opposed to a little levity. Appropriately used, it can do wonders for an otherwise dry mathematics lecture. And irony – which may be considered the most serious form of humor – is indispensable for crafting a quality lecture in the humanities or social sciences. But the "one-liner," "stand-up comedy" kind of humor that is demanded by many of today's students, draws attention to itself – its vacuous self – and greatly diminishes the effectiveness of the pedagogy.

Perhaps the worst example of this modern hegemony of humor is the role that it plays in the church. Like the teacher, the pastor is expected to be a stand-up comedian. The sermon is supposed to be entertaining, and the pastor's effectiveness is measured – subconsciously, no doubt – with the same tools used to evaluate the effectiveness of a Seinfeld or an Eddie Murphy. Beyond Sunday morning, the demand for a wealth of witticisms infects the council room as well. I invite those of you who serve as elders or deacons in your congregation to take time, at your next elders, deacons, or council meeting, to count the occasions during which time is spent on one-liners and other forms of light humor. Not only does it contribute to the obscene length of many of these meetings. More importantly, the humor has a tendency to displace or dilute the serious pondering that ought to characterize such meetings.

And that is the whole point of this essay. We live in an age of comedy where our communication with one another is normed by humor of the most trivial kind. As a result, our worldviews are losing the seriousness that is fundamental to our existence as God's image-bearing servants. We trivialize that which the Lord created, upholds, and died on the cross to redeem.

I am not arguing for a kind of puritanical sobriety that has no place for joy. But joy is far more than humor. And I am not saying we should do without humor. As I mentioned earlier, there is a kind of serious humor that we would do well to encourage – namely irony.

What I want to say can be stated simply in two words, "get serious!"

The only other solution to this pervasive plague of pernicious pleasantries is death. For the only event that has, thus far, escaped the demand for stand-up comedy, is . . . the funeral.

CHAPTER 30

COMFORTABLY NUMB

NOVEMBER 2, 1998

A few weeks ago I participated in a panel discussion of Neil Postman's book *Amusing Ourselves to Death: Public discourse in the age of show business* (New York: Penguin, 1985). I had accepted *with enthusiasm* the invitation to participate because I have read Postman's book twice, have used it as a textbook in a course for college juniors and seniors, and because I think it is one of the most important books published during my lifetime. In the book, Postman shows how western culture in the twentieth century has evidenced a transition from *typography* to *iconography*. That is, whereas at one time we depended heavily on *words* to communicate with each other, we now are much more dependent on *images* or *icons*. The best example is the way in which the majority of people acquire information about the world. At the start of the twentieth century the newspaper and the book were central to acquiring current and background information respectively. Today visual media in the form of the television and the Internet has all but replaced print media. Whether it is the nightly news or the Internet, people are doing far less reading of words and far more staring at moving images. Postman argues that this transition from print to pictures has brought with it a shift in the way we think. He suggests that print media fosters rational thought and that visual media, like television, fosters another kind of thought, one perhaps tied more closely to emotion but certainly less closely to reason. He argues further that television is an effective technology for transmitting entertainment, but that it is a cumbersome technology for transmitting complex ideas. Thus the communication technologies that have developed during this century have fostered entertainment and have inhibited rational thought.

In the panel discussion, two of my colleagues and I presented our assessments of Postman's ideas to a class of college students. We agreed that Christians need to think critically regarding the issues Postman raises. And that means being conscious of the effect that television can have in shaping not only the way we think, but also our worldview. We warned against the misshaping of young minds that occurs when the television is used as a baby-sitter. And we recommended strategies such as muting the sound during commercials when one is watching programming on commercial television stations. We compared television to alcohol, suggesting that while the technology is a part of creation that needs to be unfolded and developed in a stewardly manner, its dangerous properties need to be carefully considered. In many respects television *is* similar to alcohol, or even to certain kinds of mind-altering drugs. Used uncritically, it has the potential for dehumanizing us, for deadening our capacity to function the way the Lord created us to function – to turn us into zombies or vegetables by making us *comfortably numb* to the needs of the world around us.

While I was pleased with the way my comments and those of my two colleagues resonated with each other, I was somewhat taken aback by the response of the students in the class. There were a few who understood Postman and caught the concern that we were voicing. But there were also a few who simply could not understand Postman's thesis at all, and who could not imagine how a technology that they have known and been comfortable with from their childhood could be as perilous as we were describing television to be.

As I sat on the panel listening to these students respond with their skepticism, two threads of thought wove themselves together in my mind. The first was an almost subconscious replaying of the lines of an old song I heard recently on the radio as I was driving back to Sioux Center from Sioux Falls, late one evening.

> When I was a child / I caught a fleeting glimpse, / out of the corner of my eye. / I turned to look but it was gone. / I cannot put my finger on it now. / The child has grown. / The dream has gone. / I have become comfortably numb.[*]

The second thought I had was that *my* generation – the parent generation to these skeptical students – is largely responsible for the uncritical and anesthetic quality of their attitude vis-à-vis the perils of television, and of our godless culture in general. I can't help but think of the efforts, which churches have been making of late, to accommodate – to make comfortable – those who have been thoroughly imbued with the *comfortably numb* posture of contemporary culture. I have heard calls to keep our worship services short – which usually means truncating the sermon – and I have experienced the replacement of intellectually rich hymns by "warm 'n fuzzy" choruses that sound like 1950s love songs and

---

[*]   David Gilmour and Roger Waters, "Comfortably Numb," Disc 2, Track 12 of *Pink Floyd Pulse* (New York: Columbia Records, 1995).

simply repeat the same banal mantra over and over. All of this is supposed to attract and retain our contemporarily encultured young people. I wonder if, instead, it is simply adding further anesthetic to the seductive and numbing influences of the anti-intellectual, pleasure-seeking world of contemporary television. When the intellectual level of our worship services descends to that of the twelve-year-old, it's time for us to listen more carefully to the Word of God. From Proverbs 1:22:

> How long will you who are simple love your simple ways? How long will mockers delight in mockery and fools hate knowledge?

And from 1 Corinthians 13:11:

> When I was a child, I talked like a child, I thought like a child, I reasoned like a child. When I became a man, I put the ways of childhood behind me.

And finally, from Ephesians 4:14:

> Then we will no longer be infants, tossed back and forth by the waves, and blown here and there by every wind of teaching and by the cunning and craftiness of people in their deceitful scheming.

Perhaps it is that we in the church have confused the very necessary child-likeness of our faith with the childishness of contemporary, visual-oriented culture. Whatever the explanation, I think it's time for some hard-nosed, intellectually responsible, and prophetic calls for repentance and redirection. The alternative is to "become comfortably numb."

CHAPTER 31

DIVERSITY AND THE PURSUIT OF DIVERSION

APRIL 3, 2000

Have you ever wondered why it is that, as a culture, we value particular characteristics of persons and groups of persons? For example, bravery and honesty are characteristics valued by virtually all cultures throughout history. An appreciation for individual independence and initiative, however, is peculiar to modern western societies. Some valued characteristics, it seems, are universal, and – as Christians – we would say that they are part and parcel of the creation order. I think that bravery and honesty are like that. They are rooted in the biblical creational norms of faithfulness and selflessness. But other valued characteristics have a more mixed origin. Individual independence and initiative as values of modern western culture are actually distortions, in the direction of selfishness, of the biblical norm of fruitfulness.

In this essay I would like to examine a relatively new, valued characteristic of western society that many of us have adopted – somewhat uncritically, I suggest. That valued characteristic is diversity.

Less than a decade ago, the word diversity was a simple descriptive term that meant variety, heterogeneity, or nonuniformity. As a descriptive term, it conveys a characteristic of a group. Whether that group is a group of flowers, people, or ideas, the term diversity suggests that there are differences among the members of the group. But until recently that particular property of having differences was neither esteemed nor disparaged – it was simply recognized. Today diversity is taken by many people to be a valuable characteristic, a goal to be sought with particular vigor. The primary contexts in which diversity as a goal is operative have to do with groups of people. The sought after characteristics of that diversity are, however, somewhat nebulous. Often it is as simple as racial ethnicity. The differences among the people in the group would then be biological in nature. But almost as often the differences sought have to do with cultural ethnicity. Then the differences, often including those that are biological, are sociological as well. Economic, social, lingual, and aesthetic customs and expectations of persons within the group are varied, and that variety is prized. In some groups – the large postmodern university setting, for example – difference is sought (or perhaps I should say *tolerated*) even when it comes to moral characteristics. These include the way people think and act with regard to justice, ethics, and faith.

Rather than ask whether diversity ought to be esteemed, disparaged, or held as neutrally descriptive, I want first to recognize that diversity is esteemed by many in modern western culture, and then to ask: from where did that esteem arise? I think there are at least three answers to the question.

First, racial and cultural diversity within a society is prized in reaction to the perversity of absolutizing racial and cultural homogeneity. Two events of historical importance occurred during the middle of the last century that shaped this reaction. The first was the Nazi idolization of the Arian race, its anti-Semitism, and its persecution of European Jews. Clearly that quest for ethnic homogeneity was evil and rightfully despised. The second historically important event was the civil rights movement of the early 1960s. It was at that point recognized that while slavery had been abolished a century earlier, racial prejudice was still a serious problem in the United States. There have been other events as well, such as the battle to end apartheid in South Africa, but recognizing the injustice done to Jews in Nazi Europe and to blacks in the United States served as primary vehicles for the promotion of diversity to normative status.

A second answer to the question regarding from where diversity's esteem arose has to do with the recognition of the laws of biology. Most farmers recognize the need for diversity in their attempt to raise crops. Planting the same strain of corn year after year on the same plot of land is an invitation to disease, pest infestation, and soil depletion. It is also well known from history what happens to groups of people who intermarry within the same extended families. Genetic and other biotic problems are propagated with devastating results for future generations. More recent ecological studies have shown the importance of biodiversity in a given habitat. Artificially reducing that biodiversity usually leads to instabilities that can create ecological disasters.

But there is a third answer to the question regarding from where diversity's esteem has arisen. That can be answered by considering that the words "diversity" and "diversion" have the same root. We live in a culture that is obsessed with entertainment, with diversion, and with the quest for novelty. This has been identified by one cultural critic as a sauntering "around looking for variety,'" or as "a frivolous philandering among great diversities" (100). In any case, it is fundamental to the consumerist quest for the new and the novel. At root it has the effect of leveling us because it trivializes difference. We have developed the habit of tiring quickly of our environments and the various elements that comprise our environments. So we are a mobile society, changing jobs and changing residences with regularity. We habitually seek out different kinds of food to eat, different cars to drive, different music to listen to, and different forms of entertainment. And, unless recent divorce statistics are wrong, it seems that we are also seeking diversity in our marriage partners. Thus, when budding consumerists in their teenage years are first allowed to travel to the big city, or to other countries, they are titillated with the multiplicity of experience. The paradoxical result of all this is that true diversity is reduced to superficiality by the pursuit of diversion. We become perennial tourists, all nibbling at the same superficial smorgasbord of life. And thus our esteem for diversity leads us to a dull and trivial uniformity.

In summary then, there are good reasons for valuing diversity – reasons having to do with justice and with biotic and social health. But there is also the danger that our esteem for diversity is simply another expression of mindless consumerism. Perhaps it is time that we Christians adopt a more critical stance toward the values that western culture so readily embraces.

---

* Søren Kierkegaard, *Two Ages* (NJ: Princeton University Press, 1978), 94.

CHAPTER 32

THE MEANING OF MEANING

JUNE 1, 2000

About a week ago I was sweating over my computer keyboard trying to bang out an academic paper dealing with the concept of "meaning." "*Meaning* is the *being* of all that has been *created* and the nature even of our selfhood," wrote Herman Dooyeweerd,* a Christian philosopher whose work, in one way or another, has influenced many associated with Dordt College. What Dooyeweerd was trying to say in those words, which, at first glance, seem unnecessarily obtuse, is the same thing that Abraham Kuyper proclaimed when he said that there is not one square inch of creation that the Lord Jesus Christ does not claim as his own. Everything belongs to the Lord, and thus, in a sense, everything is always in a state of response, responding either obediently or disobediently to his Word for creation.

But I'm getting ahead of myself. I was telling you how I was laboring at the computer terminal, trying to get words to flow from my head, through my fingers, into the keyboard, and on to the 17-inch screen that dominates my desk. It was mid-morning and things were going slow. The pot of coffee that I had begun enjoying at 6:00 a.m. was empty. The stimulating effect of the caffeine on my brain had worn off. No longer was my focus fixed on what I was writing. Instead, the coffee was now urging me to get up from my desk chair in order to assist its efforts to reach its final destiny in the Sioux Center Sewage Treatment Plant. Obliging the coffee, I decided to also take a short break to read the daily newspaper. On the front page I learned that the House of Representatives had passed a bill giving China "most favored nation" trading status. On the second page I learned that gasoline prices were back up again, just in time for summer traveling. Somewhere on page 4 or 5 I learned that Bob Dylan had just turned 59 years old.

I quickly turned to the sports section: nothing of interest. Thus, before I knew it, I was reading the comic page. I enjoy reading a few of the

---

* *A New Critique of Theoretical Thought* (Philadelphia: Presbyterian and Reformed, 1969), I:4.

comic strips. *Dilbert* reminds me of what life would be like if I had stayed in industry as an engineer and never become a teacher. *Hagar the Horrible* tells what life would be like if I had married someone else – someone who would have allowed me to remain stagnated in the conservative, sexist milieu of my 1950s childhood. And *Doonesbury* reacquaints me with the life struggles of many of my college friends from the 1960s. It was Garry Trudeau that cried out to me on this particular day, "Stop! Grab the scissors and cut this out. Good stuff like this is printed less than once a year, and besides, you will probably get an essay out of it."

What was unique about the 25 May 2000 *Doonesbury* comic strip was that it dealt with precisely the topic that I had just been writing about – meaning. The two main characters in the strip were "generation-Xers," the offspring of sixties characters that have populated the strip for the last three decades. In previous strips during the week they have been fantasizing about striking it rich on the Internet. In this particular strip they are still dreaming. In their dream they are lounging on the beach of some exotic island resort, sipping cool drinks and staring at the sky. The title of the strip is "Trouble in Paradise.com." In the first frame one character says, "You know what I have a sudden longing for man? Meaning!" His buddy replies, "I hear you. Let's ask the concierge." The second frame bears the title, "Later," and shows our two generation-X'ers talking with a man in a suit who is behind a desk. He is telling them, "Sorry sir, we don't offer meaning here – only the empty pursuit of pleasure." The third frame shows the two main characters in silhouette, one of them simply saying, "Oh." In the last frame the second Generation-Xer turns to the first and says, "Yo, I can live with that." To which the second replies, "Yeah, me too. Thanks anyway." And the man behind the desk closes the strip with the words, "Rock on, sir."

That comic strip reminded me, in its own humorous way, of how our colloquial sense of the word "meaning" is a distortion in our understanding of creation. It demonstrates how even sentiments that we may wish to applaud can reinforce ways of thinking that are steeped in secularism. In the case in point, we surely would agree with Gary Trudeau in making a mockery of the "empty pursuit of pleasure." But the suggestion of the comic strip is that such pursuit is *without meaning*, or more conventionally, *meaningless*. Really? I'm afraid that the hedonists among us don't get off so easy when that hedonism is examined from a radically biblical perspective. You see, Dooyeweerd is right. Meaning *is* the being of all that has been created – including the spirit of hedonism. It's not that "the empty pursuit of pleasure" is without meaning. Rather, the meaning of "the empty pursuit of pleasure" is disobedience. Hedonism as an attitude, posture, or activity is a response to one's Creator – a disobedient response. There really is not a square inch of creation over which Christ's lordship does not extend, demanding our obedient response – not even in the comics.

DOONESBURY © 2000 G. B. Trudeau. Reprinted with permission of UNIVERSAL UCLICK. All rights reserved.

CHAPTER 33

REFORMED: AN EMPTY LABEL OR A VIBRANT WORLDVIEW?

NOVEMBER 1, 2000

Recent discussions that I have had, both with the Council of my local congregation and with my colleagues at Dordt College, have impressed me with the concern that we may be losing our understanding of what it means to be *Reformed*. It seems that this term, which is intended to – and once did – represent a vibrant worldview, is in danger of becoming an empty label. In this essay I would like to raise and attempt to answer the question, "What do we mean at Dordt College, or in the church congregations of Northwest Iowa, when we call ourselves *Reformed*?"

I think that the meaning of the word *Reformed* is captured most simply in the phrase "*all of life is religion*," and in the emphasis that "life is relig**ion**, not relig**ious**!" John Calvin makes this point in Book I Chapter 2 of his *Institutes* when he asks,

> . . . how can the idea of God enter your mind without instantly giving rise to the thought, that since you are his workmanship, you are bound, by the very law of creation, to submit to his authority? – that your life is due to him? – that *whatever you do* ought to have *reference* to him?

The same idea is expressed most poignantly by *The Heidelberg Catechism's* first question and answer, which asserts that my "only comfort in life and in death" is "That I am not my own, but belong . . . to my faithful Savior Jesus Christ." Abraham Kuyper said it most powerfully when he argued that "there is not a square inch in the whole domain of our human existence over which Christ, who is Sovereign over *all*, does not cry: 'Mine!'"* And the Dutch philosopher Herman Dooyeweerd said it rather abstrusely but very concisely when he wrote "Meaning is the being of all that has been created and the nature even of our selfhood." In other words, reference back to the Creator is the central, essential characteristic of everything that has existence.

---

* "Sphere Sovereignty" [1880] in *Abraham Kuyper: A centennial reader*, James D. Bratt, editor (Grand Rapids: Eerdmans, 1998), 488.

There are a number of **important consequences** of this basic, Reformed idea that, in short, we might call "God's sovereignty," or "the meaning character of reality." First, "life is religion" implies that "life is service." Thus we see ourselves as God's servants and seek to know his will for **every** area of our lives. There is no area of life that is neutral, that is not normed by the Word of God. There are no adiaphora – a fancy theological term that means things indifferent to God – in the Reformed vocabulary.

Second, we *Reformed* Christians have a "total" view of the Fall. We believe that every creature is touched and distorted in some way by the curse, and that what has been called "the antithesis" cuts through every human heart and every human institution.

Third, with Paul in Colossians 1, we believe that Christ is the Creator, the Sustainer, and the Redeemer of **all things**. Redemption is thus cosmic in scope. Easter has significance not only for human souls, but also for mountains and forests, for music and fluid mechanics, for theories and statesmanship, for construction projects and for baseball games.

Fourth, we take seriously what has been called "the cultural mandate." We are called to unfold, develop, and, in general, be faithful stewards of God's good creation. Our role, as expressions of the Body of Christ in history, is thus *world-formative*. Our task is to bring healing and redirection to our world, with sensitivity to the evil that is so pervasive and with an eye to the coming Kingdom, the New Jerusalem that will accompany our Lord's return. To quote a colleague of mine, "Our Christianity is either world-formative, or it is formed (that is to say, *deformed*) by the world."

Fifth, we therefore quite naturally take what some may perceive as a *counter-cultural stance* with regard to the powers and principalities that presently rule, that comprise the status quo.

And last, from a truly *Reformed* point of view, we have no interest in *integrating* "faith and science" or "religion and scholarship." Rather, if life is religion, then scholarship – as one part of life – is also religion. In other words, our science is an expression of our faith. Thus our scholarship must be authentically Christian scholarship. Our science must be biblically obedient science.

The genius and the essential distinguishing characteristic of the *Reformed* worldview is that it is cosmic in scope, including every area of human life and every created thing within the creation. If we fail to remember what it means to be *Reformed*, if we become content to call ourselves simply "Christians," or even "biblical Christians," because those are more inclusive, more seeker-friendly labels, then we and the generations that follow us will be in danger of becoming one with "Christendom," with the masses of Christians who have lost their saltiness, who are no longer a light in the world. Then science and sports, poetry and politics, economics and ecology, and the myriad areas of life that Christ suffered and died to redeem, will, for a time, be lost to the powers of darkness.

CHAPTER 34

ENTERTAINMENT AND THE PRESENT AGE

SEPTEMBER 3, 2001

It was a Thursday night during spring break of my senior year in college – April 4, 1968. My young wife – who would remain a teenager for 6 more months – had passed her driving test that morning. We were celebrating by going to the movies – escapism was the theatrical therapy against the political turmoil of those days. On January 23 of that year North Koreans had seized the USS Pueblo and its 83-man crew in the Sea of Japan. Shortly thereafter the Vietcong had launched their "Tet offensive," attacking Saigon and greatly escalating the war in Vietnam. Eugene McCarthy, the anti-war senator from Minnesota, had entered the New Hampshire primary, announcing that he could not support the president's war policies. His strong showing and the unabated anti-war protests drove President Lyndon Johnson to decide not to run for re-election.

But my wife and I were escaping all that by going to see the new Charlton Heston film – *Planet of the Apes*. Science fiction defined for us quality media that year. A new, hour-long series had debuted the previous fall; and almost every Friday night found us hustling through the grocery shopping so as to make it back to our apartment by 7:30 to watch the latest episode of *Star Trek*. It had intelligence, passion, and angst – personified respectively by Science Officer Spock, Captain James Kirk, and Chief Medical Officer McCoy. And the tension between Spock's pure rationalism and Kirk's humanist romanticism echoed tensions that were being played out in real life on college campuses across the country. So while the title of the film – *Planet of the Apes* – sounded a little silly, my wife and I had hopes that it would measure up to the high standard set by *Star Trek*.

And we were not disappointed. Charlton Heston's character in the film is a bitter, pessimistic, and self-centered astronaut. Much of the film is devoted to his expressions of anger: anger at what he considers the foolish idealism of his two fellow astronauts, anger at being treated like an animal by a culture of intelligent apes, and anger at – well, let me save *that* for the end of this essay. What we didn't appreciate about the film was its clear anti-religion message. Actually, being 1968, the message was really more anti-authoritarian than anti-religion – a message we would have embraced had it taken a different form. But in this film the authority figures are those aged apes that hold to and defend ancient, religious truths. The heroes – if there can be said to be heroes in this dystopian tale – are a trio of younger apes representing the scientific mind of their society. They are astounded at the discovery of a human – Charlton Heston – that has the capacity of speech. Questioning the ancient religious dogma, they heretically postulate that perhaps apes were descendants of an ancient race of humans, a theory suggested by one of the young ape's recent archeological discoveries in what is called "the forbidden zone." The appearance of this talking human seems to confirm the theory.

Near the end of the film the young apes help Charlton Heston to escape. What looks to be the last scene has this angry astronaut riding off on a horse – with a female human he has picked up along the way, of course – and into the sunset. But that's not quite the last scene. Instead of riding off into the sunset, our astronaut rides around the bend of a great river in what seems to be a desolate land.

Now remember, this is April of 1968. The civil rights movement is still young. The war in Vietnam is tearing the nation apart. And, of course, we were still in the midst of the cold war: less than five years after the Cuban Missile Crisis. Also consider that, while in the beginning of the film there was a lot of talk about Einstein's theory of relativity and of how these astronauts would have aged more slowly and arrived at an earth hundreds of years in the future from the time they left it, none-the-less, most of us in the theater audience simply assumed that the astronauts had crash-landed on a different planet in a different solar system. You see, the main part of the film focused on discussions of evolution rather than relativity.

So, when the last scene arrives, when Charlton Heston rides his horse around that bend in the great river and looks up, most of us in the audience responded with a huge communal gasp. For there was the leaning wreck of the Statue of Liberty. And there was the realization that our astronaut was not on another planet but was back home on Earth – thousands of years later – after twentieth century humanity had destroyed itself as a civilization. Charlton Heston's last angry words express the communal grief, anguish, and fear of a generation when he curses his fellow humanity and cries, "You did it! You really, finally did it!"

Now, I'm moved to tell this story because just a couple of weeks ago my wife and I went to the movies to see the new version of *Planet of the Apes*. What a disappointment! The story, if there can be said to have been one, was incoherent and trivial. What we experienced was two hours of violence and special effects, with a few absurd references to animal rights thrown in. I think that tells us something about our present age, starkly contrasting it with the age in which we lived a third of a century ago. I'm not suggesting that 1968 was somehow better. What I will suggest is that we – particularly we Christians – were more aware of the forces of evil that threatened to tear our world apart back then. Today those forces, no less evil, are more insidious. They lull us into the complacency of spiritual defenselessness by the narcotic effects of mindless entertainment.

So if you cannot resist going to see the new version of *Planet of the Apes* in the near future, or if you have already seen it, I urge you to view the video of the 1968 version as well. And then reflect on what the differences tell you about the society in which you live.

But wait! Don't go. I have not quite finished my story. Come back with me for just a moment to April 4 in 1968 as we exit the movie theatre, saying to ourselves, "Wow, that was quite an ending." With our aesthetically nurtured mixed feelings of awe, satisfaction, and angst, my wife and I walked to the parking lot, unlocked our car, and prepared for the short drive back to our apartment. As was my habit, I clicked on the radio. On every station was the same news announcement: Dr. Martin Luther King Jr. had just been assassinated at a motel in Memphis, Tennessee. And for the second time that evening we saw the Statue of Liberty leaning violently toward a decadent demise.

CHAPTER 35

"THE AIR CLEANER DECLARES THE GLORY OF GOD"

FEBRUARY 1, 2002

Ever since I took a shop class as a freshman in high school, I've had an interest in woodworking. A year after graduating from college, when my wife and I moved into our first house and I had a basement that I could call a workshop, I began purchasing some small power tools and making things out of wood – #2 pine to be precise. Ten years later I moved out here to Sioux Center, purchased my first floor-standing tool – a radial arm saw – and began working with oak. That was twenty-two years ago, and the sweet smell of oak sawdust has filled my basement workshop on regular occasions since then. This year I began working with cherry in addition to oak. I also invested in two woodworking machines that you likely would not find in the average hobbyist's workshop. Let me tell you about one of them.

When you spend a fair amount of time in a woodworking shop generating sawdust, you begin to feel it in your nose and lungs. I took to wearing a mask a long time ago whenever I planned to saw, sand, or rout for an extended period of time. But still, the finest sawdust particles would hang in the air for hours, would sometimes migrate from my workshop to other parts of my house, and would always find a way to get into my nose and lungs. So a few months ago I ordered a woodworking shop air cleaner. On the Saturday after it arrived I spent the whole morning installing it. I was impressed when I unpacked it. It had a simple rectangular box shape and was made of heavy gage sheet steel without any sharp edges. There were no scratches, dents, or any indications of careless finishing and assembly. Weighing in at 85 lbs, it required me to use two ladders and construct a rig made of 2x4s in order to hang it from the ceiling of my shop. Once I had it properly in place, I found it easy to make the necessary adjustments so that it was perfectly level. The adjustments complete and the power cord routed to a ceiling outlet controlled by a wall switch, I pulled the little chain to turn it on. What a sweet sound! No rattles, no vibrations, just a gentle hum that assured me the motor was working, shop air was being drawn into the front of the unit, and clean air was being discharged out the back. Later that afternoon I began cutting some 4x8 sheets of cherry plywood into 10-inch wide boards. I was generating quite a bit of sawdust and so turned the air cleaner on high. I noticed that it only took a few minutes before the air in the shop was so clean that I could remove my mask without smelling sawdust. Later that evening I returned to my shop – just to turn the air cleaner on and off a couple of times, to turn the knob that more finely adjusted the blower speed, and, in general, to admire a well-made and well-functioning machine.

The next day was Sunday, and during the morning service at my church the pastor read from Psalm 19 and preached on the way the creation reveals its Creator.

> The heavens declare the glory of God; the skies proclaim the work of his hands. Day after day they pour forth speech; night after night they reveal knowledge. They have no speech, they use no words; no sound is heard from them. Yet their voice goes out into all the earth, their words to the ends of the world. (1–4)

As I listened I could not help but think of the new air cleaner in my woodworking shop. The psalmist lived at a time long before the advent of modern technology so it would be natural for him to point at the night sky to find the most obvious examples of the creation declaring the glory of its Creator. But that air cleaner in my shop is also part of creation. It is a creature as much as are the stars. The raw materials from which it is made, like the starry host, were created by God "in the beginning," and were declared by him to be "very good." The form and function of the air cleaner are such because God's image-bearing creatures, in response to his Word to subdue and fill the earth, have unfolded and developed creation, bringing into existence things that only had potential when God rested on the seventh day of creation. So these things – these technological things – *are* creatures. And as creatures, they are servants of God. And as servants, they declare his glory for those who have eyes to see, ears to hear, and in the case of my air cleaner, a nose to smell the sawdust-free air.

Psalm 119:89–91 speaks of this universal servanthood of all things where we read:

> Your word, LORD, is eternal; it stands firm in the heavens. Your faithfulness continues through all generations; you established the earth, and it endures. Your laws endure to this day, for all things serve you.

Did you catch that last phrase: "for all things serve you?" It's true that technological things are different from the numerous creatures the psalmist lists in Psalm 148:

> . . . great sea creatures and all ocean depths, lightning and hail, snow and clouds, stormy winds that do his bidding, . . . mountains and all hills, fruit trees and all cedars, wild animals and all cattle, small creatures and flying birds. (7–10)

The big difference is that the Lord uses his fallen, image-bearing servant to bring these technological creatures into existence. Thus the fall into sin and the curse affect them in a more direct and central way than it does those creatures listed in Psalm 148 – in a way that distorts their very identities. On the other hand, technological creatures have something in common with the creatures of Psalm 148 that is bigger and more powerful than the curse. Paul writes in Colossians 1:20 that God was pleased to have all his fullness dwell in Christ, "and through him to reconcile to himself all things, whether things on earth or things in heaven, by making peace through his blood, shed on the cross." So when you come into my woodworking-shop and look around, do notice that "the air cleaner declares the glory of God and the table saw proclaims the work of his hands."

CHAPTER 36

THE NORM OF COMMUNICATION
AND THE ELASTICITY OF WORDS

MARCH 4, 2002

Have you ever wondered why, on the one hand, most people will agree that open communication is a goal that, if achieved, will facilitate peace and understanding, and yet, on the other hand, despite increasing levels of communication in our world, there is so little mutual understanding between people who hold only slightly different opinions? Recently I've come to see that a big problem with communication is the *elasticity of words*. Let me try to explain – that is, to communicate – what I mean.

Thirty years ago I began my teaching career in a Christian high school in Northern New Jersey. One of my goals was to enable chemistry and physics students to understand that when we read in the Gospel of John "For God so loved the world," we are to understand the word "world" as meaning "cosmos," or "his whole creation," and not simply "the world of human souls." But that word "world," in the context of John 3:16, had a particularly *inelastic* quality in the minds of my students and the Christian families from which they came. To them it referred only to human beings, image bearers of God, and could not be stretched to include the nonhuman creation. I ran into the same problem when I talked about Christ redeeming not just human souls, but all of creation. The word "redeem" could not be stretched to include any creature beyond humans. Despite my quoting St. Paul in Colossians 1 and in Romans 8, the word "redeem" in the vocabulary of these Northern New Jersey Christians was *inelastic*. Responding to my Scripture quotes, the students reminded me that the word in Colossians is "reconcile," not "redeem," no matter what version of Scripture you read. And, they added, the word in Romans is "deliver," "liberate," or "set free," depending on the version you chose. But in no version was the word "redeem" used. Therefore, they would argue, Christ's redemptive work – his death and resurrection – was exclusively for people; not for rocks, carbon dioxide molecules, or animals on the endangered species list.

And so I had to struggle to convince my students that such a narrow view of the atonement was unbiblical, unreformed, and a stumbling block to living an obedient life in the twentieth century. And that struggle was, in part, against the inelasticity of words. My students leaned in a literalist direction with respect to Scripture, and compounded that error by reading into the literal elements of Scripture their own very narrow and rigid definitions.

This phenomenon – this inelastic quality of the words we use – is not at all uncommon. It is simply a distortion of one of the very natural properties of language, namely, that words have distinct meanings. Even in poetry, where metaphor reigns and words have great flexibility, words always have a basic meaning, one that then gets stretched or played with. To combat this tendency toward literalist rigidity, I often deliberately use particular words in ways that, while technically correct, are playfully strange to my students. It forces them to stretch their understanding or confront the rigidity of that understanding, or both. In any case, it exercises their worldviews – and that's always good.

Recently I had occasion to ply this technique in a different context. The word, the elasticity of which I was testing, was "anthropocentrism." It's just a fancy academic term that means "human centered." In the present age the word often takes on a narrower, more pejorative meaning, referring to the belief that all of nonhuman creatures exist merely to serve humans and that humans have been given unlimited rule over the rest of creation. In other words, "anthropocentrism" has come to represent a particularly heinous distortion of the cultural mandate: one that has resulted in the environmental crisis, the energy crisis, and a host of other problems that are unique to our present age. As such, anthropocentrism is not a "pure evil," or "evil in itself" (nothing ever is!). Rather, it is a distortion of a fundamental truth found in Scripture. For example, the psalmist asks:

> . . . what are human beings that you are mindful of them, mortals that you care for them? Yet you have made them a little lower than God, and crowned them with glory and honor. You have given them dominion over the works of your hands; you have put all things under their feet, all sheep and oxen, and also the beasts of the field, the birds of the air, and the fish of the sea, whatever passes along the paths of the seas. (Psalm 8:4–8 NRSV)

Likewise, the Genesis (1:26–28) account of the sixth day of creation, we read:

> Then God said, "Let us make mankind in our image, in our likeness, so that they may rule over the fish in the sea and the birds in the air, over the livestock, over all the earth and over all the wild animals, and over all the

creatures that move along the ground." So God created mankind in his own image, in the image of God he created them; male and female he created them. God blessed them and said to them, "Be fruitful and increase in number; fill the earth and subdue it. Rule over the fish in the sea and the birds in the air and over every living creature that moves on the ground."

So, with this more basic truth in mind, I have on occasion suggested that perhaps we ought to reform that word "anthropocentrism," for there *really is* a sense in which God created humankind to play a central role in creation. Perhaps we could distinguish the truth from the distortion by using the phrase "biblical anthropocentrism" to represent the truth, and "humanist anthropocentrism" to represent the distortion. But, alas, this proposal is typically received with about the same level of befuddlement as when I suggest that we reform the word "sin," and use the phrase "biblical sin" to represent virtue.

And so here we are confronted again with the reality of "inelastic terminology" and the specter of miscommunication. In this latter case however, I am finding it more prudent to resist my inclination to playfully stretch the word "anthropocentrism," and, for the sake of peace and understanding, will try to find a different word to represent that biblical truth expressed in Genesis 1 and Psalm 8. Perhaps the word "stewardship" has some stretch left in it.

CHAPTER 37

THE TIMES ARE STILL A-CHANGIN'

JULY 4, 2002

In 1963 Bob Dylan wrote and sang a song entitled "The Times They Are A-Changin'." It may surprise some to know that he based the song on Jesus' description of the Kingdom of God in Matthew 19 and 20. The last verse of the song reads as follows:

> The line it is drawn
> The curse it is cast
> The slow one now
> Will later be fast
> As the present now
> Will later be past
> The order is
> Rapidly fadin'.
> And the first one now
> Will later be last
> For the times they are a-changin'.*

The times were indeed changing when Dylan wrote that song. In 1963 the civil rights struggle was in high gear. And in November of that year the President of the United States was assassinated – an event arguably creating even greater national trauma than occurred last September 11. Five short years later the country was deeply embroiled in the Vietnam War. The greatest civil rights leader of his time and the leading contender for the presidency were both assassinated. College campuses and inner cities were wracked by protest marches and riots. And for a short time in August the city of Chicago was transformed into a fascist police state with theatre of the absurd overtones. It wasn't until the early seventies, ten years after Dylan wrote his song, that the last of that decade's major changes took place: the resignation of Richard Nixon from the Presidency. And that last change, more than any of the others, seemed to fulfill the songwriter's prophecy.

It's been almost forty years since that song was written. And for the

---

* Bob Dylan, *Lyrics, 1962-1985* (New York: Knopf, 1992), 91.

last thirty of those years, the American people – including most Christians – have been lulled by material prosperity into a complacency of narcotic proportion. But the times continue to change and, as citizens of the United States go about celebrating Independence Day this year, the nation – and the national attitude – is different than it was only a year ago, different from what it was forty years ago, and surely different from what it was one or two centuries ago. The obvious and immediate changes brought about by the events of September 11 are significant. But significant too are more subtle changes; most occurring in the national psyche, the public consciousness – in the way we view others and ourselves.

A recent court decision that caught almost everyone's attention was when the Ninth Circuit Court declared the Pledge of Allegiance unconstitutional because it contains the phrase "under God." It reminded me of how, thirty-five years ago, I stopped reciting the Pledge – not because it contained the phrase "under God," but in part because I found unbearable the hypocrisy of reciting those last beautiful words of the Pledge in the context of all the strife of the 1960s. You know those last six words, don't you? . . . *with liberty and justice for all.* A rather fitting follow-up to the phrase, "under God," don't you think? In fact, the only words in our national lexicon that may be more fitting are those comprising the lines of the poem *The New Colossus*, which is found permanently affixed at the base of the Statue of Liberty.

> Not like the brazen giant of Greek fame,
> with conquering limbs astride from land to land;
> Here at our sea-washed, sunset gates shall stand
> a mighty woman with a torch
> whose flame is imprisoned lightning,
> and her name Mother of Exiles.

> From her beacon-hand glows
> world-wide welcome;
> her mild eyes command the air-bridged harbor
> that twin cities frame.
> "Keep ancient lands your storied pomp!"
> cries she with silent lips.

> "Give me your tired, your poor,
> Your huddled masses yearning to breathe free,
> The wretched refuse of your teeming shore.
> Send these, the homeless, tempest-tost to me,
> I lift my lamp beside the golden door!"

But the times they have indeed been changing. And I fear that a concern for "liberty and justice for all" is no longer seen as "in the national interest," much less that the nations of the world ought to send us their tired, poor, and "huddled masses yearning to breathe free." But if the United States – or any nation, for that matter – is to escape the dustbin of history, it will need to change its national character so that it sees itself, *as a nation*, as a servant to the other nations of the world. You see, that's where the Lord was going when, in Matthew 19:30, he said "But many who are first will be last, and many who are last will be first." He stated it more clearly a few verses later (20:27) by saying, "whoever wants to become great among you must be your servant, and whoever wants to be first must be your slave – just as the Son of Man did not come to be served, but to serve, and to give his life as a ransom for many." Those words are for nations, peoples, and groups, as well as individuals and are of critical importance at this time in history. For ". . . the present now *will* later be past / The order *is* rapidly fadin'."

CHAPTER 38

THE VULTURES OF OUR CULTURE

OCTOBER 1, 2003

Back in 1943 C.S. Lewis published a unique book entitled *The Screwtape Letters* (New York: MacMillan, 1962). In 1983 Os Guinness published *The Gravedigger File* (Downers Grove: Intervarsity), a book that might be seen as a take-off on *The Screwtape Letters*. What the two books share in common is an attempt to convey to the reader the point of view of a devil, one of Satan's minions, as he goes about trying to corrupt humankind. In this essay I would, similarly, like you to imagine how a slave of Satan might think, as he attempts to discharge the task of corrupting people in Northwest Iowa, in the fall of 2003.

Let's imagine that this devil is a novice, maybe a bit of a bungler, and is allowed to choose whatever segment of the population he thinks he can handle. He won't choose little children, of course. Although cultivating selfishness will work on them as well as anyone, little children are too trusting of their parents and what their parents have taught them. Corrupting them would take some creativity and some hard work. On the other hand, adults in Northwest Iowa also pose a bit of a challenge, at least when compared to adults elsewhere. Oh, they're corruptible all right. But most of them know the devil's Enemy, are strengthened by years of knowing him, and have gathered themselves into little groups that the Enemy calls his Church. That gives them a kind of resistance to corruption that a novice and bungling devil would find difficult to overcome. That leaves one group that this bungling devil just might be able to handle: a group made up of adolescents and young adults. Having put off the innocent trusting quality of children, but not having the wisdom of experience, they are the most vulnerable to that one powerful tool of evil – conformity. And in Northwest Iowa there are clusters of these vulnerable ones gathered in places called high schools and colleges.

Having solved his demographic problem, and having stumbled upon an effective basic weapon – conformity –this devil must next go about devising a plan. Now you need to remember that Satan's very reason for existing is to rebel against the Creator-Sustainer-Redeemer of the universe, to bring shame rather than honor to his name, and to increase the suffering he bore when he atoned for the corruption and rebellion of the human race. So the plan this devil devises will need to be character-ized overtly by rebellion and perversity. That could be tough . . . except for one very fortuitous (in the eyes of this devil) characteristic of the Northwest Iowa culture. Unlike any previous age, in the fall of 2003 the media has had a profound influence in Northwest Iowa. The Internet is found in almost every home. People subscribe to popular magazines and newspapers just as they do elsewhere. Movies, whether on videotape or DVD, are enormously popular among adolescents and young adults – and the cities of Sioux Center and Orange City now have their own movie theaters. And, of course, television is ubiquitous. It hadn't always been this way. In decades past, movie theaters were rare, neither video-tape nor the personal computer had been invented, and television was watched more sparingly – and more critically. Some people even talked about developing a distinctively Christian approach to media: redeem-ing the popular arts. But in 2003 things are different. The result of this rather recent media renaissance is that the citizens of Northwest Iowa are far more ready to conform to the patterns of the larger pagan culture than ever before. So even a bungling and neophyte devil should be able to devise a plan to corrupt a segment of the Northwest Iowa population.

The first step is to enlist a few of those who have already been cor-rupted, those who are so driven by greed that they are willing to prey upon their own parents, their children, or sell their own souls, just to make a buck. These are the vultures of our culture. The more success-ful ones are the producers and directors, the screenwriters and advertis-ing agents that have made the popular media what it is today. But in Northwest Iowa our junior devil need only find a couple of second-rate vultures. He will simply plant in their minds the idea that money can be made by organizing a party for adolescents and young adults. To insure corruption, this devil and his vultures will depend heavily on conformity while, ironically, creating a pretense of rebellion. But then, rebellion is very often a form of conformity. He will get his victims to think that they are rebelling, asserting their independence, when in actuality they will be conforming – surrendering – to the most corrupt elements in the wider culture.

Now this party needs to have a theme that will be both attractive to adolescent irresponsibility and offensive to the Enemy. Some sort of perversion that the larger culture has romanticized would be best. At first this bumbling devil comes up with the idea of a costume party where the young people come dressed up as sodomites and child-molesters. He figures that would really grieve the Enemy. But then he realizes that this is still only 2003 and those particular perversions have not yet been romanticized. So he spends some time watching a few dozen recent films and a more effective idea dawns upon him. In 2003 the larger culture has, for the most part, romanticized the most decadent aspects of the lifestyle of the inner city dwellers, those on the margins of society, the progeny of institutional slavery, whose broken lives manifest a more insidious kind of slavery, the selling of their own bodies and souls in order to survive. These are the pimps and prostitutes of the inner city. In 2003 the media has romanticized them, virtually legitimizing their corrupt lifestyles in its attempt to be non-condemning, inclusive, and promoting diversity of even the most salacious sort. Playing upon the hip vernacular common to rap music and "Boyz-in-the-hood" type films, this junior devil decides to call the event a "pimp and ho" party. And now he's on a roll. A few more refining touches and his scheme will be complete. He decides that there must be alcohol at this party, particularly since many of the partygoers will be legally under age. And the alcohol will serve two purposes. Not only will it potentially aid in corrupting many of the partygoers, but it will also divert the attention of adults in the community from that most powerful tool – conformity. The perverse theme of the party has the same double effect. Calling attention to itself, it calls attention away from the devil's most powerful tool: conformity.

So now our neophyte devil is all set. The vultures of the culture are ready to swoop down on their prey. The distracters of alcohol and perversity are in place. Now if only this devil could make sure that none of the clergy in Northwest Iowa decide to preach on those words of Paul in Romans 12:1–2:

> Therefore, I urge you, brothers and sisters, in view of God's mercy, to offer your bodies as a living sacrifice, holy and pleasing to God – this is your true and proper worship. Do not conform to the pattern of this world, but be transformed by the renewing of your mind. Then you will be able to test and approve what God's will is – his good, pleasing and perfect will.

CHAPTER 39

TRADITION AND OUR SENSE
OF TEMPORAL PLACE

FEBRUARY 3, 2005

Two experiences in the last few weeks have given me reason to pause and reflect on what it means to be living at the beginning of the twenty-first century. One was viewing the film adaptation of *Fiddler on the Roof,* the musical in which the head of a Russian Jewish family struggles to hold on to "tradition" as he wrestles with the forces of change that threaten his immediate family. The other experience was reading an eye opening and what I believe is a profoundly important new book from InterVarsity Press. Written by Meic Pearse, an English historian, its title is *Why the Rest Hates the West: Understanding the roots of global rage.* It attempts to understand, from a Christian perspective, the truly radical differences between Western societies and what Pearse calls "traditional societies," societies that have not been influenced significantly by the Enlightenment or the Industrial Revolution. The basic thesis of the book is that we in the West are wrong when we attribute to religion or economics, or even our government's foreign policies, the animosity that many inhabitants of the non-Western world have toward us. The people of Iraq and Iran do not hate us because we are Christian and they are Muslim, or because we are extremely rich and most of them are poor, or even because of our support for Israel or our interest in the oil under Middle Eastern soil. Those are factors, of course, but the real problem is much deeper. It has to do with a clash of cultures at the very root of what it means to be a culture.

Pearse makes the point that the cultures of the Middle East and many other non-Western nations of the world are *traditional* cultures. That is, they are cultures that have a long history. And they have a deep attachment to and unity with that history. We in the West, on the other hand, have become what we often call a *postmodern culture.* Some of the chief characteristics of a postmodern culture are individualism, relativism, intolerance for any behavior or thinking that does not fit our understanding of relativistic tolerance, and cultural imperialism. To appreciate the West's cultural imperialism, consider a recent speech given by Prime

Minister Tony Blair before the U.S. Congress where he insisted that, and I quote, "ours are not Western values; they are the universal values of the human spirit." The values that Prime Minister Blair was talking about are the values of individual freedom and relativistic tolerance.

In his book, Pearse argues that what we mean by the word "culture" includes our traditional understanding of religion, but also much more. And non-Westerners "are becoming understandably anxious about the future of" their own cultures, which they feel are being threatened by the culture of the West. More to the point, non-Westerners view Western culture as an "anti-culture." That makes sense when you realize that a culture is defined by the values it holds dear. What the postmodern West holds dear is a commitment to tolerance of all values; in other words, to the anti-value of relativizing those values that define traditional cultures. Thus a culture built upon this "anti-value" is an "anti-culture." Here is how Pearse puts it in his book:

> The truth is that we, in our hyperprosperity, may be able to live without meaning, faith or purpose, filling our threescore years and ten with a variety of entertainments – but most of the world cannot. If economics is implicated in the conflict, it is mostly in an ironic sense: only an abundance of riches such as no previous generation has known could possibly console us for the emptiness of our lives, the absence of stable families and relationships, and the lack of any overarching purpose. And even within us, the pampered babies who populate the West, something – a rather big something – keeps rebelling against the hollowness of it all. But then our next consumer goodie comes along and keeps us happy and distracted for the next five minutes. Normal people (that is, the rest of the world), however, cannot exist without real meaning, without religion anchored in something deeper than existentialism and bland niceness, without a culture rooted deep in the soil of the place where they live. Yet it is these things that globalization threatens to demolish. And we wonder that they are angry?

When I moved to Sioux Center in 1979, one of the things that impressed me most was the sense that I was now living in a place that was self-consciously Christian, and, as a sub-culture, was able to express that Christian self-consciousness in ways that were quite distinct – and unashamedly so – from the larger North American culture. And that wasn't just at Dordt College! Whether I read the *Sioux Center News*, shopped for hardware at Paul and Darlene Moerman's Coast-to-Coast store, bought gasoline at the Co-op Gas & Oil, or simply strolled Main Street on a Sunday afternoon, I had a sense of living in a culture that was different from the mainstream consumer society that I left back in Northern New

Jersey. Some, in a condescending way, would call it "quaint." Some found it narrow and restricting, a busy-body culture where everyone knew everyone else's business. I found it liberating: a wonderful environment in which to raise our three sons. But things have changed a bit in the past quarter century. Cable TV, Wal-Mart, the Internet, fast food, movie theaters, industrial growth, and numerous other artifacts of our Western culture have slowly but surely changed the face of Sioux Center and made it less distinct from that larger, surrounding culture. I don't wish to suggest that any of those artifacts are somehow intrinsically evil and ought not to have been introduced to our distinctive little hamlet. But the fact is that we have become more like the surrounding culture. And like Tevye, the Jewish father in *Fiddler on the Roof* who yearns to preserve "tradition," I lament our loss of distinctiveness. I'm not about to strap explosives to my body and express my lamentation by becoming the first suicide bomber to target Wal-Mart. But I do understand the anxiety with which the people of non-Western, traditional cultures, look upon the West's attempts to "free" them.

CHAPTER 40

DAVID AND HELMUT

SEPTEMBER 2, 2005

The Veteran's Administration nursing home in Paramus, New Jersey, is not up to the standards of the Crown Pointe facility currently under construction in Sioux Center . . . although I've witnessed much more dismal places. My wife and I have visited there regularly now for the past three years, during week-long returns to Saddle Brook, New Jersey. Saddle Brook is where we both grew up, and the town in which my mother-in-law continues to live in the house that she has lived in for the past fifty years. My father-in-law, who turns 85 in September, moved into the VA nursing home almost four years ago. His Parkinson's disease had advanced sufficiently that he was unable to be cared for at home.

For the past few years we've traveled to New Jersey in early August, and during the short week that we are there, we make a number of visits to the nursing home to see my father-in-law. Fred is a big guy and getting bigger. His usual gregariousness has been tempered by the Parkinson's disease and perhaps a bit by the immobility that pretty much confines him to a wheelchair. One thing he does enjoy is eating. And so it's common for us to visit with him at his table, at mealtime in the nursing home cafeteria.

There are three other men at Fred's table. Across from Fred sits a man whose name I do not know. He sits in silence. When his food arrives there is something within him that recognizes it and slowly moves his aged hands and lips so that he, with some difficulty, manages to feed himself. But he never communicates with the other three men at the table or with those of us who visit.

But this essay is about the other two men at Fred's table, David and Helmut. Since they sit 90 degrees from Fred around the small, round table, whenever I pull in a chair next to my father-in-law, I'm also sitting next to either David or Helmut. And I've done that with increasing delight, on a number of occasions, during the last two years. For David and Helmut, both in conversation and in their countenance, bear the image of their Creator in ways that I find worthy of esteem and respect – despite the brokenness that they bear in their bodies and their surroundings.

David is an African-American engineer, and we've talked on a couple of occasions about his career. Educated in electrical engineering, he has done a lot of work in control systems, both in private industry and in the military. His service in the Air Force is what has qualified him for a place in the V.A. nursing home, but he likes to talk more about the guidance systems on which he worked while in industry, or the short stint that he served as a college professor. He didn't enjoy the classroom very much, and I think I understand why. You see, David, beside the Ph.D. in electrical engineering, has a kind of affable but gentle dignity; something, I fear, one must forego if one is to successfully entertain college students.

But David has retained that gentle dignity, and, when I listen to him speak I can't help but wonder, "Why is he here in this nursing home?" His cordiality, verbal eloquence, and poise betray the reality of the diabetes that has taken his legs, and which hangs over his head like the sword of Damocles. It also makes it easy to forget that here is a man who grew up and was educated before the civil rights movement. Yet he shows not the slightest signs of bitterness or abasement, too often the results of struggling to live humanly in a dehumanizing, racist society. Instead, when I look at and listen to David I'm reminded of the Word of the Lord near the end of the sixth day of creation, when he created mankind in his image, looked upon all that he had made, and said, "It is very good."

Helmut sits across the table from David, on Fred's left. He's the only one of the four who can walk to the table and sit in an ordinary chair. The first thing that strikes you about Helmut is his sense of humor. His words are often seasoned with a comic irony enriched by his intelligence, education, and years as a professional musician. Born and raised in Germany, Helmut learned to become a composer of music and to gain expertise in playing the woodwind instruments. He left Germany for the United States shortly after Hitler came into power, and he served for a time in the U.S. military.

This past August I had a most fascinating conversation with Helmut. Unlike Fred, Helmut doesn't enjoy his food and eats very little of it. But motivated by some internal sense of stewardship and frugality, he hates to see it go to waste. And so he is often trying to give his food away to his tablemates or to visitors like myself. After unsuccessfully attempting to give me a half-pint of milk, he turned to me and asked about my profession. I told him that I was an engineer and an educator. And then I added, "who is very interested in philosophy." At the word "philosophy" his eyes lit up. I had connected with his experiences in the arts.

"Philosophy!" he repeated, "do you know who my favorite philosopher is?" Nineteenth century German philosophy immediately came to my mind and I responded, "I'll bet its Friedrich Nietzsche." "That's right!" he responded. And then he asked me if I had read Nietzsche's book, *Thus Spoke Zarathustra*. Well, I had. And so followed a wonderful conversation about how Hitler had misread Nietzsche, and about how Helmut thought that Nietzsche was an optimistic philosopher despite the often misanthropic character of his writings.

After that conversation with Helmut, as I turned back to my father-and-law and my wife, I couldn't help but ask myself again, why is this man here in a nursing home? The answer came a few minutes later when Helmut turned to me and asked, "What is your profession?" We repeated the same conversation about philosophy and Friedrich Nietzsche as if it had never occurred a few moments before. Helmut has lost his short term memory and therefore needs assistance in caring for himself with respect to the ordinary moment-to-moment necessities of life.

Shortly after this conversation with Helmut my wife and I were back in the outside world listening to reports of casualties in Iraq; and, just two days later, battling the madness of bumper-to-bumper traffic as we drove on Route 80, through Gary, Indiana, on our way back to Sioux Center. Reflecting on all that, I came to the conclusion that the Kingdom of God is often more present in places we least expect to find it. For although the men in the Paramus V.A. nursing home are seriously impaired in different ways, they still bear the image of their Creator. And, at times, that image shines very brightly.

Having the opportunity to see past the impairments of David and Helmut, to catch glimpses of their wisdom, humor, and dignity, I was able to see a fifth man sitting at the table with them. A man who in his first official public appearance said, "The Spirit of the Lord is on me, because he has anointed me to proclaim good news to the poor. He has sent me to proclaim freedom for the prisoners and recovery of sight for the blind, to set the oppressed free, to proclaim the year of the Lord's favor" (Luke 4:18). And then I remembered some other words of the prophet from whom he was quoting on that occasion: "He gives strength to the weary and increases the power of the weak. Even youths grow tired and weary, and young men stumble and fall; but those who hope in the LORD will renew their strength. They will soar on wings like eagles; they will run and not grow weary, they will walk and not be faint" (Isaiah 40:29–31).

CHAPTER 41

FROM THE WALL IN YORK

JANUARY 6, 2006

Early on a snowy Sunday morning a few weeks ago, I sat in the enclosed porch on the back of our house and watched the sky slowly brighten. The softly falling snow obscured the sunrise, but the view to the east and south was none the less beautiful. I had enclosed the porch last summer; and with floor to ceiling windows enveloping more than half the room, a gas-fired stove for heat, carpet on the floor, and a comfortable reading chair in which to sit, the new room provides a way of experiencing the out-of-doors while maintaining indoor comfort. Since we live on a ridge on the southeast side of Sioux Center, on a clear day I am able to see many miles toward the distant horizon, despite the presence of the surrounding houses that constitute the suburban development of which our house is one part. For example, the water tower in Orange City, twelve miles to the southeast, is often striking in its round, orange visibility.

On this particular morning, however, the obscuring quality of the lightly falling snow focused my attention on what once was the surrounding prairie. I tried imagining what the scene was like a hundred years ago, with the rolling hills dominated by prairie grass – and perhaps a herd of buffalo nearby. It was then that I was tempted to wish away the other houses, streets, lampposts, and other signs of civilization so that I could glimpse the natural, unadulterated prairie. Notice I said "tempted." I couldn't really bring myself to wish away those artifacts of civilization because I believe that houses, streets, and lampposts are just as "natural" as the prairie grass. They are simply a different kind of nature – cultural nature, if you will. You see I'm an engineer, someone who believes firmly that the Lord has called us to unfold and develop the creation, bringing forth creatures (like houses, streets, and lampposts) that exist only in potential until humankind's response to the cultural mandate brings them into being.

But on this particular morning I was torn. I developed a sense of empathy for my colleagues in the life sciences for whom, it sometimes seems, the only truly beautiful landscape is one that shows no influence of technology. I yearned to see the pristine prairie grass bending slightly under the weight of lightly falling snow, the playful scurrying of prairie dogs, and the slow moving buffalo as the snow creates a cloak of white on their woolly and dark brown backs.

But then I realized that I had seen all these things before. In fact, the imaginative longing that stirred within me that snowy Sunday morning could never have occurred had I not already been acquainted with those denizens of God's good creation. I remembered the film "The Vanishing Prairie" that was produced by Walt Disney back when I was a child in the 1950s. But I also remembered more recent real life experiences of these prairie creatures in Blue Mounds, Minnesota, and in Badlands National Park in South Dakota. It was these memories coexisting with the scene before me that helped create that yearning for a more pristine scene, a yearning that seemed somehow out of synch with my appreciation for technology.

After a few more moments of musing, however, it occurred to me that I have other memories that have the power to create other imaginative longings for very different vistas. In particular, I recalled traveling with my wife two years ago to England. One of the cities that we visited was York, a place whose medieval personality is preserved in the layout of the streets, the character of the buildings, and especially the ancient wall that surrounds the city. Many of the older cities in England were walled cities, built in early medieval times with the need for protection from less civilized neighbors. The wall in York is the best preserved of all these cities, and an ambitious visitor can walk atop the wall, almost completely around the city in just a few hours. From one vantage point on the wall there is an exquisitely beautiful view of the York Minster Cathedral of St. Peter, one of England's largest and oldest churches, built in the year 1470 after 250 years of construction. But the sight I remember best occurred when we stood atop the wall and looked outward from the city center toward the surrounding suburbs. Of course, this being England, the "suburbs" of York were developed during the nineteenth century and so represent the Victorian era, the time in which Charles Dickens lived and wrote about in his many novels and stories. What struck me most about that view were the ubiquitous earthenware chimney pots that punctuated the horizon, telling of a time when the hundreds of aged houses were heated by fireplaces that burned wood or coal. Thinking about that view from the wall in York, and contrasting it with the view from my enclosed porch, I realized that the vision of Victorian society suggested by the one and of prairie grass and buffalo suggested by the other, are both beautiful, God-glorifying, and very much natural in their own way. The tension between biotic nature on the one hand and cultural nature on the other is a false tension, and was here resolved for me by recalling the aesthetic experiences of viewing the once-upon-a-time, prairie-grass-dominated horizon from my enclosed porch, and the Victorian suburbs of Northern England from the wall in York. Truly, as the writer of Ecclesiastes has told us, God "has made everything beautiful in its time" (Ecclesiastes 3:11).

CHAPTER 42

DEATH BEFORE THE FALL?

MARCH 3, 2006

A brief article in the March issue of *The Banner* asks, "Was there death before the fall of humankind into sin?" Reverend Mark Tidd, the author of the article, recognizes that when different people ask that question, they are not always asking the same question (38). Yet for most of the article, Rev. Tidd attempts to argue for the traditional scientific view that death is as natural to the created order as life and therefore that death was present in the original good creation. I think that Rev. Tidd is going to take in on the chin in the weeks ahead as people write in to *The Banner* to express their disagreements. And although I have some sympathy for what he is trying to say, I think that he deserves the criticism he will be getting. His arguments are far too simplistic, and they seem to buy into a naturalistic rather than a biblical view of creation. Let me try to explain.

When we use the word "death" in ordinary conversation, we are usually talking about biotic death. We mean to say that a creature that was once capable of maintaining its unique existence despite its changing parts, and that was once capable of reproducing, is now capable of neither. This broad understanding of death applies, without distinction, to plants, animals, and people. On the other hand, when we use the word "death" in a restricted way, applying it only to people, we invariably are talking about far more than biotic death. In addition to the cessation of biotic functioning, we have in mind the cessation – in this world – of feeling, of thinking, of communicating, of loving, and of worshipping, just to mention of few of the functions that are unique to human beings.

Let me try to clarify this by using a couple of examples. The spinach that I eat as part of my dinner salad was once a living plant. As soon as those leaves are picked, they begin to wither and die. Usually I try to get them into my salad before they wither very much, but once they reach my stomach, they are converted into chemicals that my body can use to sustain its own life. In conventional language we would say that the spinach died. I believe that the Bible teaches that this kind of death is a natural part of the good creation order. Consider Genesis 1 verse 29:

Then God said, "I give you every seed-bearing plant on the face of the whole earth and every tree that has fruit with seed in it. They will be yours for food."

People ate seed-bearing plants before the fall. Therefore there was biotic death, in the simplest sense of that word, before the fall.

On the other hand, my father, who was born in 1915, died in 1988. But his death was not mere biotic death. His functioning as an image-bearer of God in this world ceased along with his biotic functioning. This kind of death, I believe, did not occur before the fall. Consider Paul's words in 1 Corinthians 15:21–22:

> For since death came through a man, the resurrection of the dead comes also through a man. For as in Adam all die, so in Christ all will be made alive.

Well, distinguishing between simple biotic death on the one hand and human death on the other should help us a bit. But as we will see in a moment, there are still many questions left unresolved by that distinction. Before getting to them, however, let me offer a warning about how we read God's revelation in creation and in the Bible. We twenty-first century folk have an inclination to read things – whether in the creation or in the Bible – "literally." That inclination came about primarily during the Enlightenment, when we discovered our ability to reason, and – if you will forgive the pun – let it go to our heads. We interpret all the facts of nature as empirical truth and we interpret all the words and sentences of Scripture as propositional truth. What we have done has been to el-evate our interpreting ability – which is essentially our ability to think rationally – to a position of absolute sovereignty. And that has resulted in all sorts of silliness, perhaps the worst of which is pitting God's creational revelation against his inscripturated revelation. For example, for many centuries we observed the sun rising and setting each day and took that as empirical evidence for the "truth" that the sun moves around the earth, which must be fixed at the center of the universe. Or we read in Genesis 1:5 that "there was evening, and there was morning – the first day" and, despite God not creating the sun or moon until the fourth day, we insist that those first days be 24-hour days.

Being warned against the fallacies of rationalism and literalism in science and in biblical hermeneutics does not solve all our problems with understanding death. For example, the Scriptures seem to suggest that animals fed only on plants before the fall, and from that we might con-clude that animals did not die until the fall. Listen to Genesis 1:30–31 again:

"And to all the beasts of the earth and all the birds in the sky and all the creatures that move along the ground – everything that has the breath of life in it – I give every green plant for food." And it was so. God saw all that he had made, and it was very good. And there was evening, and there was morning – the sixth day.

However, the notion that animals did not die until after the fall into sin does not square with what appears to be the fossil evidence, which strongly suggests that there were plenty of carnivorous beasts around long before the day that humankind was created. That said, there is another fallacy to beware of in these kinds of discussions. We often fail to realize that time is part of creation and that the way we experience time scientifically is only one way of looking at time – and frankly, it is one that does not square with our everyday experience of time. In that regard it's well to heed the words of Peter in his second epistle (3:8) where he writes that "With the Lord a day is like a thousand years, and a thousand years are like a day."

So, was there death before the fall? Yes and no. It depends on how we define death. But even then it's not perfectly clear because our definitions overlap and there is no clear boundary. As a scientist who is first and foremost a disciple of our Lord Jesus Christ, I put more trust in God's scriptural revelation than in his creational revelation. But then again, I recognize that it is my responsibility to interpret both. I try to do that without becoming literalistic with either, by being open to the Holy Spirit's leading, and by steering clear of the notion that my interpretation could ever be codified as an "absolute truth."

PART II
WEALTH, POWER, AND SUCCESS

"Business!" cried the Ghost, wringing its hands again. "Mankind was my business. The common welfare was my business; charity, mercy, forbearance, and benevolence, were, all, my business. The dealings of my trade were but a drop of water in the comprehensive ocean of my business!"
– Charles Dickens[*]

---

[*]   *A Christmas Carol* [1843], in Charles Dickens, *The Christmas Books, Volume I* (New York: Penguin 1971), 62.

CHAPTER 1

THE SIN OF WEALTH

NOVEMBER 24, 1982

If you read the newspaper, listen to the radio, or watch television much these days, you cannot help but be impressed with the idea that there are many people who feel that Western society, the United States included, is struggling with economic problems. For a long time I allowed these reports to go in one ear and out the other, figuring that the media needed something to report, and without the war in Vietnam, a Iranian hostage crisis, or the Prince Charles-Lady Diana wedding, the state of the economy would just have to provide the filler. But lately I've come to the conclusion that Western culture, particularly the United States, really is in deep trouble. We are suffering from the sin of wealth.

Oh, I know what the media is telling us: high unemployment, high cost of living, spiraling inflation, high interest rates. But only a fool would believe one of the fat-cat media people, standing there in their $200 suit, suffering the deprivation of a $60,000 a year salary, and trying to talk about economic suffering. Or Ronald Reagan getting off the helicopter after returning from his millionaire's ranch, trying to encourage us through the "hard times" because there's an economic heaven on the other side of his budgetary programs. Even when the media people interview the supposedly typical working family, it turns out to be a couple or a family of three or four who are "having difficulty making ends meet" because their family income is "only" between $20,000 and $30,000 per year.*

I guess we should expect that. Every sin has its peculiar manifestations, and one of the manifestations of the sin of wealth is that we believe we are poor.

In numerous discussions I've had with other Christians, after considering this problem, someone always asks me, "Is it a sin to be rich?"

---

* According to http://www2.census.gov/prod2/popscan/p60-142.pdf [accessed 1 April 2014] "The 1982 median income for all married-couple families was $26,020 and $30,340 for those with wives in the paid labor force, both down 2.2 percent from the previous year."

Up till now I have always hedged at this point, suggesting that while wealth is a powerful corrupter of men, it still cannot be evil in itself, and so it is theoretically possible to live an obedient Christian life and be wealthy. But that argument is terribly weak and of no useful purpose in our situation. Besides, I don't believe it any more. One cannot live obediently before the Lord in our world and be wealthy. The two are mutually exclusive. An obedient Christian who suddenly, without seeking it, comes upon great wealth, would find need to share it with those all over the world who truly are in need. He would thereafter cease to be wealthy.

Jesus said in Luke 6:20–21:

> Blessed are you who are poor, for yours is the kingdom of God. Blessed are you who hunger now, for you will be satisfied. Blessed are you who weep now, for you will laugh.

He adds in verses 24–25:

> But woe to you who are rich, for you have already received your comfort. Woe to you who are well fed now, for you will go hungry.

And later on in Luke 18:24–25, we read him saying,

> How hard it is for the rich to enter the kingdom of God! Indeed, it is easier for a camel to go through the eye of a needle than for someone who is rich to enter the kingdom of God.

What's wrong with being wealthy? I'd like briefly to identify some concrete ways in which wealth evidences itself as sin. Much of this is taken from a really fascinating book that I read a number of years ago, by Arthur Gish, titled *Beyond the Rat Race*.

First, wealth in the form of material possessions has addictive power. The more we have the more we want. And wealth is basically non-satisfying in that we tire of our possessions and at the same time are always looking to improve them. On the other hand, even though they are non-satisfying, we become dependent on them. Once we get the color TV, we can't imagine settling for only a black and white, much less a radio, or a good book. If we've had two cars in our family we can't imagine going back to just one. And who thinks they can live without air conditioning in the summer once they've spent a summer with it? We become slaves to our possessions to such an extent that we even pay people who con us into thinking that these possessions should be insured.

Early Communist leaders said "religion is the opiate of the people." They're wrong. Affluence is the opiate of the people. Consumer affluence keeps us subservient to the will of the system. The fat cats in Washington and on Wall Street have us in the palm of their hand. Yearly cost-of-living raises pacify us. These are the trinkets thrown to us by the great white fathers who use us to enslave the rest of the world so that they can make their millions.

This point is worth expanding on. Affluence, even when it's not used as a tool of the power brokers in Washington and on Wall Street, still distracts us from the important aspects of life. We cannot be truly busy seeking God's kingdom if we at the same time are seeking material possessions or higher wages.

Wealth creates an illusion of security. We feel somehow more at ease with $10,000 in the savings account than without any savings at all. We put our trust in wealth, especially when we earn it ourselves. The Lord cautioned the Israelites about this when he told them in Deuteronomy 8:11–17:

> Be careful that you do not forget the LORD your God, failing to observe his commands, his laws and his decrees that I am giving you this day. Otherwise, when you eat and are satisfied, when you build fine houses and settle down, and when your herds and flocks grow large and your silver and gold increase and all you have is multiplied, then your heart will become proud and you will forget the LORD your God, who brought you out of Egypt, out of the land of slavery. He led you through the vast and dreadful wilderness, that thirsty and waterless land, with its venomous snakes and scorpions. He brought you water out of a hard rock. He gave you manna to eat in the wilderness, something your ancestors had never known, to humble and test you so that in the end it might go well with you. You may say to yourself, "My power and the strength of my hands have produced this wealth for me."

Wealth encourages false values. This is so obvious that we are forced to concoct euphemisms in defense of these false values. For example, we use the term "profit-motive" because it sounds so much better than what it really means – greed. We use the phrase "enlightened self-interest" when we really mean selfishness. If you have watched commercial television anytime over the last six months you have no doubt seen the "thank you Paine-Webber" commercials. One of the many false values promoted here is that of laziness – letting someone else use your money to earn more money for you, while you sit back and relax.

There are numerous other false values encouraged by wealth. Paternalism, arrogance, vanity, and pride are just a few. One of the most deadly is the elevation of property rights and the corresponding apathy to human rights. Some militarists among us recommend the use of chemical weaponry because it would kill people without destroying property. The infamous neutron bomb is an example of this way of thinking. If your house were on fire, would you run in first to save your child or your stereo system? Our trouble is that our very world is on fire, and we are wrapped up in saving our trinkets while God's children are dying.

There are many other ways in which wealth manifests itself as sin. Its creation of barriers between peoples (the haves and the have nots) and its power to make us fearful and defensive (the truly poor don't invest much in locks and burglar alarms) are just two more manifestations we often overlook.

Excessive wealth is the basic cause of most violence in this world. The truly poor don't start wars, nor do they stockpile nuclear weaponry in an attempt to deter others from stealing their precious standard of living. Virtually all violence, both individual and institutional, is related to the acquisition or protection of material goods. Even from a techno-logical point of view, it can be shown that violence directed against the creation, resulting in energy and environmental crises, finds its root in the increased affluence of our Western lifestyles.

There is much more that can be said, but I would like to close with a word from the Scriptures, which we hear all too infrequently; from James 5:1–6:

> Now listen, you rich people, weep and wail because of the misery that is coming on you. Your wealth has rotted, and moths have eaten your clothes. Your gold and silver are corroded. Their corrosion will testify against you and eat your flesh like fire. You have hoarded wealth in the last days. Look! The wages you failed to pay the workers who mowed your fields are crying out against you. The cries of the harvesters have reached the ears of the Lord Almighty. You have lived on earth in luxury and self-indulgence. You have fattened yourselves in the day of slaughter. You have condemned and murdered the innocent one, who was not opposing you.

CHAPTER 2

WHY HAVEN'T YOU TURNED IN YOUR SWEEPSTAKES CERTIFICATE YET?

JANUARY 6, 1984

"Is that your Publishers Clearing House Sweepstakes Certificate lying on your desk? How come you haven't turned it in yet?" How many times during the last few weeks have you heard that question? Even if you watch TV very sparingly, you can hardly escape the media confrontation with a salesman who seems to be offering you something for nothing – a possible two million dollars, for doing nothing more than mailing an envelope.

Well, what about it? How do you answer that repetitious question asked by the man on TV? I must confess that I usually throw the whole envelope, unopened, into the garbage with all the other junk mail. But this time, thinking it might be an appropriate subject for an essay, I kept it, and even opened it. Inside I'm told I have multiple chances to win a $365,000 "dream home," a $125,000 "luxury houseboat," or a $100,000 "vacation hideaway." And if I'm just perverse enough not to be seduced into a frenzy at the prospect of such "goodies," I'm told I can choose instead to win cold, hard cash. As I turn over to the back of the certificate, I'm told what all that money can do for me:

> Your superprize options are: $365,000 cash . . . imagine you with $1,000 a day, everyday for a year, to spend on cars, trips, furs . . . the choice is yours. Or $40,000 yearly for 25 years to spice up a long happy life.

> Your giveaway #1 prize options are: $125,000 cash to spend any way you want! Invest in real estate, stocks, bonds or any valuable securities you choose. Or $30,000 a year for five years . . . to make your life happier and more secure.

> Your giveaway #2 prize options are: $75,000 cash to pay off bills, splurge on the luxuries you always wanted, keep in the bank to give you ease of mind. Or $20,000 a year for five years on top of your regular income to make your dreams come true.

So what do we say to the question? Why do we throw away unopened the chance to get a million dollars? Is it because we believe it's impossible to win? No, there is some ridiculously small chance that our number will be chosen. Is it because we think that there must be some strings attached? Well, perhaps, but not the kinds of strings that most people would be concerned about. We know they're trying to sell us magazines that we neither want nor need. But we also know that we don't have to buy any. The only tangible string is a 20 cent stamp.

Maybe we throw it away because we're insulted by the gimmickry of the sales pitch and consider it beneath our dignity. I think we should feel offended, but pride is never a very good reason for any kind of action, so there must be another answer.

I'd like to suggest a very concrete way for us to answer the question why we throw away the sweepstakes certificate – a way that will not only stick in your mind, but also speak authoritatively and undeniably to your children. The next time the commercial for the sweepstakes comes on the TV, wait until the question as to why you haven't returned your certificate is asked for the first time. Then quickly turn down the sound to your TV set and move to the back of the room. With all of your friends, relatives, and children sitting there staring at the mute images on the tube, read with a loud voice the following quotes from the Scriptures. First from Ecclesiastes 5:10–11:

> Whoever loves money never has money enough; whoever loves wealth is never satisfied with his income. This too is meaningless. As goods increase, so do those who consume them. And what benefit are they to the owners except to feast his eyes on them?

Then quickly turn to 1 Timothy 6:6–9:

> But godliness with contentment is great gain. For we brought nothing into the world, and we can take nothing out of it. But if we have food and clothing, we will be content with that. Those who want to get rich fall into temptation and a trap and into many foolish and harmful desires that plunge people into ruin and destruction.

Finally turn to the words of our Lord in Luke 12:15 and 29–31:

> Then he said to them, "Watch out! Be on your guard against all kinds of greed; life does not consist in an abundance of possessions. . . . And do not set your heart on what you will eat or drink; do not worry about it. For the pagan world runs after all such things, and your Father knows that you need them. But seek his kingdom, and these things will be given to you as well."

The Scriptures should open our eyes to what is really going on here and provide a solid foundation for a course of action to take so that we are not duped like the rest of the pagan world. Perhaps the very best reason to dump the sweepstakes envelope is simply that we don't need what it has to offer. All of us have plenty of food and clothing and what we need is not more, but rather the grace to be satisfied with what we do have.

However, even if we were in need, there is a principled reason to dump the envelope. The whole idea of getting rich quick, of suddenly acquiring a large sum of money simply by chance, without having earned it in any way, is perverse and contrary to the way the Lord structured his creation. And seeking after that perversity can only occur when our hearts are turned away from service to God and neighbor and greedily turned toward self-service.

Although the Lord gives great wealth to many people, to most wealth is a curse, not a blessing. People sometimes point to Job as an example of a man who had great wealth and was nonetheless considered righteous before God. But listen to what he says about his wealth in Job 31:24–25 and 28:

> "If I have put my trust in gold
>   or said to pure gold, 'You are my security,'
> if I have rejoiced over my great wealth,
>   the fortune my hands had gained . . .
> then these also would be sins to be judged,
>   for I would have been unfaithful to God on high."

Those people who do "win" large amounts of money in sweepstakes usually have their lives wrecked by it because they are not equipped to use it properly.

Our society is one that is based on the faith that money is the route to happiness, and that one can never have too much money. The Publishers Clearing House Sweepstakes is a gimmick concocted by some very greedy advertising agents. It is designed to appeal to the greed of the American masses and get them to buy what are for the most part worthless magazines, which we neither want nor need. When a person sends back the sweepstakes certificate, that person has allowed himself to be manipulated by the greedy publishers and advertising agents, even if he buys no magazines. For he has in effect said to them, "Yes, I agree with you that money means everything, and brings happiness."

There are some interesting verses in the second chapter of the second epistle of Peter that, I believe, speak directly to what is going on here. I quote from the King James Version:

But there were false prophets also among the people, even as there shall be false teachers among you. . . . And many shall follow their pernicious ways. . . . **And through covetousness shall they with feigned words make merchandise of you**. . . . (2 Peter 2:1a, 2a, and 3a)

The next time you hear the question about why you didn't send back the sweepstakes certificate, remember that the true answer is found throughout the Scriptures. It is not the lot of God's people to be made into merchandise by the false prophets of this world.

Chapter 3

Exposing Your Income

October 7, 1985

"Say, what is your yearly income, anyway? Oh! Pardon me. I didn't mean to be offensive. I guess it is none of my business, if you say so."

Have you ever noticed how reluctant people are to reveal how much money they earn? Do you ever wonder why this is so?

When I was a very young child my mother taught me that there are parts of the body should always to be covered when in public. Some of us learned that lesson so well that we actually became embarrassed when in sixth grade gym class we had to take off our pants to do our exercises – even though we had our gym shorts on underneath. But by the time we became adults we understood that there was something very proper and right about maintaining that area of personal privacy.

No one ever taught me, however, that the amount of income I receive is also a part of my personal life that I ought to be ashamed about revealing in public. So I am continually amazed when I encounter a person, especially a confessing Christian, who is reluctant to talk about how much money they earn.

Money is a funny thing. It can be a blessing or it can be a curse. We read in Deuteronomy 8:18 that we must "remember the LORD your God, for it is he who gives you the ability to produce wealth, and so confirms his covenant, which he swore to your ancestors, as it is today." This and numerous other passages indicate that wealth is a blessing from God that is to be used in his service. On the other hand, we also read passages like the one in James 5:1, which starts out, "Now listen, you rich people, weep and wail because of the misery that is coming on you." And this is only one of many passages which warn us against trusting in wealth or in selfishly using it.

In our modern world there is a perspective on money completely different from that found in the Scriptures. It is seen as a source of power, a symbol of status, and a measure of success. People who have a great deal of money are judged to be "successful." High income means high social status, while low income means low social status. Because of this view of money, it is understandable that some people are ashamed to reveal their income in public. If it is low in the eyes of society, then revealing it would be revealing one's failure and impotency. And since our income, at least in the eyes of those who buy into this means of evaluation, places us on a scale of success, and since there will always be someone above us on the scale, then revealing our income will always be an ego-debilitating experience, since it fixes our position on the scale and is a confession that we are somehow less worthy than those above us.

There is another reason people may be ashamed to reveal their income. Most of us do not want to be looked upon as selfish, greedy, and materialistic, even though those are precisely the characteristics that our hedonistic society attempts to cultivate in us. Even the non-Christian has been created in the image of God, and while he may attempt to stifle it, he has some awareness of God's law. Virtually everyone is moved to compassion by scenes of starving children in the third world. But not too many people are moved to do anything about it, since doing so would mean giving up part of what provides them with their sense of security and self-worth – their money. There are social mechanisms in place that allow us to cover over this obvious tension. We become compassionate for others only when we don't have to think about how much money we have. And when we are involved in financial dealings, we make sure that compassion for others is no longer a factor in our thinking.

By making our income a matter of public knowledge we also reveal our hypocrisy. We can't have feelings of compassion for others when our income is dangling in front of our face, for then we would be in danger of doing something about it, like giving our money away. But we can't give our money away, because we find our security in it. To be caught in this dilemma is one thing, but to be caught in it in public is quite another. So to avoid such difficulties we keep our income a secret and whenever necessary pretend that we are not quite wealthy enough to do what needs to be done to exercise our compassion.

For a Christian this problem is multiplied. We know that the Lord requires us to be good stewards of the wealth he has entrusted to us. Yet, unless we are properly exercising that stewardship, to reveal how much money we have is to reveal the extent of our disobedience. The Lord is tough on those who are in love with wealth. Listen to the words of Paul in 1 Timothy 6:9–10 and 17–19:

> Those who want to get rich fall into temptation and a trap and into many foolish and harmful desires that plunge people into ruin and destruction. For the love of money is a root of all kinds of evil. Some people, eager for

money, have wandered from the faith and pierced themselves with many griefs. . . . Command those who are rich in this present world not to be arrogant nor to put their hope in wealth, which is so uncertain, but to put their hope in God, who richly provides us with everything for our enjoyment. Command them to do good, to be rich in good deeds, and to be generous and willing to share. In this way they will lay up treasure for themselves as a firm foundation for the coming age, so that they may take hold of the life that is truly life.

In summary, there are two reasons* why we might be hesitant to reveal our income in public. The first is that we believe that the money we make is an indicator of personal worth and success and we are ashamed to admit that we are less valuable or successful than anyone else. The second is that we have been entrusted with a fair amount of money but are ashamed to admit how very selfish we are in our handling of it.

Both of these reasons are rooted in disobedience and ought never to be used by a faithful Christian. Therefore, the next time you are asked to reveal your income to someone, do so freely. If you hesitate to do so, it will indicate far more about you than your income does. Of course, if it is indeed true that we are rich and selfishly misappropriating what the Lord has entrusted to us, then we better rectify that quickly to avoid potential embarrassment. But we ought to do that anyway, for the day is coming when it will be the Lord who asks for an accounting. And we will not be able to tell him that our income is a personal matter and is none of his business.

---

* Over the years, many Dordt students have made the valid point that there is a third and proper reason for not readily revealing your income: doing so may lead a "weaker brother" into the sin of covetousness and an attitude of jealousy. (2005, Editor)

CHAPTER 4

LIVING VIBRANTLY IN A SLUGGISH ECONOMY

DECEMBER 3, 1991

It seems like every time we sit down to read the newspaper or to watch a news program on TV these days, we are confronted with lamentation upon lamentation regarding the health of what is referred to as "the economy." For almost a year now, we have been told that the economy is sluggish, that consumer spending is down, that unemployment is on the rise, and – oh, the horror! the horror! – that the average middle-class family income, adjusted for inflation, is in decline. Things have gotten so bad in Washington that President Bush may have to take drastic measures soon or his enormous popularity will be eroded away just at the time when he needs it most – the November elections. This morning I heard a hint of what those drastic measures might be as I listened to a report on National Public Radio. No, it's not a cut in taxes, not the establishment of a national health care system, and certainly not any cuts in government spending. No, it seems instead that consideration is being given to invading North Korea, making what are called pre-emptive air strikes against what are believed to be North Korean nuclear arms facilities. This would, of course, start the second Korean war. And since the North Koreans are much better prepared, militarily, than the Iraqis were, there would be (as they say) greater competition. The war might go on for six months rather than just a few days. The average middle-class American would be back in his lazy-boy in front of the TV set with his bucket of popcorn, cheering on the troops, with commentary and instant replay supplied by CNN. Not only would that take people's minds off the sluggish economy, it would put President Bush back in the role he played so well last winter – the conquering hero. And not only that, it would very likely stimulate people to go out and buy bigger TV sets – ones with surround sound so the fighter jets would seem to be coming from behind you as their computers guide them to North Korean targets. And, of course that would help the economy, which is the reason to have the war in the first place, right?

Well, let's hope that scenario is wrong. We have enough violence, carnage, and genocide here in the United States. We don't need to go around the world, forcing it on other nations, shoving it down their throats with an F-14.

But now, what about this sluggish economy, this recession, we are supposedly suffering from? In this essay I want to argue that the problem the United States is facing is far more serious than economic recession. In fact, phrases like "economic recession" and "sluggish economy" are simply euphemisms that gloss over the real problem: the fact that our gods are failing us.

Since the industrial revolution the United States and other western nations have worshipped at the altar of the god of self-centered greed. We have written its commandants on our hearts. "Thou shalt love thyself with all the material possessions possible to acquire. This is the first and great commandment. And the second is like unto it: thou shalt do everything in thy power to get ahead of thy neighbor. On these two commandants hang the free enterprise system and the hope of a better world."

Well, we've faithfully followed those commandments, and where has it gotten us? We have a growing class of politically powerful fatcats, the John D. Rockefellers, Michael Milkens, Lee Iaccocas, and Ivan Boeskys, who will do anything to make their billions, and who are looked up to by the masses as demigods – divinity being equated with success, which is measured, naturally, by the size of one's bank account. We have a shrinking class of politically impotent, pathetic wimps, who convince themselves that they have it tough because their family income is a mere $50,000 a year. This, of course, is the great middle class. Made up of skilled laborers, professionals, and middle managers, these people work in order to make money. These are the "true believers," whose

> . . . faith is built on nothing less,
>   than social status and business success;
> Who dare not trust the sweetest frame,
>   but wholly lean on profit's name.

And lastly, we have the growing class of the disenfranchised: the homeless and the hopeless; those who have lost their faith as they are trampled underfoot by those climbing the ladder of success over their backs. They inhabit our large cities, our mental institutions, and, of course, our prisons. They have no political voice whatsoever because they have no economic power. The god of material greed has swept them away like dandruff from the shoulders of its gray, pinstriped suit. They have become the willing subjects of a lesser god, the god of sensual pleasure, and their short, miserable lives are anesthetized by alcohol, drugs, and sexual promiscuity, until they die of cirrhosis of the liver, a heroin overdose, or AIDS.

We might ask where the Church is amidst this menagerie of monetary miscreants. Sadly, there are Christians in each of the three classes, although the vast majority inhabits the middle class. On Sunday they worship the one true God, but during the week they give themselves up to the god of self-centered greed. They want lower taxes, less government interference in their lives, a higher standard of living, and the security that supposedly comes with a fat retirement package. They admire the powerful fat cats (unless, of course, they're exposed, like Milken and Boesky), despise the disenfranchised (who are too lazy to work for a living), and pity the plight of the poor middle class, who have such trouble these days "making ends meet."

Just yesterday an incident occurred that confirmed to me the schizophrenic quality of the mindset of many middle class Christians. My daughter-in-law received a chain letter in the mail. It wasn't the standard kind of chain letter where you are supposed to send money to a number of people on a list, and then send the letter on to a greater number of new people – all in the hopes that they will send even more money to you. Those kinds of chain letters are illegal. This chain letter is equally nefarious, however. On first glance it seemed to be appealing to the religious character of the person to whom it is sent. Reading with a little more care, however, reveals that its appeal is to superstition and greed. The letter was entitled "With Love, All Things Are Possible." Sounds all right so far. The letter is cryptically signed, "St. Jude," and then, "It works!!" (yes, with two exclamation points). Let me read just the first paragraph:

> This paper has been sent to you for good luck. The original is in England. It has been around the world nine times. The luck has now been sent to you. You will receive good luck within four days of receiving this letter - providing you, in turn, send it on.

If you are like me you will react to the phrase "good luck," judging that the sender is some sad and seriously misled soul who is both gullible and superstitious. But what's really interesting is the form that this "good luck" is supposed to take. The other five paragraphs describe reputed accounts of people who received the letter and either sent it on as instructed or ignored it. Those who sent it on allegedly received "good luck." That good luck, however, was always in the form of money, except in one case when a person supposedly received a new car. So in spite of the fact that this chain letter was not originated and propagated with the intention of immediately producing money, its superstitious purpose is to produce good luck – which usually comes in the form of money. The god of self-centered greed is often wrapped in a sacred (or in this case, superstitious) shroud to make its worship more palatable to Christians.

So what are we to do about all this? How can we free ourselves from the power of this miserable god? How can we communicate to our brothers and sisters in Christ that worship of the god of self-centered greed is hatred of the one true God, the God of Abraham, the Father of our Lord Jesus Christ? Here are three suggestions to consider. First, stop working for money. Work ought to be our response to the Lord's call that we serve our neighbors. The money we earn only serves to enable us to continue that work. People who work in response to the Lord's call never have serious money problems, no matter how much or how little they earn. If you find yourself working for money, consider quitting your job and finding one that is meaningful, regardless of the pay. If you can't find meaningful work that will enable you to be who the Lord calls you to be, then continue in your meaningless work, but find some volunteer project to get involved in. The time and energy commitment that you will gladly make will serve to ward off the god of self-centered greed.

My second suggestion is to ignore the ranting of the Republicans and Democrats. Their gods are political power and self-centered greed. You should have nothing to do with such polytheistic organizations unless it's to infiltrate and subvert them. Everything they preach is based on the gospel of materialism.

Finally, when you and your family are watching the news and you hear about the lamentable condition of the economy and the sorry economic situation of middle class Americans, laugh! Laugh hard. It will teach your kids a healthy contempt for the god of self-centered greed. And it will annoy the hell out of the devil.

CHAPTER 5

POVERTY AND PUBLIC JUSTICE

MARCH 3, 1992

The preoccupation of today's presidential campaign rhetoric with respect to the so-called economic crisis has, like a skilled magician's slight-of-hand in a circus side show, tricked its audience into thinking that a woman has been sawn in half, when in fact something far worse but less visible has occurred: the woman has been degraded – reduced to a stage prop, an economic object to assuage the greed of the show's owners. The Sununu-esque magic of 1990's Reaganomics has brought the public to believe that having to raise a family of five on a $30,000 per year salary is akin to suffering the slings and arrows of outrageous fortune, and that having to depend on unemployment payments for a few months while searching for a new job is a virtual death sentence. The reality is that while the public's eyes have been selfishly focused on its own pocketbook, Ebenezer Scrooge has continued enriching his coffers at the expense of Bob Cratchit – and the prognosis for Tiny Tim's survival is bleak.

Now we ought not to deny that losing one's job can be a traumatic experience. And raising a family of five on $30,000 per year no doubt requires some effort at economic self-discipline – especially if the children in that family are teenagers with hopes of a college education. The difficulty is that preoccupation with the minor discomforts of the American middle class has blinded the public's eyes to the enduring problem of poverty in our nation. On the streets of our major cities, in the hills of Appalachia, and on the reservations of the Midwest, there are children suffering from malnutrition; there are men and women with spirits so crushed by despair that they give themselves over to alcohol and drugs; and there are infants yet to be born who will never live to see their first birthday because their mothers can't afford prenatal care. Assuming that we can wrench our concentration free of the magic show, what should be our attitude toward those who are truly suffering economic deprivation in our society?

When Ebenezer Scrooge was asked to make a charitable contribution in consideration of the "hundreds of thousands [who were] in want of common comforts," he replied, "Are there no prisons?. . . Are there no workhouses?" And when told that "many can't go there; and many would rather die," he replied, "If they would rather die, they had better do it, and decrease the surplus population." Today's Scrooges ask, "Is there no welfare? Is there no public school system? Are there no equal opportunity laws? Isn't this America, where everyone has the ability to pull himself up by his own bootstraps?"

Such lack of compassion and outright ignorance is rightfully disdained by a majority of American citizens and is denounced as sin by virtually all who call themselves Christian. But there are other attitudes that, while not as ugly on the surface, are equally destructive. One such attitude, rooted in a faith in individualist capitalism, is often referred to as "the trickle-down theory." According to this theory, as the rich get richer, the poor will benefit from those riches as well. But the facts blatantly contradict the argument. There are more people suffering and dying today from poverty-related causes than there ever was. A better response to the trickle-down theory, however, was given by Abraham Kuyper just over a century ago in the Netherlands. Speaking to a group of Christian citizens concerned about *The Problem of Poverty* in their country ([1891] Sioux Center: Dordt College Press, 2011), he said:

> The tremendous love springing up from God within you displays its radiance not in the fact that you allow poor Lazarus to quiet his hunger with the crumbs that fall from your overburdened table. All such charity is more like an insult to . . . the poor man. Rather, the love within you displays its radiance in this: Just as rich and poor sit down with each other at the communion table, so also you feel for the poor man as for a member of the body, which is all that you are as well. . . . You, too, must suffer with your suffering brothers. Only then will the holy music of consolation vibrate in your speech. Then driven by this sympathy of compassion, you will naturally conform your action to your speech.
>
> For *deeds* of love are indispensable. (68–69)

A second argument that often surfaces to justify our less than Christ-like concern for the plight of the impoverished is grounded in the words of Jesus in Matthew 26, where he says, "The poor you will have always with you. . . ." Listen to how Abraham Kuyper responded to that:

> If we have food and clothing, then the holy apostle demands that we should content ourselves with that. But where our Father in heaven wills with divine generosity that an abundance of food grows from the ground, we are without excuse if, through our fault, this rich bounty is divided so unequally that one is surfeited with bread while another goes with an

empty stomach to his pallet, and sometimes must even go without a pallet. If there are still some who, God forgive them, try to defend such abuse by an appeal to the words of Jesus, "For ye have the poor always with you" (Matthew 26:11), then out of respect of God's holy Word I must register my protest against such a misuse of the Scriptures.'

'The[se] words . . . give no rule, but merely state a fact. There is no implication that it *should* be so, but at most that such will be the case. Second, it will not do to conclude from this statement that Jesus is giving a prophecy about *later* ages. In the third place, we often completely overlook the reproach that hides in these words. The Greek actually means not "with you," but "in life as you are patterning it, you will always have the poor." This was said to Judas and his like, men who carry the purse and use it like Judas. (55–56)

Finally, there is an attitude prevalent among many Christians that seems to suggest that our proper response to poverty ought to be simply to preach the gospel of the coming Kingdom, where all sorrow, pain, and suffering is banished. After all, it's argued, you can't change people's lifestyles without changing their heart direction. Well, that may be true. But the assumption here is that it is the poor who are in the greatest need of change, and that the change hoped for is that they will be content with their lot. Such pie-in-the-sky evangelism is foreign to the witness of Jesus' life, and outrageous hypocrisy to most Christians and non-Christians alike. Kuyper spoke eloquently to this point when he said:

Every man's inner sense of truth rebels against a theory of eternal happiness that serves only to keep Lazarus at a distance here on earth.

There cannot be two different faiths – one for you and one for the poor. The question on which the whole social problem really pivots is whether you recognize in the less fortunate, even in the poorest, not merely a creature, a person in wretched circumstances, but one of your own flesh and blood: for the sake of Christ, *your brother*. It is exactly this noble sentiment that, sad to say, has been weakened and dulled in such a provoking manner by the materialism of this century. (67–68)

I've been quoting Abraham Kuyper from a book titled *The Problem of Poverty* edited by James Skillen. It's a translation of his opening address at the First Christian Social Congress in the Netherlands on November 9, 1891. I whole-heartedly recommend it as highly relevant to our present situation. To close this essay, however, I wish to quote from the words of our Lord (Luke 4:16–21) at what was perhaps his first public speaking opportunity.

He went to Nazareth, where he had been brought up, and on the Sabbath day he went into the synagogue, as was his custom. He stood up to read, and the scroll of the prophet Isaiah was handed to him. Unrolling it, he found the place where it is written: "The Spirit of the Lord is on me, because he has anointed me to proclaim good news to the poor. He has sent me to proclaim freedom for the prisoners and recovery of sight for the blind, to set the oppressed free, to proclaim the year of the Lord's favor." Then he rolled up the scroll, gave it back to the attendant and sat down. The eyes of everyone in the synagogue were fastened on him. He began by saying to them, "Today this scripture is fulfilled in your hearing."

As Christ's disciples we are not called to merely read these words, and certainly not to spiritualize them away. We are called to live them.

CHAPTER 6

WORDS AND BOTTLES

FEBRUARY 2, 1999

Words are like empty bottles. We pour into them the meaning that we as a culture – or as a subculture – agree they should contain. Thus, when we use a word, it's like drinking from a bottle that has been filled by a particular company, like Coca-Cola, Blue Bunny, or Anheuser-Busch. It's always important to read the label before taking a sip.

In a little book titled *On Being Human* (Burlington, ON: Welch, 1988), Calvin Seerveld, at one point, reflects on the royal priesthood (1 Peter 2:9) we are called to as Christians. With reference to the task we have been given of reforming science and communication he says,

> Christ's ambassadors will fight the secularized pragmatic quantification of knowledge and speech not by being "tops in the field" but by redirecting knowledge-gathering and by selecting differently the news for distribution within a labour-intensive, face-to-face teaching and telling that keep knowledge and information open to a homely wisdom. (83)

Each year, in a class on technology and society, I read this quote to a group of senior engineering students. And each year I get a similar response: "What? Not strive to be *tops in the field*! What's he talking about? Of course we want to be tops in our field!" Then we have a long discussion in which I try to convince them that "tops in your field" is a phrase whose meaning is very specifically defined by our secular culture. It's rooted in the social Darwinist notion of "survival of the fittest." It suggests competition – cutthroat competition – if not to biotic death, than certainly to professional death, for only one can be tops in her field. And it demonstrates a concern for one's professional self that is placed ahead of a concern for the wellbeing of others. As such it is diametrically opposed to the biblical notion of servanthood. It runs completely counter to Christ's words (Mark 10:42–45) where he says,

> You know that those who are regarded as rulers of the Gentiles lord it over them, and their high officials exercise authority over them. Not so with you. Instead, whoever wants to become great among you must be your servant, and whoever wants to be first must be slave of all. For even the

Son of Man did not come to be served, but to serve, and to give his life as a ransom for many.

There are many other words and phrases that, like bottles, have their content poured into them by the secular culture in which we live. And the extent to which we readily use those words and phrases demonstrates the extent to which we have appropriated a non-Christian worldview and compromised our witness. Imagine that you were in the process of starting a biblically obedient business enterprise. There would be a host of commonly used words and phrases that you would want to avoid. How about words like "boss" or "employee"? Or what about the title "director of human resources"? Each of these has poured into them the secular notion that a worker is nothing more than a commodity to be bought and sold. True, it's not quite like the slavery that stained the reputation of this nation two centuries ago. At least it's the worker who sells his or her self. Thus it's more akin to prostitution than slavery. But neither slavery nor prostitution is what the Lord has in mind when he calls us to be servants, one of another.

The examples I've used in this essay have all been related to the spirit of individual self-promotion. There are other spirits of our age that pour content into common words and phrases. Consider the spirit of technicism, which suggests that more and more technology is the solution to all of our problems. Some words that we ought to use very carefully if we are to be sensitive to this particular evil are "efficiency," "productivity," and "maximize."

The Lord says "by their fruit you will recognize them" (Matthew 7:16). Included among our differentiating fruits are the words and phrases that we use. So it may be helpful to keep in mind the bottle metaphor. Words *are* like empty bottles. It's not the bottle. It's what is poured into it that can either nourish or poison us. For the sake of the Kingdom of God we need to be discerning about the contents of the bottles we use.

CHAPTER 7

SERVICE AS COMMODITY?

FEBRUARY 3, 2003

One of the gravest evils that humankind has inflicted on itself through-out history is slavery. Even in the twenty-first century, the memories and repercussions from the days when America treated the majority of its black citizens as "commercial wares to be bought and sold" haunt the nation. But as evil as slavery is, it is only a symptom of a more basic sin. That sin is greed, avarice, the self-serving seeking after wealth. Certainly slavery adds to that the sin of cruelty, lack of compassion, or lack of love toward one's neighbor. But the origin or starting point of slavery is not lack of love; it is the self-serving seeking after wealth. And that is surely confirmed by Paul's assertion in 1 Timothy 6:10 that "the love of money is a root of all kinds of evil." Slavery treats people as commodities.

Now commodities are merchandise. They are commercial wares: things that are bought and sold. Today in North America we do not treat people directly as commodities. That is, we do not practice slavery in the classic tradition of buying and selling human beings. But – in perhaps a less overtly cruel, but more insidious manner – we do sustain the spirit of slavery by treating certain essential human services as "merchandise to be sold at a profit."

If one concedes that, because of sin, a kind of modified free en-terprise system – a system built upon competition – is best for the de-veloped nations at this point in history, then there certainly must exist services that legitimately may be treated as commodities. I would guess that trading in stocks – that is, the work of a stockbroker – would be considered a legitimate service to sell for a profit. After all, the purpose of trading in stocks is to gain a profit in the first place. Likewise many forms of entertainment might be considered legitimate commodities since they are discretionary services, services that people don't really need. But what about other, more essential services? We certainly don't think of church services or the pastoral services of the clergy as merchandise to be bought and sold. That's why there is no admission charge when you go to wor-ship on Sunday. Sure there's an offering. But no one is compelled to put anything in the offering basket. The services of the church are not commodities. Despite the distortions of TV evangelists, the church does not exist to make a monetary profit.

Other forms of essential services have been more ambiguous regarding their status as commodities to be bought and sold. Consider the supplying of water or the removal of sewage and garbage. These services are usually either performed by or contracted by a local community that will charge its citizens for the services, but will not make a profit. That community may work with a provider that *is* making a profit, but the community serves as an economic buffer, isolating its citizens from the vagaries of the marketplace, protecting them from being plundered by profit-seekers.

But now consider one of the most important, essential services of our modern age – that of health care. In most counties of the developed world people are provided health care by their government in ways that protect them from profit-seekers. But not in the United States. Here health care is a commodity to be bought and sold. There are some safety nets, so to speak, like Medicaid, that serve a very small portion of the population. But that's the exception. Consider specifically health insurance. Unless you are blessed with an exceptional health care plan where you work, health insurance is both very expensive and very elusive. If you have a health problem, for example, the health insurance companies will deny you insurance. Think about that. The people who most need the service are denied it. And why? Because the health care insurance provider is seeking to make a profit – and it is not profitable to insure sick people. It's only profitable to insure healthy people who will pay the premiums but rarely require the services.

The result of all this is that increasing numbers of people are being made subject to the avarice of health care executives who seek first not to provide an essential service, but to make a monetary profit. In doing so, they ignore the needy and make merchandise of out healthy people. And that is our modern way of practicing slavery – more subtle, to be sure, but just as evil as classic, pre-civil war slavery. It makes me think of those words of Paul in 2 Peter 2:1–3, where we read about the only legitimate form of buying human beings: our redemption by Jesus' blood. And we also get a strong sense of Paul's (and God's) contempt for the self-centered, profit-seeking enslavers of this world. Listen; this is God's Word:

> But there were false prophets also among the people, even as there shall be false teachers among you, who privily shall bring in damnable heresies, even denying the Lord that bought them, and bring upon themselves swift

destruction. And many shall follow their pernicious ways; by reason of whom the way of truth shall be evil spoken of. And through covetousness shall they with feigned words make merchandise of you. . . .

Don't you just love that old King James translation?

CHAPTER 8

POLISHING IMAGES AND HIRING CONSULTANTS

AUGUST 2, 2004

Earlier this summer I read two books, the messages of which are being acted out with uncanny precision in these days of political conventions, news from Iraq, and other less disturbing forms of entertainment. The first book, *The Image: A guide to pseudo-events in America*, was written by the historian Daniel J. Boorstin in 1961 (New York: Random House). In it, Boorstin argues that our society is in the process of replacing substance with the illusion of fabricated images based on extravagant expectations. The second book *Life: The Movie – How entertainment conquered reality*, written by Neal Gabler and published in 1998 (New York: Random House), is a more recent work. But its theme is similar. Gabler argues that our society has become entertainment-driven and celebrity-oriented, replacing the substance of real life with the inauthentic fluff of theatrical impressions. The insights gleaned from these two books have helped sensitize my Christian worldview so that I retain a healthy skepticism regarding images communicated by the media. I'm thus very much aware that what I read in the papers, see on television, or hear on the radio regarding the presidential candidates, the turmoil in the Middle East, and the state of the economy are distortions of reality, images fabricated and shaped for some purpose other than communicating the truth.

But reading these two books has also sensitized me to illusions that are formed close to home. For example, a small part of my work at Dordt College is to help communicate to high school age young people that a scientific and technological education is best acquired at a Christian college. By reading these books, however, I'm reminded that college advertising and recruitment has become the very big business of fabricating and selling illusions. Create in the mind of the potential student the fantasy of college life filled with fun and frolic followed by a career of personal advancement and money-making – that is, create such a fantasy better than your competitors can create – and you will win the hearts and minds, as well as the tuition dollars, of many of today's young people. But of course for me or anyone else at Dordt to do that would be to sell our institutional soul and to make a sure reservation for the fitting room where we might be measured for millstones, in anticipation of our being thrown into the depths of the sea (Matthew 18:6).

So I'm thankful for Boorstin's and Gabler's books, and for the insight and prophetic sensitivity they provide to those of us who want to live obediently in a culture that incessantly invites us to full participation in Pinocchio's "Pleasure Island." But there are more subtle ways in which this cultural problem of fabricating images rather than dealing with reality can affect us; and in the remainder of this essay I wish to point out just one of them.

It seems to me that living in a culture shaped by fabricated images rather than by substantive reality amplifies the apparent distance between those who have achieved a modicum of success in any endeavor, and those who have not. You see; it's very easy to create an image once you achieve any level of success – you simply polish your success story. Whether it's in business or politics, education or technology, if you've achieved any amount of success, you can easily shape that story of success to create a rather attractive image. Attractive to whom, you might ask? Well, attractive to those who are involved in similar endeavors but who have not yet achieved success. As a result of this, two interesting kinds of phenomena have arisen in our society. First is a general lack of confidence that people acquire when they compare themselves to images of success. For example, let's suppose I wanted to start an engineering program at a small Christian college. If I spend a lot of time with the images of engineering schools portrayed by the MIT's, Cal Tech's, or even the Iowa State's of the world, I might develop a bit of an institutional inferiority complex. But if I'm relatively ignorant of those images, knowing only the substance of a few engineering programs that I've carefully researched or of which I've actually been a part, then I'll have much greater confidence in the possibility of achieving my goal.

The second phenomenon is related to the first. If I become sufficiently intimidated by the image of success portrayed by those who have come before me, I may feel compelled to run to one or more of them for help. Knowing this, many of those in our society who have had a little success, and who have created an impressive image based on that success, are ready to offer their advice and services – at a price, of course; they have become consultants. And sadly, many of us in the Christian community are eager to pay whatever it costs to get that advice and those services. And the saddest thing of all is not the money we are wasting in doing something we might very easily do ourselves, rather it's the nature of the advice and services that we accept. You see, the vast majority of consultants – whether in education, business, engineering, or whatever – work from out of a secular perspective and will be of little help to us in our efforts to further the Kingdom of God. In fact, their advice will likely be counter to those efforts.

So *my* advice – which, by the way, is free, except for the time you've spent reading this essay – is to avoid consultants, don't be overly impressed by the images of success that come to us via the media, and consider taking seriously that old saying, "if you want something done right, do it yourself." If that sounds a little arrogant to you, consider another saying you may have heard before: "you are not your own" (*The Heidelberg Catechism*, Lord's Day 1: Q&A #1). Ironically, there's a kind of legitimate confidence that arises from the knowledge that we are completely dependent creatures, called to a task "for such a time as this" (Esther 4:14) . . . by a Creator who, more often than not, equips us well for the tasks to which he calls us.

CHAPTER 9

TOWARD A MODEL OF
CORPORATE RIGHTEOUSNESS

MAY 4, 2005

A few weeks back I had the opportunity to sit in on a panel discussion with three businessmen, one from a multinational chemical company, one from a local biotechnology company, and the other – the moderator of the panel – from the Dordt College Business Department. The discussion was billed as "Christians in Corporate Community" and the topics of discussion were to range broadly from the nature of business and the role of the Christian in business, to the effect of globalization and multinational corporations on society. I may have been asked to serve on the panel because I teach a course to senior engineering and computer science students that asks them to design – on paper, of course – an ideal business that would be pleasing to the Lord: a "Christian business," for want of a better descriptor. I've also had the opportunity during the last year and a half, in a very small way, to provide some advice and assistance to one of my sons – who was also one of my engineering students – as he found himself in the situation of being able to take over and give direction to an engineering company here in Siouxland.

Well, as it turned out the panel discussion was all about globalization and I was left with two pages of notes on a Christian approach to business that I never had chance to put into words for people to hear – until now. So if you, dear reader, will indulge me, I'd like to suggest a few things about what a model of corporate righteousness might mean.

What is the purpose of a company and what is the nature of work? Both of those questions can be answered with the same word: service. The Lord calls all his creatures into existence for service. We individuals are called to serve our fellow humans as well as the nonhuman creation. Companies are entities that we create in order to serve more effectively; they might be called communal institutions for service. They exist to produce goods and services, where "goods" are simply embodied forms of service. Therefore individual members of a company – often referred to as "employees" – are "servants in community." There is always, of course, an economic dimension to any company whether it is classified as "profit" or "nonprofit." The company must "earn" sufficient monetary resources to enable it to perform its service. And the individual members of the company must be given sufficient monetary resources to enable them to do their work as well as discharge the other responsibilities that they have in connection with family, church, community, etc.

Is there such a thing as a "Christian company?" I don't think so. Christians are people, image bearers of God who are called to responsible selfhood and who identify themselves as disciples of the Lord Jesus Christ. Companies – like buildings, cars, and telephones – are technological artifacts. They serve as a communal response to the Lord's call to service. They will always demonstrate elements of obedience and disobedience, righteousness and unrighteousness, regardless of whether they are run by Christians or not. Some companies may want to identify themselves as "Christian." What that may mean for a particular company seems to me to be uncertain. What it, perhaps, ought to mean is that the company is run from a distinctively Christian perspective, it is assumed that everyone in the company is a self-declared Christian, and the company seeks to be a "corporate witness to the truth," a "light shining in the darkness of global capitalism," a model of corporate righteousness that consciously and deliberately seeks to bring shalom to the world of business. The same thing might be said for any college that dares to call itself "Christian."

Well this all sounds very nice in theory, but what does it mean specifically for a modern, technological company? How might such a company serve as a model of corporate righteousness and obedience? Well, let me try to answer that with my son's company in mind, Adams Thermal Systems of Canton, South Dakota. It does not identify itself as a "Christian company"; a number of its employees would likely not identify themselves as Christians. But the intent of the owner is to run it in obedience to the Word of the Lord for technological corporations. And by the way, these are the words of an enthusiastic father and professor. I can't claim to speak in any detailed way for my son or his company.

First of all a model company will provide a valuable, high-quality, and needed product, based on its intellectual and infrastructural resources. In this specific case the product is heat exchangers that are based on a solid understanding of modern heat transfer technology. But a model company will also be looking to the future and anticipating societal needs that it may be called to meet. For example, when your corporate expertise is in heat transfer technology, you will be very conscious of the need to be good thermal energy stewards. You will be aware of how the nature of thermal energy resources are and will be changing in the future. Right now you may be supplying the major agricultural implement manufacturers with oil coolers for such things as tractors and combines. But as the need to conserve petrochemicals gets increasingly critical, you may be called to use your expertise in a wider range of thermal energy technology; for example, space-heating, general transportation technology, and innovative ways of improving the thermal energy stewardship of American industry in general.

Secondly, a model company will aim to structure its own organization in ways that respect the expertise and the responsibility of all persons who work for the company. Mind-numbing, repetitive work will be eliminated or performed by machines. And everyone, from the manual laborer on the assembly floor to the plant engineer will play a responsible role in enabling the company to serve effectively.

Third, the company will seek to meet the needs of all those it employs, starting with the basic needs of nutrition, housing, and health; including the need for meaningful and responsible work; but also maintaining a concern for higher level needs, such as aesthetic, ethical, and faith needs, by insuring that all persons have sufficient time, resources, and encouragement to address them. One way one does that is by imagining, and then developing, a radically biblical salary and benefits program.

Fourth, a model company will create relationships with vendors and customers that are based on mutual trust, honesty, and a desire for mutual service – even if some of its vendors and customers are initially hostile to anything but dog-eat-dog competition. After all, it will take a model of corporate obedience to help them change their ways.

Fifth, a model company will relate to its community – both local and global – in ways characterized by justice and righteousness, in ways that foster shalom.

Finally, in all of its activities, a model company will demonstrate a Christ-like alternative to adversarial competitiveness. In terms of the current standard for American businesses, it will be counter-cultural.

Impossible? Wildly idealistic, you say? I don't think so. Keep your eye on Canton, South Dakota.

Chapter 10

The Price of Gasoline

November 4, 2005

For the past few months weather events, hurricanes mainly, and the price of fuel, mainly gasoline and natural gas, have been prominent in the news. And if you believe what you hear and read, then you are led to see a necessary correlation between those "natural disasters" and the rise in the price of fuel. The explanation is given as "supply and demand." Natural disasters reduce the supply of refined oil, the demand for it remains unquenched, and therefore the price "naturally" rises. I say "naturally" because most people seem to believe that the law of supply and demand is as natural as the law of gravity.

But other things have been happening as well during the past few months. One is that we are getting closer the Christmas. And as Christmas approaches I can't help but think of Charles Dickens. This Christmas I've promised myself that I will read *A Christmas Carol* (1843) again. And I eagerly look forward to once more reading that part, early in the story, where Ebenezer Scrooge is confronted by the ghost of his recently deceased business partner, Jacob Marley. Marley is trying to convince Scrooge to change his greedy and self-centered ways. But Scrooge doesn't seem to understand, and responds fearfully to the apparition by telling him what a good businessman he was during when he was alive. Then we get my favorite line in the whole book. In frustration, the ghost raises his arms, rattles his chains, and we read...

> "Business!" cried the Ghost, wringing its hands again. "Mankind was my business. The common welfare was my business; charity, mercy, forbearance, and benevolence, were, all, my business. The dealings of my trade were but a drop of water in the comprehensive ocean of my business!"

Poor Jacob Marley. He was in life, as Scrooge said, a "good businessman." And in the nineteenth century, which was when Dickens was writing and which was the setting for most of his novels, the spirit of the age was naturalism. Naturalism is the belief – I dare say, the faith – that everything that happens, happens as a result of natural causes. In other words, there are laws of nature that govern all events. Some of those laws we already, pretty well, understand, like the law of gravity. We are working on under-standing the rest. And, one day, because the laws of nature are expressible in the language of mathematics, which is rooted in "iron-clad" logic, we will be able to understand everything.

One of the "laws of nature" that got a lot of attention in the late nineteenth century was that of "natural selection." It was based on the observation that in the animal and plant kingdoms, the strong survive and the weak die out. Charles Darwin used that as at least partial ground for his theory of evolution: earlier forms of life evolve into later forms by a process in which random changes give a competitive advantage to certain members of a species. Those changes are passed on to the descen-dants of a privileged few, and that line of descent survives while other lines die out. In more colloquial language it became known as "survival of the fittest."

The nineteenth century was also the time of the Industrial Revolution in Western culture. In addition, the secularizing forces of the Enlightenment were holding sway; humankind was seen as its own master, with no need for belief in a Creator to whom humankind owed service, allegiance, and worship. And so was born, by the combination of all of these forces, the myth of "free enterprise." That's the idea that if you simply leave people alone to exercise their absolute freedom and their natural inclination to survive and progress, the most energetic among them will create great industrial enterprises. And as those enterprises struggle against each other, the best – that is, the most "fit" – will sur-vive and the weaker ones will die out. This complex ideological system, resting as it is on a curious conflation of belief in naturalism and human autonomy, came to be known, at times, as social Darwinism. That term is not used very often today, of course. The reason being that many of those who hold to the naturalistic ideology of free enterprise call themselves Christians, and Christians, of course, have problems with the atheistic aspects of Darwinism.

And so here we are at the beginning of the twenty-first century, with the price of fuel escalating and the poor facing a cold winter. And it's all due to natural disasters like hurricanes and natural laws like supply and demand, right? No, no, no, no, no; it's all due to an ideology grounded in the belief that the law of supply and demand is a natural law. It's due to the belief on the part of entrepreneurs that they should charge whatever the market will bear, because that's good economics, that's good business, and the human species will somehow improve if they do so.

But Christmas is coming. And hopefully a few of those entrepreneurs who are committed Christians, despite their bowing the knee to the god of naturalism, will read Dickens' *A Christmas Carol*, and will understand what the ghost of Jacob Marley was trying to tell Ebenezer Scrooge. And if their hearts are softened they may repent, turn to the Word of God, and hear afresh Paul's words to Timothy where he writes about business and economics, saying

> But godliness with contentment is great gain. For we brought nothing into the world, and we can take nothing out of it. But if we have food and clothing, we will be content with that. Those who want to get rich fall into temptation and a trap and into many foolish and harmful desires that plunge people into ruin and destruction. (1 Timothy 6:6–9)

CHAPTER 11

AUTHENTICITY, HYPOCRISY, AND APING THE WORLD

APRIL 7, 2006

Each year in the spring, as part of my course in Technology & Society, I give a lecture on the reactions of romanticism and existentialism to the effects of technology on our lives. The core of the lecture contrasts the sin of hypocrisy with the virtue of authenticity, and near the end of the 90 minute session I turn to those words of our Lord in Matthew 23, where he rails against the hypocrisy of the Pharisees. Although I read the whole of that chapter, the key verses are the first few where Jesus says,

> "The teachers of the law and the Pharisees sit in Moses' seat. So you must be careful to do everything they tell you. But do not do what they do, for they do not practice what they preach. . . . Everything they do is done for men to see." (Matthew 23:2–3, 5a)

Reading those verses to my class often stimulates me to write something on the kinds of hypocrisy that too often creep into our lives, especially our institutional lives. This year I am doubly motivated to write because, shortly after that class where I referenced Matthew 23, I read the last chapter of Nancy Pearcey's recent book *Total Truth: Liberating Christianity from its cultural captivity* (Wheaton, IL: Crossway, 2005). And in that last chapter was a section that dealt with the problem of institutional hypocrisy and authenticity. It is so good, so faithful to the spirit of Matthew 23, that I will cite her at length.

Pearcey is concerned with the way in which Christians have a tendency to separate what we believe from how we live, categorizing our beliefs as "spiritual," belonging to what she calls the "higher story" of our two-tiered existences, and categorizing our everyday lives as somehow "neutral," belonging to the "lower story" of our two-tiered existence. In doing that, we too often find ourselves "aping the world." She writes:

> Looking back over the history of evangelicalism, we can understand better why there has been a strong temptation to split belief from practice – to do the Lord's work but in the world's way. . . . [In] the late nineteenth century,

evangelical scholars adopted methodological naturalism in dealing with the subjects in the lower story, treating them as religiously neutral – as merely technical subjects where biblical truth did not apply in any integral way. As a result they tended to accept a largely functional and utilitarian approach to areas like science, engineering, politics, business, management, and marketing.

In the late nineteenth century, evangelicals even stopped sending their children to Christian liberal arts colleges, where the classics were still taught (they were suspicious of those pagan Greeks!). Instead they sent their children in droves to the newly founded state universities, to receive the technical training required to succeed in an increasingly technological society. Studies show a steady decline in church-related colleges, while the numbers in state institutions boomed. And the students attending those state colleges were predominantly evangelicals – Methodists, Baptists, Disciples of Christ, Presbyterians. "Ironically," says historian Franklin Littell, "it was the misguided piety of revivalist Protestantism which . . . gave the first great impetus to the state colleges and universities."

Littell calls it "misguided" precisely because it was shaped by the two-story division of knowledge. Christian students were avoiding fields like philosophy and literature and the classics, where they would have to deal with ideas, while avidly seeking technical and vocational training in fields that they thought were safely neutral. They were willing to accept an exclusively technological and utilitarian concept of knowledge in the technical fields (the lower story), as long as they were allowed to supplement their studies with campus religious activities designed to nurture the spiritual life (the upper story).

This explains why many Christian churches and ministries today continue to treat areas like business, marketing, and management as essentially neutral – technical fields where the latest techniques can simply be plugged into their own programs, without subjecting them to critique from a Christian worldview perspective. Start the business meeting with prayer, by all means, but then employ all the up-to-date strategies learned in secular graduate schools. Douglas Sloan calls this "the inner modernization of evangelicalism." That is, we have resisted modernism in our *theology* but have largely accepted modernism in our *practices*. We want to employ the latest techniques and quantitative methods, where the results can be calculated and predicted.

For example, a Christian ministry once hired a young man who had just received his master's degree in marketing to head up its fundraising department. Immediately he set about implementing the standard techniques he had learned in his courses, including a sharp increase in the number of fundraising letters sent out. When other staff members questioned the new strategy, asking whether increased mailings were a good use of funds given sacrificially to the ministry, his response was, *but this works.*

Brandishing graphs and studies, he said: "Statistics show that if you send out X number of letters, you will get Y rate of return – guaranteed."

But if any secular organization can achieve the same results using the same "guaranteed" methods, where is the witness to God's existence? How does relying on statistically reliable patterns persuade a watching world that God is at work?

Doing the Lord's work in the Lord's way means forging a biblical perspective even on the practical aspects of running an organization, instead of relying on mechanical formulas derived from naturalistic assumptions. We may reject naturalism as a philosophy, but if our work is driven by the rationalized methods we have learned from the world, then we are naturalists *in practice*, no matter what we claim to believe.

"The *central* problem of our age is not liberalism or modernism," [Francis] Schaeffer writes – or even hot-button social issues like evolution, abortion, radical feminism, or homosexual rights. The primary threat to the church is the "tendency to do the Lord's work in the power of the flesh rather than the Spirit." Many church leaders crave a "big name," he continues: They "stand on the backs of others" in order to achieve power, influence, and reputation – instead of exhibiting the humility of the Master who washed His disciples' feet. They "ape the world" in its publicity and marketing techniques, manipulating people's emotions to induce them to give more money. No wonder outsiders see little in the church that cannot be explained by ordinary sociological forces and principles of business management. And no wonder they find our message unconvincing. (364–366)

The only thing I want to add to what Nancy Pearcey says in that section is to quote our Lord's words where he said, "Whoever has ears to hear, let them hear" (Mark 4:9).

CHAPTER 12

THE RED FLAG OF WEALTH

NOVEMBER 10, 2006

My grandson Justus likes to show people his wallet and all its contents. And that's pretty normal for a six-year-old boy. His dad has a wallet, so he's proud to have one too. His dad has some credit cards and some membership cards in his wallet, and so does Justus – not quite as "official" as those of his dad, but . . . they look the same. And Justus's dad carries some cash in his wallet. So does Justus.

A short time ago his grandmother and I took Justus and his older sister to a house-warming party for one of my nieces. Surrounded by relatives and friends, Justus pulled out his wallet and began showing everyone its contents. My brother and I watched in amazement. Back when we were six years old, we felt rich if we had a quarter in our pocket. That was wealth to us. A penny could get you a bubble gum with a cartoon wrapper. A nickel could get you a good-sized candy bar and a dime could buy you a comic book. But a quarter! Wow, that was more than you could spend in one place. Justus had a few quarters. But he also had a ten dollar bill, a couple a fives, and good number of singles. He actually had more cash in his wallet than I had in mine.

Now Justus is just an innocent six-year-old, knowing nothing about the value or vice of money. To him it's simply something he can count – which *is* something he is learning to do right now – and something that enables him to relate to his father. OK, he's a rich kid. There's no getting around that. But as yet he hasn't been corrupted – in the twenty-first century, North American sense – by avarice or mammon.

None-the-less, there was something disconcerting to his grandmother and I in watching our six-year-old grandson show off the cash in his wallet. Perhaps it was related to the sermon we had heard on the previous Sunday. That sermon dealt with the passage from the Gospels where Jesus tells the rich young man to sell all that he has, give the proceeds to the poor, and follow him. Shortly thereafter, when explaining his words to his disciples, Jesus says, "It is easier for a camel to go through the eye of a needle than for someone who is rich to enter the kingdom of

God" (Mark 10:25). These words, and many others like them in both the Old and New Testaments, have made it clear to me that wealth is something of a "red flag." If Jesus says that it is difficult for a rich man to enter the Kingdom of God, then there must be something about riches that should give us pause, that should make us think twice when evidence of riches is displayed before us. I think that's part of why good old Grandma and I felt a bit uncomfortable at our otherwise cute grandson's displaying the cash in his wallet.

So here's the point I wish to make: wealth is first and foremost a "red flag." It is not first and foremost a sign of God's blessing, or an indicator of success in this life. If you are introduced to a stranger and the first thing you learn about the stranger is that he is exceedingly wealthy, you ought to take pause; the same way you should take pause if you are introduced to a stranger who is coughing and blowing his nose. The cough and Kleenex are "red flags" that alert you to the fact that this person may be sick, and that the sickness may be contagious. Likewise, wealth is a "red flag" that alerts you to the fact that a person might be sick in a different way – a way that also may be contagious.

Nancy Pearcey, an increasingly well-known Christian author, has written an excellent book entitled *Total Truth*. In it she describes Christian organizations that have conformed to the materialistic methods of the world with regard to marketing and fund-raising. One example of such conformity is for a Christian organization to appoint persons to its board of directors who are known to be wealthy, expecting that wealthy board members will automatically contribute big bucks to the organization. Can you imagine that? Having as a criterion for appointing persons to your board of directors that they are among those for whom it is easier to pass through the eye of needle than to enter the Kingdom of God?

I've warned my sons, including Justus's father, to beware of organizations like that. And my grandson Justus, when he grows up and has to deal with the wealth that his father, through no fault of his own, has had thrust upon him; will have learned to be very careful about opening his wallet.

CHAPTER 13

GUIDING PRINCIPLES FOR SALARY AND BENEFITS

JUNE 4, 2007

Two institutions with which I am involved are asking questions such as "what is a proper salary to pay those who work for the institution" and "what kind a benefit package ought we to have in place for our workers?" Those two institutions are Dordt College and Adams Thermal Systems – ATS, for short. Dordt's Salary Committee will be working this summer to develop a set of guiding principles for faculty salary and benefits. So I would like to spend a few moments asking the question, "What *does* the Lord require of us, when it comes to paying salary and benefits to the workers in our institutions?" What norms, what principles should guide us as we develop salary schedules and benefit packages for different kinds of institutions – institutions that we want to honor the Lord and be models of Christian service?

It's important that any specific principles we develop be grounded in general biblical principles to which we Christians have committed ourselves. A first such principle is that God is sovereign over all of creation and over every area of life. Our world – including all of our institutions – belongs to God. A second biblical principle that follows from the first is that all things – including human beings – are God's servants. Thus the primary reason that any of us, or any of our institutions exist, is to serve. That service is, of course, first to our Creator. But we render that service to our Creator by serving his creation, both humankind and the nonhuman creation. A third general, biblical principle is the recognition that our world is stained and broken by sin. Until the Lord returns we will have to contend with distortions, both in the way that human beings live and even in the behavior of the nonhuman creation that surrounds us.

Now, given these basic principles, what specific principles can we develop regarding the nature of work, the place of institutions in our world, and, finally and specifically, how an institution ought to develop salary schedules and benefit packages for its workers?

With respect to work, we can say that work is as natural and as necessary to a human being as is eating and sleeping. Since we are created as servants, we need to have ways to serve. An institution, therefore, besides providing a product or service to those it serves directly, serves its own workers by providing meaningful work for them. The institution is the agent through which a person may render meaningful service to the Lord, to fellow humans, and to the nonhuman creation.

Thinking ahead to salary schedules and benefit packages, we can therefore say that these are secondary ways in which an institution serves the worker – secondary to the institution's *enabling the worker to serve*. In fact, salary and benefits exist to *enable* the worker to serve. Just as food and clothing are not central to life, but exist to enable human beings to live their lives of service, so salaries and benefits are not central to the institution or the worker, but exist to enable workers to provide service, both within the institution and within the other spheres of life such as family, church, school, and community.

We have just zeroed in on a very important principle for any institution that desires to honor the Lord and his creation: we do not pay workers *for* their work; we pay them *to enable* them to work. This principle will influence heavily the way we consider salaries and benefits and will be key to the way in which our Christian institutions provide a distinctive witness to the secular world. But because of the brokenness due to sin, this principal will be applied differently in different institutions. For example, at Dordt College the faculty see themselves as fellow-servants one of another, working for a distinctive Kingdom cause. What motivates them is not money but the satisfaction of knowing that they are making a significant contribution to the Kingdom of God by serving students. To suggest that we ought to pay them for services rendered is to insult them. What will attract them to Dordt College and motivate them will not be the self-centered desire for a high salary, but rather a Kingdom-centered vision for obedient service. Thus they will be looking for a salary and benefit package that is a model of distinctive, Christian obedience, just as they seek to make their classes models of distinctive Christian obedience.

On the other hand, an institution like ATS must work with a wide variety of people, some who are not Christians, and many of whom are motivated by the need to support a family, and who would not necessarily choose to work at ATS otherwise. Here *justice* demands that salary and benefits correspond more closely to the amount of work – usually measured in hours – that the worker performs.

Actually, one finds both kinds of workers at both kinds of institutions. At Dordt College there are persons who are interested in advancing the cause of Christian higher education, but who see their efforts at Dordt first as a way of helping to support their families. Conversely, there are workers at ATS – hopefully among the management team – who see their work first in terms of advancing the Kingdom of God, and are therefore relatively disinterested in their own salary level, but want to see the company put together a salary and benefit package that is first and foremost just.

For now, I will focus only on those management personnel at ATS and the faculty at Dordt – persons who see their salary and benefits as enabling them to work rather than as payment for services rendered. What specific principals ought to shape their salary schedules and benefit packages?

Well, regardless of nature of the worker, the first basic principle for the development of a salary and benefits package will be justice: enabling the workers to be who they are called by God to be. That means focusing first on needs rather than merit. What do workers need in order to make it possible for them to serve within the institution? Clearly different people have different needs. But we cannot pay people salaries based on their needs because we have laws that require "equal pay for equal work." And those laws are a reaction to injustice that has occurred in the past. So a focus on what the worker needs, to be him or herself, will require an institution to minimize the salary schedule and develop a robust benefits package. A benefits package that covers basic living, educational, health, retirement, and similar needs will be just because it will meet the needs of workers, in their own particular situations, thereby enabling them to be the unique servants the Lord calls them to be.

We now have at least an outline for a set of guiding principles for salary and benefits in an institution that desires to honor the Lord and be a model of Christian service. Let's review them:

> **First principal**: God is sovereign and we are his servants called to live lives of mutual service.
>
> **Second principal**: We live in a sin-filled world, requiring us to proactively seek justice.
>
> **Third principal**: Salary and benefits ought to meet needs, enabling workers to be who they are called to be.
>
> **Fourth principal**: Salaries ought to be de-emphasized: they ought to be relatively low and they ought to be relatively equal across the institution.
>
> **Fifth principal**: Benefits packages ought to be robust: they ought to be the primary vehicle whereby the institution meets workers' needs and enables them to be who they are called to be.

PART III
CHRISTIAN EDUCATION

What the teacher "teaches" is by no means chiefly in the words he speaks. It is at least in part in what he is, in what he does, in what he seems to wish to be. The secret curriculum is the teacher's own lived values and convictions, in the lineaments of his expression and in the biography of passion or self-exile which is written in his eyes.

– Jonathan Kozol[*]

---

[*]   *The Night is Dark and I am Far From Home* (New York: Bantam, 1975), 115.

CHAPTER 1

CHRISTIAN TEACHING:
HIRING FOR PERSPECTIVE

APRIL 7, 1989

One of the greatest problems with living in a Christian communi-ty, growing up in a Christian home, or attending a Christian school, is the tendency to take for granted the defining characteristics of that community, home, or school – the qualities and attributes that set it apart as distinctively Christian. For example, what does it mean to be a Christian teacher? If you had the responsibility for hiring a teacher for your Christian school, what would you look for?

In the past few years, on a number of occasions, I've been involved in the process of hiring a Christian teacher. That occurred at the college level, here at Dordt, when we were searching for engineering professors. It also occurred on the high school level when I served as a member of a Board of Trustees and was involved in interviewing prospective teach-ers in areas such as English, music, art, and Spanish. What struck me on these and other occasions was, in contradiction to our profession as reformed Christians, the ease with which it was taken for granted that a teacher who is a Christian is a Christian teacher.

Let's imagine that we are on the board of a Christian school. It could be a grade school, high school, or college. Imagine that we have the re-sponsibility for interviewing and approving or disapproving a candidate for a mathematics teaching position in our school. It doesn't matter that we are not experts in mathematics or in math education. Determining the candidate's technical qualifications is not the job of board members, it's the job of the professionals on the school staff. As board members we have the responsibility to determine whether or not the candidate will be a truly Christian teacher. So what specifically will we be looking for? What kinds of questions ought we to ask?

Unfortunately, the easy thing to do is to ask about the candidate's faith life and faith commitment. By listening to this Christian describe her faith in Christ, her church activities, and her commitment to the community of believers, we can easily be fooled into thinking that the candidate is a Christian teacher, when all we have really ascertained is that she is a Christian. I don't mean to imply that the candidate is consciously fooling us or that we as board members are consciously trying to avoid the real question. It's just that it's so easy to take for granted that a Christian who is trained as a teacher is a Christian teacher.

What kinds of questions ought we to ask our candidate for the position of math teacher in our Christian school? Let me suggest three.

First: "How will your mathematics class be different from the mathematics classes of your secular counterparts?"

Second: "How do you expect the lives of your students to be different from the lives of their contemporaries, because they were in your class?"

And third: "What people and what books have been most influential in forming your ideas about a Christian approach to mathematics and to math teaching?"

The first question attempts to determine if the candidate has an understanding of what it means to teach Christianly. The second question gives the candidate an opportunity to tell you about her vision of what teaching math means for the Kingdom of God. The third question is, in a sense, a check. People don't suddenly become Christian teachers overnight. They have to develop. That development takes time and the influence of other Christians. Some of that occurs in school. But most of it – the most important part, anyhow – occurs when the potential student interacts with other Christian teachers – either directly or through books. Thus, a genuine Christian teacher is going to be enthusiastic about telling you who her teachers were, and what they taught. She will also want to tell you about the books that have greatly influenced her. And if the candidate is a really sharp Christian teacher she will be able to tell you about books she is reading now and plans to read in the future. A serious Christian teacher is one who is immersed in and excited about his or her calling.

But let's go back to the first question: "How will your mathematics class be different from the mathematics classes of your secular counterparts?"

If your candidate tells you that $2 + 2 = 4$ is the same for Christians and non-Christians and therefore the "facts" that he teaches will be no different from what is taught in the secular schools, then I think you ought to politely thank him for his time and go on to your next candidate.

When I ask that question to a prospective Christian math teacher

I listen to hear three things. Those three things can be categorized as "content, purpose, and methodology," or phrased in even simpler terms, "what, why, and how."

First I want to hear that in this teacher's class students will learn that numbers, equations, and the like are creatures of the Lord. They are not mere neutral facts that simply exist or that man has created. Rather they, like stars, songs, elephants, and people, are servants of the Lord Jesus Christ, some of whom have been here from the beginning and some of whom have been unfolded in time by God's image-bearers. Perhaps she will even quote Psalm 119:89–91, where we read:

> Your word, Lord, is eternal;
>> it stands firm in the heavens.
> Your faithfulness continues through all generations;
>> you established the earth, and it endures.
> Your laws endure to this day,
>> for all things serve you.

That brings me to the second thing I want to hear. In this teacher's class, students will be confronted with the reality that they are servants of God called to develop their insights and talents in obedience to his Word – and called to claim the world for the cause of his Kingdom. That means the purpose of studying math is not merely "to get a job" or "become well-rounded persons," but to redeem a part of creation from the clutches of the evil one. In studying math with this Christian teacher, they will come to see themselves as "subversives-in-training" – being prepared to go out and turn the world upside down in the name of Jesus.

Finally, I want to hear how the teacher plans to do this. I look to see the teacher's eyes light up as she tells me how she would avoid the pitfalls of radically student-centered education where the individual is all-important, and of subject-centered education where the student is made the slave of the curriculum. And if the teacher is really good she will be telling us how to combat the effect of the TV on the students by using the media to deepen the Christian perspective of the students, and how she will integrate the learning of mathematics with the other learning experiences that the students will be undergoing.

If I hear all that, I will attempt to hire her on the spot. If I hear what sounds like the beginnings of that perspective, I will rephrase my questions in an attempt to enable the teacher to convey that perspective in her own language.

To close, let me say that there's only one thing worse than hearing no perspective at all from the candidate. That is hearing my colleagues or fellow board members say "she may be a little weak on perspective now, but she is a solid Christian and is open to learning." For a Christian teacher, perspective is not something you learn while "on the job." Our responsibility to Christian students will not allow it.

CHAPTER 2

CONFORMING TO A PATTERN

NOVEMBER 7, 1986

A few weeks ago I was asked what I thought of the possibility of a Christian high school sponsoring a dance for its students. That's not a new question, of course. I've been involved in Christian schools during the last fifteen years – as a teacher, parent, and board member – and the question of dance has a habit of arising with amazing regularity. There are those who are in favor of it – primarily young people who see dance as a legitimate form of recreation and a vehicle for self-expression. You will often hear them telling of how in the Old Testament, David danced before the Ark of the Covenant. Then there are those who are totally opposed to it. They see it as corrupt and immoral, and would view the idea of a Christian school sponsoring a dance with the same horror that they might view a local church congregation opening a mini-gambling casino in order to meet its budget.

Well, I don't fall into either group – or perhaps I fall a little bit into both. In any event, I think that the question of a Christian school sponsoring a dance is an interesting one, because it can help us exercise our Christian minds. You see, there really is no simple yes or no answer. Those who would uncritically accept a proposal that our Christian schools sponsor dances lay themselves open to the charge of conformity to the world. And those who simply dismiss the dance out of hand as inherently evil lay themselves open to the charge of legalism, and maybe even of pharisaism. Let's spend a moment or two considering the question.

First of all, I think we have to agree with at least one point made in the recent Synod of the Christian Reformed Church decision on the dance – there is nothing inherently sinful with expressing oneself by moving the body in a way that is somehow coordinated or in harmony with music. Motion and music are both basic components of God's good creation. As reformed Christians we believe that God calls us to be his agents in reconciling the creation back to him. It is our task to claim every area of life for Christ. That would imply that at least some of us ought to be busy in the aesthetic area of life – including music and dance – seeking out the will of the Lord. And if this is happening, we ought certainly to encourage our young people to learn from those who have developed insight into God's will for music and dance, and develop abilities themselves in those areas.

But there is another side to the question. The Christian community has a very nasty habit of uncritically accepting what the world has to offer, without consideration for our tasks as reformers. How much serious deliberation goes on in our homes these days with regard to television watching or the role of the automobile, the question of stewardship of income, or how we dress, and so on? For the most part, we simply accept the customs and norms that are thrust upon us by the media and surrounding society. And we all know what would happen if we simply said "yes" to our teenage children with regard to the question of having a dance. They have learned from us to be uncritically accepting of what society offers them. So, of course, we should not be unduly surprised if a dance organized at our Christian school would be no different from a dance at a state school, or worse yet, a dance like those portrayed in some of the latest movies or television programs. We haven't taught our children what the will of the Lord is for dance, but we have taught them how to conform to the will of our pagan society in many other areas of life.

I empathize with many in the Christian community who say, "No! The dance as it is known in today's hedonistic culture has no place in a Christian school." But we cannot say only that. If we do, our children will have every right (and duty) to respectfully say to us, "Dear father, mother, teacher, pastor, board member, we submit to your decision, but we must point out that in denying us the privilege to dance while allowing yourself the privilege of uncritically watching television, running your business, driving your car, and choosing what clothes to wear, you act as a hypocrite and Pharisee and call down on yourself the judgment of Matthew 23."

So perhaps you see why I consider the question of dance to be an important one. It raises much bigger questions having to do with conformity to the surrounding culture, Christian distinctiveness, our ability as a community to be a light in this world, and our task as representatives of our King, claiming every area of life for him.

To the question of whether or not a Christian school should sponsor a dance for its students: I say, sure, as soon as we begin to understand what it means to dance obediently before the face of our Lord, thereby reclaiming that area of life for him – but not before. The same ought to be said about every other activity in life. Having a dance that conforms to the pagan culture around us is not any different than what we do now with our "junior-senior banquets." How many parents will look on with ignorant admiration this spring when their sons and daughters, dressed in tuxedos and long dresses, conform to the world's initiation rites for its little socialites and debutantes?

Let's play that questions like the one about dance will help wake up God's people to what it means to live holy and distinctive lives. And it is parents and community leaders who must first do this themselves if they expect their children to ever learn to do the same. I close with the word of the Lord from Romans 12:1–2:

> Therefore, I urge you brothers and sisters, in view of God's mercy, to offer your bodies as a living sacrifice, holy and pleasing to God – this is your true and proper worship. Do not conform to the patterns of this world, but be transformed by the renewing of your mind. Then you will be able to test and approve what God's will is – his good, pleasing and perfect will.

CHAPTER 3

CONFESSIONAL CONVICTIONS AND "THINGS"

SEPTEMBER 1, 1989

One of my heroes is Andrew Kuyvenhoven, the soon-to-be retired Editor of *The Banner*. During his tenure as editor he has brought a distinctively biblical and reformed worldview to the pages of that magazine. More than that, he has, in his editorials, demonstrated the kind of joyful poise that those of us who find our only comfort in belonging to the Lord Jesus Christ ought to have a great deal more of. He hasn't been afraid to write candidly and biblically on controversial issues. And because of that he has earned the hot displeasure of many who are not blessed with the same degree of mature Christian poise that he demonstrates week after week. It is perhaps in responding to that hot displeasure where Kuyvenhoven demonstrates that poise most succinctly.

But I learned long ago that even heroes are human. And to be human is to be finite. And to be finite is to occasionally make mistakes, or, in the case of the Editor of *The Banner*, misstatements. I'm thinking of his editorial in the August 28 issue. There he writes an open letter to a consistory that criticized him for a previous editorial he wrote on the topic of Christian schools; an editorial entitled "Christian Schools: Reaffirm the Principle; Rethink the System." He made a number of interesting points in that editorial. But the one that bothers me has to do with the idea that some subjects can be taught without concern for the confessional convictions of the teacher. I quote from Kuyvenhoven's "Letter to Consistory."

> The editorial says that it does not matter whether a Buddhist or a Christian teaches us the skill to handle computers and word processors. You say that is not true because computers can be servants or idols.
>
> You are right about the way we might use the whole realm of knowledge represented by the computer. But I was talking about learning how to handle the thing. Does it really matter what the confessional convictions are of the person who teaches us how to drive a car? Of course not. Yet the way in which we drive in terms of speed, regard for law, regard for other drivers and pedestrians, and the role of the automobile in our scales of values have much to do with our faith.

As an engineer and as an engineering educator I have a great deal of concern for "things" and for "learning how to handle" things. And I believe that it is impossible to separate the heart direction, the "confessional convictions" of a person, from the relationships that that person has with things.

Let's use as an example the (now archaic) typewriter. It would seem that Kuyvenhoven, and probably many other faithful Christians, believe that our children could be taught the skill of typing as well from a Buddhist as from a reformational Christian. But that belief rests on the prior belief that the skill of typing – that is, the purely mechanical relationship between human fingers and typewriter keys – can be separated from the totality of the context in which that skill is exercised. I would maintain that that separation occurs only in our mind. In order that we might understand this world God created, he gave us the ability to *abstract*, that is, to separate in our mind one concept from another. So what is separable is not the actual skill of typing and the total context in which it occurs, but rather our conceptualization of the skill of typing, and our conceptualization of the total context in which it occurs. That's why it's very important that our children have biblically solid and reformationally alive typing teachers. Because in the real world, where our children learn, nothing is totally separable from anything else. A typing teacher is going to, by necessity, teach far more to our children than "the skill of typing." That teacher cannot help but teach all sorts of attitudes about typing because s/he cannot help but teach from out of a total worldview.

The same thing goes for driving a car. In reality one cannot separate such things as "the way we drive in terms of speed, regard for law, regard for other drivers, and the role of the automobile in our scales of values" from the simple skills of steering, braking, and raising or lowering the windows. And Kuyvenhoven admits that those former attitudes "have much to do with our faith."

But there is a second argument that can be made. It seems to me that Kuyvenhoven, while maintaining that things can be used for good or for evil, is saying that they are otherwise neutral. I would argue that automobiles, word processors, typewriters, computers, and every other artifact designed by man, have values embedded in them by virtue of the process by which they were created. Designing a computer necessarily involves the designer in the consideration of economic norms, aesthetic norms, ergonomic norms (that has to do with what is often called the "user-friendliness" of the computer), legal norms, and, while perhaps not as obvious, ethical and faith norms as well. So when a person uses a computer, they will be to some extent directed by the response of the designer of that computer to those norms.

To use a different example, if I go out and buy a fancy BMW automobile, I will be strongly influenced by the values – such as mechanical integrity, but also status consciousness – that are necessarily a part of that car (in fact, the mere purchase indicates that I have already been so influenced).

In conclusion then, Andrew Kuyvenhoven is wrong when he says that the confessional convictions of the person who teaches us to drive a car do not matter. And it is therefore critical that Christian schools look for biblical, reformational drivers-ed instructors, keyboard teachers, and soccer coaches, with the same sense of urgency as they do for Bible teachers. To do otherwise is to fall into the very western and very unbiblical trap of thinking that reality can be separated into parts – and that some of those parts can be considered neutral with respect to the lordship of Christ.

CHAPTER 4

GRADES AND THE HEIDELBERG CATECHISM

DECEMBER 2, 1988

This semester at Dordt College a disturbing phenomenon is occurring with a particularly high frequency. That phenomenon is "course dropping," quitting a course halfway through the semester. It's occurring at especially high frequency among freshmen.

In this essay I would like to first consider why students drop a course halfway through, then suggest some good and bad reasons for dropping a course, and finally suggest an explanation for the increased frequency of course drops that we are seeing.

So why do students drop courses halfway through? One good reason for doing so is the realization on the part of the student that s/he does not have the prerequisite knowledge to benefit from the course. Theoretically, if the student has been advised properly, this should not happen. But advising is a tricky business. Imagine a freshman who has had some math in high school and who wants to take Introductory Calculus. On paper it might look like the student is ready for the course. But after a month of struggling in the class it becomes obvious to both the student and the professor that there is just too much background that is missing. So the student correctly drops the course and spends more time on the remaining courses.

Another possible justification for dropping a course halfway through is the realization that the student's interests and talents lie in a different direction. While the student may be capable of succeeding in the course, little benefit is gained by staying with it. And having more time to spend on the remaining courses usually helps the student considerably. Many freshmen come to college without a firm idea about the area in which they want to major. Thus students might find themselves in an introductory engineering course and part way through the semester discover that their real interests and abilities are in history and philosophy. Dropping the engineering course might be a wise thing to do, depending on the time the students have to spend on other courses and activities.

Those are two legitimate reasons for dropping a course part way through. Other legitimate reasons, usually over which the student has no control, are sickness, and problems at home that require the student to leave campus.

But from what I've observed, especially this semester, there is another reason. Many students are dropping courses because they believe they are getting low grades. Most students come to college from high school with fairly good grade point averages. Although they may have gotten an occasional "C" here and there, they are used to receiving A's and B's. It's very likely that all through grade school and high school their parents praised them when they brought home A's on their report cards, and gave them grief when they brought home C's. This is reinforced at school by such things as honor rolls, honor societies, and potential college scholarships – all of which place a heavy emphasis on grades. Unfortunately the student fails to see that grades are merely a device used to measure how much s/he has learned, and instead begins to see grades as a measure of self-worth.

This misplaced emphasis on grades is somewhat understandable. It's difficult for parents to tell whether their children are benefiting from their high school education without having some kind of simple measuring stick such as grades. I can ask my children questions like, "Why did Hamlet say 'the time is out of joint'?" or "Why does silver have only one electron in its outermost orbital?" or "Why did Abraham Kuyper start the Free University?" But I would have to ask a tremendous number of questions before gaining assurance that they have benefited from their high school education. And asking all those questions would probably do more to alienate them than simply harping on grades.

But this is no excuse. Each student is created different, blessed with unique talents, and called to responsible service in ways that differ from each other student. Thus we ought not expect our students to get all A's, whether in high school or in college. In fact, I have real concern for those students who do maintain "straight A" or "4.0" averages. My guess is that such a student is playing the system, probably learning a lot of things that are unimportant to him or her, and probably failing to grow in some very crucial other areas of life that are not measured by grades.

But now back to the main point. I think that students are dropping courses halfway through because they believe they are getting low grades (low grades for most students are "C" or less), and their self-image can't handle the psychic blow that they attach to getting those low grades. Students with talents and interests in a particular area, but whose less than perfect high school preparation in that area mean that a few low grades are inevitable at first, are dropping out not just of courses, but also career fields, due to their hang-up with grades.

That's why we have to work harder to teach our students to better understand the meaning of the first question and answer to the Heidelberg Catechism, "What is your only comfort in life and in death?" The answer has nothing to do with high grades, high salaries, secure jobs, or comfortable homes. Instead the answer is "That I am not my own. . . ." Our students need to learn that they are not their own, that they belong to their Lord, and therefore their worth is not measured by grades. And they certainly are not all called to get straight A's.

I would urge any student who is ready this essay to not worry about grades. If you work to develop the talents and interests that the Lord has created in you, the good grades will come – maybe not right away, but in good time. I am not suggesting that parents and teachers give up on grades. But we need to convey to our students that grades are only a measure of what has been learned and ought not to be pursued for their own sake. Remind them that their high school and college grade point averages are of no significance when they decide to marry, when they raise a family, or even when making important decisions about their work twenty years in the future. And when the Lord finally brings them into his Kingdom, saying, "Well done, good and faithful servant," it will not be because of their grade point average.

CHAPTER 5

GRADUATION AND BAPTISM

MAY 7, 1985

Well it's that time of year again. This Friday, Dordt students will graduate. In about two weeks seniors at the high schools and eighth graders in the grade schools will go through a similar exercise. There's a great deal of "hype" associated with these events. Grandparents travel many miles to attend the ceremonies. Students, even college students who have been through it all twice before, psychically pump themselves up to a state not unlike that just preceding their seventh birthday, when they might have anticipated receiving their first two-wheel bicycle. Parents, throwing off all inhibitions, arm themselves with cameras and flashcubes, and in the midst of the ceremony itself, march conspicuously down the aisle, take careful aim, and at the consummate moment drown themselves and everyone else in a brilliant flash of light.

So why all the fuss? What is it about graduation that causes us to anticipate it so eagerly? I remember my own graduations. I specifically remember graduating from eighth grade because it was one of the biggest letdowns in my life. A number of my relatives showed up for the big occasion. I still have a couple of those silly pictures of my good friend and I dressed up in white jackets and red ties and our hair combed carefully and solidified in place with Odell's hair trainer. And I remember receiving a fantastic gift from my grandparents. It was an 8 mm movie outfit that belonged to my great-uncle and had been passed on down to me. That was really something for 1960! Nevertheless, I remember going home after the graduation exercise, playing the radio, and feeling very let down, as if something big was supposed to have happened but never did. High school graduation was much the same, except that we had the traditional graduation party to numb us a bit. But afterward the feeling of being let down persisted.

This year I'm going through graduation again. Not as a student but as the parent of a student. In fact this is the third time in four years that I'm a parent of a graduating eighth grader, and next year I look forward for the first time to being the parent of a high school graduate. But these graduations have been, and are, very much different than my own. There still is the ceremony, the visiting relatives, and the photographs. But all this is part of an event that now has genuine meaning. There is a connectedness to other events that have occurred and to events that will occur.

When I think about it, the reason for the difference is quite obvious. My sons have, are, and will be graduating from Christian grade and high schools in northwest Iowa. I graduated from a public school in northern New Jersey, just outside New York City. These schools represent two different worlds. For the public school student in New Jersey, graduating from grade school meant anticipating having a "blast" in high school and getting a little bit closer to being "grown up." Graduating from high school meant being able to go to college. And graduating from college meant getting a good job – "good" being defined in New Jersey public school language as "high paying." Graduating always meant breaking ties with the past and looking ahead to a life that would somehow be better. Why it would be better we were never really sure of. We assumed it would be better because it seemed the older you got the more "things" you could acquire. It never sunk into our young heads that the grownups, either as represented on the TV or in the neighborhood, were always trying to regain their lost youth.

But graduating from a Christian school is different because the focus is not exclusively on the future. In fact, I'd like to suggest that if our graduation ceremonies are carried out properly, the focus is as much on the past as it is on the future. And here is that connectedness with other events that I mentioned earlier.

Perhaps the most important of those events connected or related to graduation is baptism. Before our children were born, we as parents prayed for their physical and spiritual wellbeing, and in our hearts we knew the Lord was hearing our prayer and was looking upon our yet unborn child as one of his covenant people. Shortly after the child was born we sealed that covenant with the sacrament of baptism. During the baptism ceremony both we as parents and we as members of the Christian community made certain promises. Remember the questions? "Do you promise, in reliance on the Holy Spirit, to do all in your power to instruct this child in the Christian faith and to lead him by your example into the life of Christian discipleship?" To which we answered, "We do, God helping us." And then the question to the congregation: "Do you, the people of the Lord, promise to receive this child in love, pray for him, help care for his instruction in the faith, and encourage and sustain him in the fellowship of believers?" To which we again answered, "We do, God helping us."

In the Reformed faith we believe that "instruction in the faith" means more than learning the Scriptures and catechism. "The faith" means the "way of life of the redeemed in Christ." And instruction in the faith means learning about history, mathematics, biology, penmanship, singing, and all other possible subjects, not because they are ways of getting us a good job, but because they are areas in our Lord's creation to which he calls us in order to develop them, enjoy them, and redeem them back from the curse, so that they may be used to honor him. So graduation from a Christian school means at least the partial fulfillment of those covenantal vows we made at the time of our child's baptism. It is an indication that the Lord has been faithful to his covenant, and has upheld us as members of his body in our attempt to remain faithful by establishing, running, and sending our children to Christian schools.

In light of this, I hope that the speeches made at graduation ceremonies might be just a little different this year than the usual. Often we hear the student graduation speaker expressing thanks to parents, teachers, board members, the supporting community, and so on. But really, do we as teachers, parents, board members, and supporting community deserve all that thanks? After all, we were only doing what we promise to do every time a child is baptized. In fact, if we failed to establish, support, and send our children to Christian schools, then we would have failed in our covenant obligations – we would have reneged on our promise.

So this year let's enjoy the graduation ceremonies by seeing them for what they really are: landmarks along the road to the Kingdom of God – landmarks that surely point ahead to a life to be lived in service to our Creator-Redeemer – but also landmarks pointing back to such events as baptism, and reminding us of who we are, and of the promises we made.

In the next few weeks not only will there be graduations taking place, but also professions of faith and baptisms. It would really be neat to go to a graduation ceremony on Friday night, participate in a worship service involving a profession of faith on Sunday morning, and then have a baptism during the worship service on Sunday evening. The beauty of the Reformed faith is its awareness of God's presence in each of these ceremonies and in their relationship to each other.

And one last point. Parents, when you get ready to sneak up the aisle with your camera and flash unit, remember, the fact that a Christian school graduation ceremony is not a sacrament does not make it any less holy an event than baptism. I think cameras can certainly be a part of a holy celebration. But perhaps a more sensitive awareness of God's presence will make us just a wee bit more discrete.

CHAPTER 6

COMMENCEMENT, THE TROJAN HORSE,
AND CHRISTIAN SUBVERSIVES

MAY 6, 1988

Today at 10 a.m., Dordt College will hold its commencement exercises. In reflecting on what I'd like to believe is the significance of this event, it struck me that perhaps the Trojan Horse is as apt a metaphor as any. Even if you have never read Homer, I'm sure that most of you know something about the Trojan War, and in particular, the Trojan Horse.

You may recall that the Greeks, led by the super-heroes Achilles and Hercules, had sailed to Troy with the intent of invading the city and re-claiming Helen, the queen who had run away to Troy with Paris, the son of the Trojan King. After nine years of waging war, the Greeks devised a plan to gain access to the city. They built a huge wooden horse that was ostensibly a propitiatory offering to Minerva, the goddess of wisdom. In reality the horse was to be filled with Greek soldiers. After building the horse and arming it with the soldiers, the rest of the Greek army got into their ships and sailed away, leaving the Trojans to believe that they had finally abandoned the siege. The curious Trojans came out of the city to inspect the horse and believed the story that the horse was an offering to Minerva, made especially big just so that it would be difficult to capture and bring within the city.

According to the planted story the Greeks believed that if the horse were captured and brought within the city of Troy, the Trojans would certainly win the war. The gullible Trojans believed the story, dragged the horse into the city, and partied all night, celebrating what they thought to be their victory. In the early morning, the Greeks, who had supposedly sailed away, were back outside the city gates. The soldiers inside the horse got out, opened the city gates to let in their comrades, and the Trojans, drunk and sleepy from partying all night, were defeated.

Now, what has all this to do with a Dordt graduation exercise? Well, on the surface, four years of a Dordt College education, capped off with a commencement exercise and a diploma, looks pretty much the same as does its counterparts at the state universities or private liberal arts colleges across the nation: a propitiatory offering to the goddess of success. Students in business administration, mathematics, education, journalism, and engineering, just to mention a few, pick up their bachelor's degrees and thereby gain entrance into the career fields that will supposedly yield up to them the rewards of the American dream: money and prestige. Leaders in American society and industry, seeing only the surface of the Dordt students' diplomas, welcome them into the system. Little do they know that these former students are really subversives – revolutionaries who have spent the last four years in a kind of guerrilla training camp learning how to infiltrate the system so that they can turn it upside down, claiming it for its rightful King.

That's the cornerstone of Christian education, at least reformed Christian education: reclaiming culture for the Lord Jesus Christ – the Creator, Sustainer, and Redeemer of culture. But that process of reclaiming requires the training of subversives, nonconformists, and prophets who are aware of the presence of evil and who see their task as exposing and correcting it, and proclaiming the good news of the Kingdom.

Whereas the status quo educational system trains students to conform to society, Christian education trains students to reform society. Whereas the status quo educational system teaches students to be concerned about evils such as poverty, injustice, and racism, Christian education convicts students of their responsibility to do something about the same evils. Status quo education is really sixteen years of moral desensitizing and social indoctrination. This results in nice people who will be productive, won't rock the boat, and will never feel guilty because they have so much while so many in the world suffer.

Christian education is training in moral outrage, it is the nurturing of disciples. The result is Kingdom citizens who won't buy into the system, who reject the American dream as self-centered materialism, and who aren't concerned with living productive lives as long as they live obedient lives. These are people who are not afraid to feel guilt, because they have learned how to repent, and who consider it their duty to right the wrongs for which they bear a communal responsibility simply by being citizens of the richest nation on earth.

Does all of this sound a bit idealistic to you? I guess I have to admit that Christian education isn't perfect. We have our failures too, many of them. But then again I've had some of these Dordt students in classes this past year. And I know that the fields of engineering, psychology, communication, social work, and education are in for some world-shaking in the days ahead because of them.

So if you happen to see the graduates in their gowns spilling forth from the Dordt Chapel today, or even if you only see them leaving town in their cars, look a second time. If you look with prophetic eyes you just may get a glimpse of the Trojan horse of the Kingdom as its belly gives up the cultural subversives and reformers of the next five decades.

CHAPTER 7

FUNDING THE KINGDOM: I

NOVEMBER 8, 1985

In this, and in several subsequent essays, I would like to speak to an issue we have all heard about, and will continue to hear about in the future – the funding of Kingdom causes. Very specifically, I want to address the problem of funding Christian education.

To say this is a problem is not new. Throughout the history of the Church, and certainly since the Christian school movement in North America began, raising the necessary funds to educate the Lord's covenant children has always been a struggle. At times the struggle has been a minor concern but at others has been most critical. Here in the Midwest we are moving into a period when this struggle will become increasingly critical. If you live in Sioux Center, for example, you are faced with at least four institutions of Christian education: one grade school, two high schools, and a college, all of whom are beginning to seriously feel the pinch of a depressed farm economy.

I'm not going to suggest any solutions here. What I'd like to do is consider some principles that I believe can be defended on the basis of Scripture, and might help us as we search for solutions.

The first principle has to do with the priorities we assign as we organize our giving. Many Christians believe that the Old Testament concept of the tithe remains in effect today – the first ten percent of your income must be given for the work of the church. I think one danger of this is that we "get off the hook" too easily. The Lord has revealed to us in his Word, and this is especially clear in the New Testament, that everything we have, including our very selves, belongs to him. In that sense the norm for giving is that we must give 100 percent of our resources for the work of his Kingdom. There is nothing that is ours alone, because, as the first answer to the Heidelberg Catechism so clearly states it, we are not our own.

On the other hand, assuming that we recognize that our eating and drinking, our work and our play, our worshipping and our celebrating are all Kingdom activities, it is probably very good self-discipline to decide that the first 10, 20, or 99 percent of our income – whichever is appropriate to our situation – ought to be reserved for Kingdom work outside of our immediate home environment.

But, there is yet another problem. In spite of our reformed emphasis on the fact that all of life is religion, and that the Church of Jesus Christ is the body of believers everywhere, we still fall into the dualistic trap of considering the Lord's work to be primarily the work of the institutional church, and only secondarily the work of such institutions as the school. The point I wish to make, therefore, is this: the Christian school is an expression of the Church as much as the institutional church is an expression of the Church. Neither institution deserves to be placed on a pedestal above the other. I'm sure we would admit that to have Christian schools while allowing our Sunday worship practices to disappear would be hypocritical. But I would like to suggest that maintaining healthy local congregations with their buildings, choirs, and various societies, while allowing the Christian school to suffer, is just as hypocritical.

When we set aside a portion of our income at the beginning of each month, we ought to consider which Kingdom causes have the greatest need. Then we ought to distribute our giving among them. I don't think anyone can tell you how to distribute your giving. But that is exactly what we have allowed to happen. The dualistic influences of pietism and fundamentalism have somehow duped us into unconsciously thinking that the institutional church is the most holy expression of Christ's body and therefore must receive our full share before we even begin to think about other Kingdom causes.

Please don't misunderstand me. I believe we ought to support our local congregations. In fact I think it would be wrong to give our full "fair share" to the Christian schools and then have nothing left over for the institutional church. But likewise, it is wrong to give all to the institutional church and nothing to the Christian schools.

Often we are exhorted by our pastors or deacons to remember that we agreed to a certain budget at our congregational meeting, and we ought to give according to the terms of that budget: so much per family per week. But I would remind us that every time a child is baptized, we promise the Lord that we will do all in our power to teach that child to know his heavenly father and how to respond as a true servant of the Lord in every area of life. We need Christian schools to do that. Therefore, if we do not support the Christian schools with our giving, we are reneging on the promises we make at each baptism.

In these days of economic hardship a few families can simply not afford to give to all of the Kingdom causes to the extent that has been agreed upon by such things as the church budget and the Christian school fair share program. But that does not mean that they ought to give only to one cause and not to any others. We must be wise in the distribution of our giving, being faithful to support each expression of the Church according to our ability to give and according to the need of each cause.

As I mentioned at the outset, Christian schools – grade schools, high schools, colleges, and graduate school – will be facing some rough financial times in the next few years. It's important that those who can afford to give liberally, do so without waiting to be asked to do so. Try to find out where the needs are greatest and then try to address those needs. And it's important that those who can only give modestly because of financial hardships do so wisely, not allowing themselves to be manipulated by a dualistic worldview that divides sacred from secular and declares tithes as appropriate only to so-called sacred causes. If we had three children and only enough food to feed one of them completely, would we allow the other two to starve? Of course not. We would use all of our wisdom to divide the food, giving the most to the child who had the greatest need, and generally splitting the food up while praying for the day when we would have sufficient for all.

In terms of our Kingdom institutions, that day of plenty will come if we are faithful in our stewardship now. But being faithful requires wisdom. So let us pray for wisdom, let us pray for the ability to meet the needs of our Kingdom institutions, and let us carefully divide our resources so that no part of the body of Christ goes without nourishment.

CHAPTER 8

FUNDING THE KINGDOM: II

DECEMBER 9, 1985

In my previous essay I raised the issue of how we as a Christian community ought to fund Kingdom causes. I dealt with the problem of how we assign priorities as we organize our giving. Now I would like to address a parallel issue: the question of how those funds ought to be distributed by a Christian organization as it pays the salaries of those who serve the organization on a full-time basis. Very specifically, I want to examine the ruling of the Supreme Court that equal pay ought to be given for equal work, and to inquire as to its relevancy for Christian organizations.

Let me first, however, make clear what I mean by a Christian organization or a Kingdom cause. Many Christians use the phrase "full-time Christian service" to apply only to the institutional church, and occasionally to the Christian school also. But this betrays an unhealthy dualism that breaks life into two parts – the "religious" and the "secular." The Scriptures and the reformed faith in its attempt to remain faithful to the Scriptures tolerate no dualism of this sort. Rather, emphasis is placed on the wholeness of life and the fact that every area of life must be lived as service to our Lord.

This means that there exist many Kingdom causes other than churches and schools. Your own family is a Kingdom cause because it has been called into existence by God and its primary purpose is to respond obediently to his Word for the family. Your role in that family as father, mother, or child is normed by the Word of God. To the extent that your family is faithful in living according to those norms it is an important force in furthering God's Kingdom here on earth. And not only churches, schools, and families, but also institutions of mercy such as Pine Rest Christian Hospital or Bethany Christian Services are Kingdom causes. And it would be wrong to stop there. Christian political organizations like the Association for Public Justice and Bread for the World, and Christian journalistic enterprises like *The Reformed Journal* or *The Other Side*: all of these are Kingdom causes.

But now let's get back to our problem. In a Christian organization of even a modest size there will be a number of full-time servants. The question is: How are those servants to be financially supported so that they are able to carry out the service that they have been called to? In the world of secular American economics, these servants are referred to as "employees." This is first because the secular mind has no biblical understanding of the nature of servanthood, and second because secular institutions are set up, for the most part, on the assumption that there exist two types of persons in the institution – employers and employees – and that there exists an adversarial relationship between these two types: the employer wants to get the most work for the least money from the employee, and the employee wants to get the most money for the least work from the employer. This adversarial relationship often results in one side dominating the other, and the government has had to step in to prevent gross injustices from occurring. Such laws dealing with nondiscrimination, minimum wage, and "equal work for equal pay" are examples of the government's attempt to prevent injustice.

The problem is that such laws, while necessary to prevent injustice, bring with them a spirit that is foreign to a Kingdom spirit. And since these laws are applied to Kingdom as well as secular institutions, they have the effect, if obeyed uncritically, of warping the financial policies of an institution whose purpose is to function according to the norms of the Word of God.

A good example is the "equal pay for equal work" law. This law originated when it was recognized that women were being dealt an injustice by organizations that would pay them far less than they would a man for doing the same work. And to a certain extent, Christian organizations have been among those most guilty of that practice. But the law, in its attempt to curb injustice, is written from an individualist, materialist perspective. As such, it makes no exceptions whatsoever. Each person is looked upon as a totally independent individual. To the extent that Christian institutions simply obey the law, without addressing the underlying perspective, they become like secular institutions.

Let me give a concrete example. Christian organizations in Northwest Iowa area all struggle with finances. None has an overabundance of money, and in fact, none has even enough funds to operate in a manner such that they can completely fulfill their callings. Salaries of full-time servants in such organizations are often kept as low as possible. However, because of the "equal pay for equal work" law, and because of our often uncritical acceptance of the spirit behind that law, some crazy results occur. First, there are a number of families where one parent must be present in the home full time to take care of pre-school children while the other parent serves the Christian organization full-time. Such families struggle financially. Then there are a number of families where the children are partially grown, attend school, and do not require a parent to be home full time. In such situations, very often both parents are serving in Christian organizations full time, sometimes even in the same organization. These families usually have no financial struggle whatsoever; in fact, as a family they are paid far too much (at least when there exists in the same community serious financial struggles on the part of other families and the institutions). But the courts declare adjustments to salaries that might address these severe imbalances unconstitutional. And worse yet, some Christians have begun to use the same secular arguments against such adjustments. So what do we do?

I would like to suggest that we first ought to recognize the validity of the "equal pay for equal work" law in the context of our secular society. Even though the foundations of the law are corrupt, the application of the law does help curb injustice. Second, I think we ought to obey the letter of the law simply because it does not, by its very nature, force us to do something that would be disobedient to the Word of God. But third, I think we ought to reject the secular, individualistic spirit of the law. This we can do by developing salary packages that, while holding to the letter of the law, go far beyond it – and even violate that secular spirit – by helping families who have only one income.

CHAPTER 9

FUNDING THE KINGDOM: III

JANUARY 8, 1986

In my last two essays I have dealt with the issue of how the Christian community ought to fund Kingdom causes. I first addressed the problem of how we assign priorities as we organize our giving and then considered the question of how those funds ought to be distributed by a Christian organization as it pays the salaries of those who serve the organization on a full time basis. This time I want to consider a salary package for a Christian organization that addresses some of the injustices present in many of the salary packages we find today, and which I believe can be defended by appealing to basic biblical principles.

There is one basic principle that must guide us as we attempt to formulate any salary package. That principle is that the people of God are created to work, that we are created in such a way that our work ought to be a joyful experience, and that a sense of accomplishment and the knowledge that we have been a good and faithful servants is sufficient motivation to continue our work. To suggest to the person of God that we ought to be paid for our labor is the same as suggesting to a husband or wife that one ought to be paid for loving one's spouse. No, it is worse than that. To tell the person of God that you want to pay for his or her service is to say that you consider him or her to be a common prostitute – a person who takes what God has given and uses it to make money. The only payment that the Christian community owes its servants, specifically in exchange for their labors, is the gratitude of the community.

"Ah", but you say, "this is wildly idealistic. Come back to the real world." Yes, I'm afraid it's true. In the real world people work for money. And they are not all prostitutes either. In this fallen world there are many people who have no means to supply the necessary food, shelter, and clothing for themselves and their dependents other than to go out and work for money. No, they are not prostitutes. But neither are they truly "workers" in the Kingdom sense of that word. If you doubt the validity of what I am saying, try this little test. Go to the person who you believe is a true Kingdom worker and remind him that in the parable of the talents, Jesus suggested that those of us who have faithfully served him in a small way here on earth will be put in charge of much more in the Kingdom to come. Then ask the person what he expects to be doing for all eternity after the Lord returns. If he says, "Why, just what I'm doing now, of course; only without the encumbrances of sin," then I will agree with you, that person is a true worker.

But just to make sure, try this second test on the person. Tell him that your rich uncle died and left you a six million dollar estate and you want to share half of it with him. Then ask the person to describe in detail how he will use the money. If the person will quit his job to do something else (whether it is to lounge on the beach in Acapulco, or become a missionary to Siberia), you will know his commitment as a worker. On the other hand, if he plans to use a portion of the money to increase the effectiveness of his work, then you have encountered one of this world's most rare persons – a true worker.

Let's look at a possible salary package for full time servants of Christian organizations. You may say "That's easy! Based on the principle that a true worker does not work for money, we have a very simple salary package, each worker will get one big thank-you per year – and that's all." You silly idealist, come back to the real world! It's true we shouldn't pay our servants for their work. But given the realities of life in the twentieth century, we will need to pay them to enable them to work. And it is very important, as we prepare our salary schedule, to realize that the purpose of the salary is to enable our workers to fulfill all the various responsibilities they have in the community.

These may include family responsibilities (such as food, shelter, clothing, and the education of their children), church responsibilities (such as serving on the consistory or as a Sunday school teacher), school responsibilities (such as serving on the school board or on a support organization like a booster club), as well as the responsibility of a worker. For that reason the salary schedule must try to address needs, even though these needs will be different for different workers. To best accomplish this, I think it will be helpful to break the salary schedule into four parts: a basic wage, a fringe benefit package, a scholarship program, and a grant program.

If the other three components of the salary package are carefully developed, the basic wage can be a simple scale. If we assume that the average person begins work and likely begins a family at the age of about twenty-five, we expect that needs are lowest for the first five to ten years, greatest during the following ten to fifteen years, and then decrease again until retirement. Let me suggest a $12,000 starting salary with $400 increases each year for twenty years leading to a maximum annual salary of $20,000. [Keep in mind: this was written in 1986, in Northwest Iowa.]

The fringe benefit package should be designed to meet the needs of both single workers and families. A major medical plan such as Blue Cross/Blue Shield should be in place to take care of serious medical problems and hospitalization. In addition, a dental, optometric, and minor medical plan should be available that provides up to a fixed amount (let's say $1000) per year that a worker can draw upon if needed. A second component of the fringe benefit package ought to be a retirement plan. This should provide at least half salary to any worker retiring after age 65. The third component of the fringe benefit package ought to be a comprehensive insurance program. It should provide a substantial term life insurance benefit and total disability benefit to the worker, but also should provide up to a fixed amount for automotive and home insurance (let's say $1000, and $300 per year respectively).

The scholarship program is an important part of the salary package that would meet the needs of large families. It should be based on need; thus, a worker would apply for it by submitting evidence of his or her yearly family income. For a worker with no other family income but the basic wage mentioned earlier, it ought to pay 100 percent of Christian grade and high school tuition and 50 percent of Christian college tuition. In addition, it ought to pay up to a fixed amount of any approved tuition expense incurred by the worker in connection with his work.

The last component of the salary package would provide a series of grants to meet specific needs. First, and perhaps most obvious, would be a housing grant. This would be in the form of a low interest loan (the interest rate ought to be set at the inflation rate) of up to $30,000, which would enable a worker who had no other means to purchase a house. Second, a "professional expenses" grant ought to be made available to the worker, let's say of up to $2,000 per year. This would be applied for once during the year, and the worker would then be free to draw upon the amount granted as it is needed. It would be intended to cover such things as tools for mechanics, books and periodicals for educators, conference fees and conference travel expenses, etc. The amount of the grant would be based on need and determined by total family income. As with every other part of the salary package, it is intended to enable the worker to be an effective steward of his or her talents.

Obviously the details of such a salary package would need to be worked out very carefully; realizing that some elements of the package might not be possible for every Christian organization. It might be costly in some cases. For example, a worker with a family of six children in Christian school and with twenty years of experience could cost the organization $36,000 each year, even though his wage is only $20,000. Two requirements would need to be met for such a salary package to work. First, the organization would have to have sufficient funds to set it up. That's the easy part. The second requirement is that each member of the Christian community would have a heart that seeks the good of his neighbor, be committed to enabling his neighbor to develop the talents the Lord has given, and be totally disinterested in making comparisons between the material state of himself and his neighbor. If this second requirement were achieved, the necessary funds would come naturally.

CHAPTER 10

THE OUTRAGEOUS IDEA OF
RELATING FAITH TO LEARNING

AUGUST 1, 1997

George Marsden, a former Calvin College professor who is now teaching at the University of Notre Dame, has written a new book titled *The Outrageous Idea of Christian Scholarship* (New York: Oxford University Press). It is a brief follow-up to his earlier book *The Soul of the American University* (New York: Oxford University Press, 1994), in which he says, "I explored how and why American university culture, which was constructed largely by Protestants, has come to provide so little encouragement for academically rigorous perspectives explicitly shaped by Christian or other religious faith" (i). The follow-up book was a necessity of sorts because Marsden is a respected historian and he seemed to be suggesting, in that first book, the possibility of a distinctively Christian approach to academics – not just in theology, mind you, but in all academic fields of inquiry. Well, he *was* suggesting that. And most scholars who are part of the American university culture – Christians included – *do* think that the idea is outrageous. And so Marsden has written a new book titled *The Outrageous Idea of Christian Scholarship*.

It's an excellent book. It's one of the best in recent years to attempt a description and defense of the reformed position that all of life is religion, scholarship included, and that if one claims to be a Christian, one ought to be pursuing scholarship that is uniquely characterized by one's Christian faith. I whole-heartedly recommend that everyone interested in education get a copy and read it before the year is out.

But this essay is not a book review. What I want to do in these few lines is to pick at what I consider to be a little "bug" in this otherwise excellent book. It is a bug that has been a part of our reformed language for so long that we are not even aware it is there.

Marsden begins his book by writing "Contemporary university culture is hollow at its core" (3). In fact that is the first sentence of the introduction on the very first page of the book. He goes on to describe how even though the most prominent academics are preoccupied with politics, they proffer no basis for preferring one set of political principles over another. They do, on the other hand, he suggests, tend to be dogmatic moralists. But, Marsden goes on to say, the theories that they espouse tend to undermine not only traditional moral norms but their own moral norms as well.

So, before the first page is over, Marsden has gotten to the heart of the issue. Contemporary university culture has no core, or in other words, no foundation, no starting point. Why this is so has to do with the fact that any core, basis, or starting point is beyond proof or demonstration. It's where you stand in order to prove or demonstrate other things – things less basic, less foundational. Therefore it must be accepted on faith. And this is something the contemporary university culture refuses to do (or better, it is something the contemporary university cultures *pretend* that they will not do).

After making this opening point however, and before we are even off that first page, Marsden writes the following, "American higher education should be more open to explicit discussion of the relationship of faith to learning." Sounds good, right? It even sounds consistent with what he previously wrote. Instead of pretending that faith commitments play no role, American higher education ought to recognize that faith commitments, Christian and otherwise, do exist, and that they provide the foundation for our academic work.

But listen again to the last few words in Marsden's sentence. "American higher education should be more open to explicit discussion of *the relationship of faith to learning*" (emphasis added). Now what I am going to say may initially sound very picky, but I believe it is important. Consider what is meant by the phrase, "the relationship of faith to learning." It's clear that what is aimed at is a discussion of how these two things, faith and learning, are related to each other. But notice that before we can discuss their relationship, we must presuppose their separate existence. That is to say, we must assume that there exist these two categories that are clearly distinguishable from each other in our minds.

Here we have the old enigma of integration. You see, to integrate is to unite, to blend, to combine, to synthesize, or to incorporate into a unified whole. But to do any of these things, one must identify at least two distinguishable entities, that is, one must presuppose a kind of duality or plurality. The process of integration is then either an obliteration of the characteristics that distinguish the two or more entities, or a rendering of those characteristics insignificant in the context of the unified whole. The enigma is the stubbornness with which the original distinction resists that obliteration or that rendering as insignificant. Consider the example of racial integration. It is possible to integrate a school, a bus, or a lunch counter by passing laws that prohibit segregation. But unless the racial distinction at the root of that segregation is rendered insignificant in the context of the school, bus, or lunch counter, the goal of achieving a truly unified whole, that is, true integration, will not be achieved. It is likewise with faith and learning.

So I think we ought to avoid language that speaks of "integrating faith and learning" or, as Marsden and many others put it, "the relationship of faith to learning." In either case we have presupposed – at least semantically – the very disunion we seek to overcome. From a reformed Christian perspective, learning is an expression of one's faith. It is one of the ways that we respond, either in obedience or disobedience, to our Creator. So our task is not bringing together, not integrating, not relating faith and learning. Rather, it is to make others aware that one's learning activities are always directed by one's faith – whatever that faith might be.

And this, despite my semantic pickiness, George Marsden does a superb job of accomplishing in his book *The Outrageous Idea of Christian Scholarship*.

CHAPTER 11

WHY GO TO A CHRISTIAN COLLEGE?

MARCH 8, 1996

Are there two kinds of knowledge; one dealing with facts, the other dealing with opinions? I raise that question because it seems like our culture has presupposed an affirmative answer to it. But I believe that accepting that answer has led many Christians to unchristian conclusions on a variety of topics.

In a recent discussion with my Engineering Department colleagues at Dordt College, we were trying to develop solid, reasonable arguments for why a student who is interested in engineering ought to study at a Christian College. That's usually not a very difficult task for us, since we believe that there are many reasons and we are involved on a daily basis in working them out. But we wanted to be sure that we had a common, well-articulated rationale for the many high school students and their parents who visit us each year while deciding where they should go to college.

In the course of our discussion, we asked, "Ought *every* appropriately talented Christian, who is interested in pursuing engineering as a vocation, go to a Christian college for such study?" In particular, we considered that unique Christian young person who has already determined, while in high school, her specific area of interest, who feels called to a vocation in that field, and who is eager to begin the kind of specialized learning needed for that kind of service. Is it possible that such a person may benefit more from going to a state university where she can immediately get into that area of specialization? You see, at a Christian college the undergraduate engineering program will usually be more general than at a state university. There will be required courses in the humanities and social sciences that are not required at the university. Consequently, the Christian college student will not have as much room in her four-year schedule for highly specialized, technical electives.

Well, as you might imagine, I have an answer to that question. And, also as you might imagine, the answer is that the Christian college offers a better engineering education than the state university, and that *every*

Christian ought to choose to go to a Christian college. But I won't spend the rest of this essay talking about how the general engineering education found at a Christian college is more holistic, more effective, more human, and consequently more obedient than the specialized engineering education of the state university. What I want to consider is how it is that such a question can arise in the first place. That is, what is it that might make us think that a specialized university education could be as good or better for a particular Christian student than a more general Christian college education?

Recall that at the beginning of this essay I asked, "Are there two kinds of knowledge; one dealing with facts and the other dealing with opinions?" If we believe that there *are* two kinds of knowledge, then it might be possible to reason as follows: A student who has grown up in a Christian family, had the benefit of a church community, Sunday school, and Christian education, whose faith is strong and who feels called by God to serve in a very technical area of engineering, and, finally, who has a great deal of interest in math and science but not much interest in courses like English and history . . . such a student would benefit from what the large university has to offer. After all, two plus two equals four is the same for the Christian and the non-Christian, and therefore the engineering courses at the Christian college cannot be very much different from those at the university. And the university has a much greater variety of technical courses, with millions of dollars' worth of high-tech equipment as well. So the technically talented Christian might even be guilty of hiding those talents by choosing to go to a Christian college instead of the university.

Well, let me say unequivocally that such reasoning is seriously flawed in more ways than I can possibly address in this essay. But I do want to point out one serious flaw that is not often or easily seen, and it has to do with this notion of there being two kinds of knowledge.

If you are familiar with the old TV series, *Star Trek*, you will remember that the two main characters, Captain Kirk and Mr. Spock, are almost exact opposites in their ways of thinking. Mr. Spock represents the cold logic that supposedly characterizes science. Captain Kirk, on the other hand, represents the fallible, emotional, sometimes tragic, but often heroic thinking that seems to depend more on impulse than logic. At times it is the cold logic of Spock that saves the day. At other times, the passion or compassion of Kirk triumphs where logic fails. What is consistently portrayed episode after episode, however, is the notion that there are two ways of thinking. One is based on cold, hard, logic, the other is based on softer stuff: feelings, opinions, beliefs. And the majority of Americans who watch the show, accept it all, without so much as a blink.

That is why it is possible for some Christians to consider studying engineering at a state university. They tend to think that the experience will be mostly logical, without much in the way of feelings, opinions, or beliefs. But that is terribly false. The Lord created us as whole persons. No one works, learns, or thinks only in the realm of logic. Mr. Spock is fictional. And there are not only two kinds of knowledge, rational and emotional, there are many other kinds. And when a real live human being engages in thought, all the different ways the Lord gave us to think are active.

So here are two concluding thoughts that may be helpful whenever you hear of a Christian contemplating going to a state university to study science or engineering. First, no matter what you study, science or philosophy, engineering or poetry, every aspect of your thinking will be active and you will grow logically, socially, ethically, aesthetically, emotionally, and so on. Second, to study in an environment that you believe only teaches you to think logically is to close your eyes to how that environment will/must necessarily also teach you to think regarding other aspects of your life. It's like Daniel getting his education in Babylon but remaining oblivious to the fact that King Nebuchadnezzar had a penchant for building golden images.

CHAPTER 12

YEARBOOKS, THE NEW SCHOOL YEAR, AND CHRISTIAN DISTINCTIVENESS

SEPTEMBER 9, 1996

At the beginning of a new school year, it is hoped that many a teacher in a Christian school – grade school, high school, or college – will ask the question, "Will my work during this year be truly and distinctively Christian?" It's the kind of question, of course, that every Christian, not only Christian school teachers, ought to ask. But it is particularly meaningful for the teacher. After all, if there is little distinction between Christian schools and state or private schools, then one could hardly justify the resources spent to support Christian education. In this age, however, when *inclusiveness* is a dominant cultural theme, Christian distinctiveness comes to bat with two strikes against it. It sounds too much like *ex*clusiveness, or even discrimination, two words that have evoked concern and taken on deeply pejorative meaning of late. And Christians ought to be able to appreciate those concerns, despite their associations with pagan postmodernity. After all, the message of Acts 10:35 is very clear. "God does not show favoritism, but accepts from every nation the one who fears him and does what is right." Those are the words of the Apostle Peter. So what *does* it mean to teach in a truly and distinctively Christian manner?

A few weeks ago, near the end of the summer, I spent a weekend at my brother's house near Chicago. It was a family reunion of sorts and, since my family has grown to number thirty people spread over four generations, there were a fair variety of persons present. Despite the variety, however, more than half the individual families have one or more members who are teachers. One of those teachers – a young, Christian high school, English teacher – was eager to show the rest of us the *video yearbook* that he had helped produce during the past school year. Its technical quality surprised me. Fancy fades and wipes and the careful coordination of audio and video made it clear that this young English teacher knew his stuff, enjoyed his work, and enjoyed showing it off. When it was over, however, one nagging thought remained – a thought quietly verbalized by one of my brothers, even as it lingered in my own mind. The video yearbook was for a Christian high school. Yet, except for a few references to chapel services and the Bible teacher, it could have been the video yearbook for any North American high school. There was nothing distinctively Christian about it.

A yearbook, in general, is a pretty good indicator of the practical, communal worldview of a given school. Both its form and it content reveal the kinds of things valued by that school community. That's why, whenever I look at a Christian school yearbook, I try to find evidence of resistance to the surrounding culture. Or, put more positively, I look for a kind of Martin Luther-like "Here we stand" impression to permeate the pages from the opening statements and photos to the index. I've got twenty-five yearbooks on a shelf in my living room, and some of them really do evidence, at least in part, that Christian distinctiveness. Sadly, most do not. Oh, I know it's easy to carelessly scatter Bible quotes amidst the pages and insure that there are well-distributed pictures of chapel services and Christian service clubs doing their things. But I'm not talking about that. I'm talking about something deeper; an indication that the day-to-day life of the school is self-consciously Kingdom-oriented. Some of my yearbooks, however, don't even have the more superficial distinctiveness.

What makes something distinctively Christian is not only references to the Bible or to church-related service organizations. Whether it is the work of a Christian school teacher or the yearbook of a Christian school, the distinctiveness ought to be found in the way in which its most mundane characteristics evidence a Christ-centeredness, and stand thereby in stark contrast to mundane characteristics manifest in the surrounding culture. For example, there are times when almost any high school or college student, while studying science, experiences the thrill of learning something new. I don't mean just learning some factual information. I mean seeing for the first time how things work together, understanding the explanation for some ordinary natural phenomenon. It might be the explanation for how soap works to clean your hands, or why it is that diluted vinegar is great for getting the hard-water streaks off the windshield of your newly washed car. What distinguishes the Christian student from the North American herd of science students is the immediate awareness that it is God's good creation that is better understood, that the understanding itself and the pleasure that it brings are gifts from God, that the understanding and pleasure are also a valid way of responding to God, responding to his call to be a servant bearing his image, and that God shares the pleasure experienced by his human creatures when they respond as he intended. This awareness moves the Christian student to gratitude and praise. That gratitude may be expressed enthusiastically and verbally, or simply as a thought – or better, a prayer quietly but consciously uttered within the closet of the mind. This sort of thing should be happening all the time in a Christian community.

When someone asks me, "What makes Christian education Christian?" and I have opportunity only to give a brief answer, I tell them it is in the *what*, the *why*, and the *how*, of everyday life in the classroom. By *what*, I mean that there is a conscious awareness that what we study are God's creatures. By *why*, I mean that the purpose of every activity is to enable God's children to be the servants he calls them to be. And by *how*, I mean that the methods used to realize that purpose are rooted in a view of the student as image bearer and servant of God. So the pedagogy, too, bears a characteristic Kingdom distinctiveness. The subject matter, the purpose, and the pedagogy. It seems to me that covers all of the bases. And to borrow a phrase from Abraham Kuyper, there is not one square inch of the Christian school experience that Christ in his sovereignty does not claim as his own, and not one square inch, yearbooks included, that ought not to bear the visible mark of Christian distinctiveness.

Chapter 13

Poetry for Technophiles

October 3, 1997

As an engineering professor, I encounter many students who consider themselves nerds or technophiles; that is, people who are busy with, and appreciative of, the technical areas of modern life – mathematics, the natural sciences, and the products of modern technology like the computer – to the virtual exclusion of other areas of life, such as the aesthetic and the social. One of the goals of the program in which I teach is to enable such students to grow beyond that narrow technical flower-pot in which they find themselves and to blossom into the whole, vital, and contributing persons that the Lord created them to be.

Part of the problem that we face with such a task is that an engineering education, by its very nature, tends to be narrow, technical, and constricting. Thus powerful tools and a concerted effort are required to move the students in the opposite direction – the direction of academic breadth, aesthetic and moral sensitivity, and a passion for living life in its Kingdom fullness.

About ten years ago I had a student in class during the fall semester of that student's senior year. I remember specifically an event that occurred midway through that semester. The class was taking an exam, and apparently this particular student, who ordinarily did quite well on exams, had run stuck on one problem. I wasn't aware of that until the sound of a desk crashing to the floor assaulted my ears, and I watched as the student stormed out the door in otherwise quiet frustration. After the exam I reflected to myself, with some sadness, on how the student had no vehicle for releasing that frustration other than violently throwing his desk to the floor. At that point I noticed a little paperback book staring at me from the corner of my desk. The multiple applications of cellophane tape had long ago lost their capacity for keeping the first ten pages secure within its binding. On opening it, my eyes were greeted with chestnut gradations on the periphery of each page and my nose tried to tell me that I was rummaging through ancient stacks at a university library. The real give-away, however, was the price innocently printed on the front cover: 60 cents. Cheap, even for an engineering freshman in 1964, it disguised the influential role that the book, *Immortal Poems of the English Language*, played in the education of this engineer. There sat an answer to my reflective questions regarding my student's frustrations.

Like many engineering students, along with enjoying "the existential pleasures," I too suffered "the slings and arrows of outrageous fortune." For me, it seems the latter were all concentrated in my freshman year. I recall the angst that accompanied the return of my first physics midterm; and, following an "all-nighter," my frustration when realizing at 5 a.m. that I had just thrown away alertness, my best defense against a monster calculus test. But I found a vent for my distress in the unlikely form of my English class assignments. At first it was one of Shakespeare's sonnets (no. 66):

Tired with all these, for restful death I cry, –

Then we were assigned to read the book of Ecclesiastes:

"Vanity of vanities," says the Preacher,
"Vanity of vanities! All is vanity."

What advantage does man have in all his work
Which he does under the sun? (1:2–3 NASB)

The poetry resonated with my psychic state. Not literally, of course (I never considered quitting engineering, much less life!), but the quality and intensity of the metaphors were so deeply satisfying that I found crying with Hamlet,

The time is out of joint: O cursed spite,
That ever I was born to set it right! (I, 5: 205-206)

to be far more heartening than slamming my books down on my desk.

There have been many occasions since my freshman year in college when reciting the lines of an appropriate poem or song – usually in my head – has expanded my narrow little world and been a providential vehicle for supplying the solace, encouragement, and perspective needed to work through a period of frustration, pain, or even delight.

Recently, the fabric of our usually quiet and orderly Dordt College lives was ripped apart by the sudden and seemingly senseless death of one of the more prominent members of our community. There is no way to explain or rationalize such occurrences. But the Lord provides, in aesthetic expression, a valuable component of our creatureliness that allows us to transcend the finitude of our day-to-day lives. Once before, many years ago, I was confronted head-on with horror, tragedy, and grief. At that time I took refuge in reciting, over and over in my head, the third and fourth verses of Martin Luther's great Reformation hymn *A Mighty Fortress Is Our God*. Those verses begin: "And though this world, with devils filled, should threaten to undo us, we will not fear, for God has willed his truth to triumph through us." If you don't know the remaining lines, I suggest you find a hymnal and read through them. They have visited my often-wandering mind on numerous occasions during the past two weeks.

This semester I once again have at least one student, a senior, who, when looking into the mirror, concludes that the image looking back is that of a "techno-nerd." There is an irony in this particular case, however. The irony is that anyone observing this student in or out of class, for more than a superficial moment, is soon impressed by the student's potential for contributing far more to the twenty-first century and the Kingdom of God than a few well-cranked-out integrals, computer programs, or techno-gadgets. But it's only October. Graduation is seven long months away.

My colleague Jim Mahaffy is a biologist. He studies plants and gets to watch them blossom into flowers that bring beauty to this fallen, and at times ugly, creation. But I'm happy being an engineering educator. I get to watch techno-nerds as they blossom into culture-forming, world-healing, God-glorifying, whole persons.

CHAPTER 14

DEEP MAGIC FROM BEFORE
THE DAWN OF TIME

APRIL 1, 1998

The stereotypical engineer is a person who is logical, who looks to reason as a guide in decision making, and who is highly skeptical of any element of thought that might suggest equivocation, mystery, paradox, or inconsistency. The engineer isn't alone in this, of course. A bias against thinking that is not purely rational is shared by analytic philosophers, lawyers, and business executives as well. In fact, it is the central feature of what we call modernism, the attitude of thought that has dominated Western culture since the Enlightenment. Chief Science Officer Spock, one of the main characters of the original Star Trek television series, epitomized it in the late sixties and early seventies. And, you may be surprised to learn that modernism is one of the primary obstacles against which Christian engineering educators have to struggle in their attempts to educate students from out of a distinctively Christian perspective. You see, engineering that is obedient – Christ-centered engineering – requires the development of wisdom. And wisdom is far more than the tight logic of the part-Vulcan science officer Mr. Spock or the Enlightenment philosopher and mathematician René Descartes. That's why, of late, I have been playfully using the phrase *deep magic* with my engineering students here at Dordt College.

But before saying more about *deep magic*, allow me to give a biblical explanation to those who may be wondering how it is that a professor of engineering can have difficulties with his students' affinity for rational thought. In the first few chapters of 1 Corinthians, the Apostle Paul deals with quarrels and divisions in the church. His way of dealing with the argumentation is to associate it with what he calls "cleverness of speech," "persuasive words," and "the wisdom of this world." These are pejorative phrases that Paul contrasts with a different kind of knowledge and wisdom, which in 1 Corinthians 1:18 he calls "the message of the cross." In chapter 2 Paul argues that he does not come to the Corinthian Christians "with eloquence or human wisdom," but "in weakness with great fear and trembling." Then, in verses 6–8 he says the following:

We do, however, speak a message of wisdom among the mature; but not the wisdom of this age or of the rulers of this age, who are coming to nothing. No, we declare God's wisdom, a mystery that has been hidden and that God destined for our glory before time began. None of the rulers of this age understood it, for if they had, they would not have crucified the Lord of glory.

So what is this "wisdom" that is "not of this age," that has associations with "mystery," that might be contrasted with mere cleverness of speech or persuasive words? Some answers to that question can be found in the book of Proverbs, particularly in chapter 8 where wisdom is associated with prudence, righteousness, truth, discretion, and life. Those associations suggest strongly that wisdom includes far more than just reason. And that is something particularly important for future engineers to understand. For solving societal problems by the unfolding and developing of creation requires more than just logic and mathematics. It requires insight into areas such as economics, sociology, language, ethics, justice, and aesthetics. More than that, it requires a kind of vision that can perceive the tapestry of the creation in its multifaceted complexity. I like to use that tapestry metaphor with my engineering students. The creation is perceived as a great tapestry woven together of innumerable fine threads and in which is embedded an uncountable assortment of jewels. Science, mathematics, and reason are among the tools that enable us to glimpse some of those jewels and threads and thereby allow us some appreciation of the creation tapestry. But to see the interconnectedness of the threads, to gain some vision of the whole, we need much more than what those rational tools can provide. We need the wisdom of Proverbs, the wisdom Paul is talking about in 1 Corinthians when he uses the word "mystery."

C.S. Lewis describes this wisdom in a particularly appropriate way in his gospel allegory *The Lion, the Witch and the Wardrobe*. In the story, four English children find themselves magically transported to the land of Narnia. Narnia is a fascinating world that was created by a great and noble lion named Aslan, where animals talk and contrive schemes just like people. When the children arrive, however, Narnia is under the control of evil, in the form of the wicked White Witch. Edmund, one of the children, is tempted by the Witch and joins up with the forces of evil. In order to save Edmund's life, Aslan gives himself to be killed in Edmund's stead. He is tied to a great Stone Table and slain. At dawn, on the following day, two of the children, Susan and Lucy cannot believe their eyes as they see the Stone Table cracked in half and empty. At that point, Aslan, alive again, appears to the children.

"But what does it all mean?" asked Susan when they were somewhat calmer.

"It means," said Aslan, "that though the Witch knew the Deep Magic, there is a magic deeper still which she did not know. Her knowledge goes back only to the dawn of Time. But if she could have looked a little further back, into the stillness and the darkness before Time dawned, she would have read there a different incantation. She would have known that when a willing victim who had committed no treachery was killed in a traitor's stead, the Table would crack and Death itself would start working backward."

It seems to me that *deep magic* is a particularly fruitful phrase that helps describe the wisdom that Paul is talking about in 1 Corinthians. It's fruitful in a least two ways. First, it suggests that true wisdom involves mystery, something beyond mere logic. Secondly, it's iconoclastic; that is, it flies in the face of the modernist notion that the only valid knowledge is rational or factual knowledge. Using it in a class filled with engineering students inevitably creates a healthy kind of discord, a tension that causes the student to sit up and ask, "What's going on here?" And any time you can get an engineering student to ask that kind of a question, a nontechnical question, you have achieved something significant as an educator.

But, alas, *deep magic*, or wisdom is not something one can teach like one teaches solar collector design. There are techniques by which an engineering educator can pretty much insure that the student will learn the mechanics of solar collector design. But all a teacher can do is offer pointers and suggestions when it comes to wisdom. The student must grasp it for herself. And often the teacher must endure the kind of anguish that the Christian philosopher Søren Kierkegaard is known for describing. Kierkegaard wrote two kinds of books. One kind involved deep philosophical probing that was very academic in nature. His most famous treatise *Either/Or* best represents it. The other kind of writing he described as "religious." These are more like sermons, reflections on his Christian faith, and according to Kierkegaard, these are the "things of most vital importance" and foundational to his philosophical writing. His book titled *Two Edifying Discourses* best represents these. About these two kinds of writings, Kierkegaard said the following:

> I held out *Either/Or* to the world in my left hand, and in my right the *Two Edifying Discourses*; but all, or as good as all, grasped with their right what I held in my left.*

It's like that in engineering education too. It takes a deeper magic to instill deep magic in the hearts of students; a deeper magic that one can only pray for.

---

* Søren Kierkegaard [1848/1859], *The Point of View for My Work as An Author* (New York: Harper Torchbooks, 1962), 20.

CHAPTER 15

THE HEAVENS DECLARE THE GLORY OF GOD – AND THE DIFFERENTIAL EQUATION PROCLAIMS THE WORK OF HIS HANDS

JULY 2, 2003

In the summer, in Sioux Center, the sun is a night owl. Perhaps that's because the "heavenly tent," which the Lord has "pitched" for the sun, is so huge out here on the plains of Northwest Iowa. From horizon to horizon that "champion rejoicing to run his course" is like a long distance runner during the latter days of June, beaming to us mortals to within two hours of midnight. I'll always remember my family's first night in Sioux Center back in late June of 1979. Our three young sons wanted to explore the town after supper and so, giving them permission, we told them to be back home by sundown. When 10:00 p.m. rolled around and the boys were not yet home, my wife and I became a bit concerned – except for that fact that there was still a fair amount of light in the sky. We found them in the Open Space Park, along with a crowd of other people, watching the conclusion of a baseball game. Our three young city-slickers from New Jersey were genuinely surprised to find out what time it was.

This summer I'm again enjoying the late June sunsets and hearing the words of Psalm 19:1–6 as I do so:

> The heavens declare the glory of God; the skies proclaim the work of his hands. Day after day they pour forth speech; night after night they reveal knowledge. They have no speech, they use no words; no sound is heard from them. Yet their voice goes out into all the earth, their words to the ends of the world. In the heavens God has pitched a tent for the sun. It is like a bridegroom coming out of his chamber, like a champion rejoicing to run his course. It rises at one end of the heavens and makes its circuit to the other; nothing is deprived of its warmth.

But this summer those words have additional meaning for me because I'm preparing to teach a course in the fall that I've never taught before: a mathematics course. And mathematics is one of those subjects that can be especially difficult to teach from a distinctively Christian perspective. "After all," people ask, "isn't math the same for everyone?" Well sure, in the most obvious ways it is. But then, so is the sun! And yet the sun has such an important role in the Scriptures – particularly in the Psalms. In these verses from Psalm 19 it's clear that the sun is a servant of God. And it's clear that the sun is not a silent servant. Rather it "pours forth speech," declaring "the glory of God." Just as "nothing is deprived of its warmth," so all are without excuse that observe the sun and fail to see the glory of its Creator evidenced in its warmth, its light, and its faithful daily passage across the sky. But the psalmist chose the sun *only* as a lucid example. *All* of God's creatures evidence his handiwork. And so an important theme of the math course I will be teaching will be adopted from those first lines of Psalm 19: "The differential equation declares the glory of God, the Laplace transform proclaims the work of his hands."

If that sounds a little strange to you, consider the next five verses of the Psalm:

> The law of the LORD is perfect, refreshing the soul. The statutes of the LORD are trustworthy, making wise the simple. The precepts of the LORD are right, giving joy to the heart. The commands of the LORD are radiant, giving light to the eyes. The fear of the LORD is pure, enduring forever. The decrees of the LORD are firm, and all of them are righteous. They are more precious than gold, than much pure gold; they are sweeter than honey, than honey from the honeycomb. By them is your servant warned; in keeping them there is great reward.

In this second part of the Psalm, the emphasis shifts from God's creatures to his relationship with his creatures. That relationship – which includes God's bringing his creatures into being and sustaining them – is what we often refer to as the "Word of God," or the "will of God" for creation. The psalmist takes pains to insure that his reader understands this by using a number of different terms that mean much the same thing as "Word" or "will." He talks of "the law of the Lord," "the statutes of the Lord," "the precepts of the Lord," "the commands of the Lord," and "the decrees of the Lord." These certainly include the Word of the Lord for how we human creatures ought to live our lives. That's why the psalmist mentions that "in keeping them there is great reward." But they are also intended to include God's governing Word for the rest of creation – which includes mathematics. And when we study mathematics, we study numerically structured parts of that creation. In the course I am going to teach, we are not so much interested in numbers as in abstract things that can be related by numbers and the changes in numbers over time and space. For example, if a lake is fed by a mountain stream that keeps the lake full and clean, but is also having pollution dumped into it by a factory, we can represent the rate at which pollution is building up in the lake, or the amount of pollution in the lake at any point in time, by writing and solving a differential equation. And it's the law of the Lord for the numerical, the spatial, and the logical aspects of his creation that enables us to write such an equation and to have confidence that the results of the equation will give us insight regarding the situation with the lake.

Having confidence presupposes a kind of order or regularity. We find that kind of orderliness in creation that we cannot help but recognize with our minds. It's true that our human minds construct a specific kind of order, and therefore we can discern differences in the concepts of order between different cultures throughout history. But underlying all those concepts of order – or even disorder – are those ordinances, commands, precepts, and statutes about which the psalmist writes. And that underlying order, rooted in the Word of the Lord, is what makes mathematics possible. It is also what makes the late evening summer sunshine possible. And thus it is that differential equations, as well as the heavens above, declare the glory of God.

CHAPTER 16

THE TACIT DIMENSION OF KNOWLEDGE . . . AND DISTANCE LEARNING

JANUARY 1, 2004

Have you ever learned to ride a bicycle? I'm sure that most readers of this book have. Do you remember what it was like? For me it was mostly trial and error after observing how the "big kids" rode their bicycles and after riding the sidewalk with a little tricycle for a year or so. Some of us had "training wheels" on our first bicycle, but I went from the tricycle on the sidewalk right to the bicycle on the street. It wasn't a very big bicycle, but it was a brand new experience. My father jogged behind me at first, steadying my balance and getting me going fast enough so that the angular momentum was sufficient. He might have taught me about conservation of angular momentum – the basic physical principle that explains how and why a moving bicycle doesn't just fall over like it does when it's standing still – except for three factors. First, *he* never learned it. Second, at age five, *I* wouldn't have understood it. And third, even the most profound knowledge of physical theories will not help you very much when it comes to learning how to ride a bicycle.

In this essay, I would like to explore briefly that last point. Let me put it a little differently. There are kinds of knowledge, and therefore kinds of learning, that cannot be gained in the conventional way (e.g., by reading a book, attending a lecture, using a computer, or otherwise working only with your mind). On one level, we might call these other kinds of knowing/learning experiential knowledge or experiential learning. Certainly that fits the example of learning how to ride a bicycle. No one can learn how to ride a bicycle unless they actually experience sitting in the saddle with hands on the handlebars, moving across the road with wheels turning – and, of course, occasionally falling down. But there's something deeper going on here than mere experience, and so the term experiential knowledge is not quite adequate. You see my youngest grandchildren – a set of sixteen month-old twins – already have a fair amount of knowledge. They know their parents. They know how to eat solid food (although they still make quite a joyful mess when doing it). And a few months ago they learned how to walk – something they are doing with increased agility each time I see them. But they haven't learned to ride a bicycle. And no matter how much experience we may give them of sitting in a saddle, holding on to the handlebars, and so on, they are simply incapable, at this stage of their development, of learning to ride. That's because learning to ride a bicycle takes more than mere experience. It requires a kind of knowledge that is based on that experience.

The scientist and philosopher Michael Polanyi – a Christian, I might add – did most of his work in the latter half of the twentieth century and developed a theory that nicely explains what I have been describing. He identified what he called "the tacit dimension" of knowledge. Tacit knowing is non-rational knowing. It's not *irrational*. It is simply all the types of knowledge that lie outside that narrow definition of knowledge that we westerners have grown up with since the Enlightenment. Do you know the smell of a rose? That's sensory knowledge, one type of tacit knowing. Would you know the difference between a beautifully arranged bouquet of flowers and a dull, almost ugly arrangement? That's aesthetic knowing, another kind of tacit knowing. Do you get into your car on a regular basis with the knowledge that the brakes will work when needed, even though you haven't had them checked recently? That's a simple form of faith knowledge. Do you believe that what we call the law of gravity will hold you to the earth ten years from now just as it does today? That's a more profound form of faith knowledge. Do you truly know the One who created you, who sustains you moment by moment, and who has delivered you from sin and death? That's a most profound form of faith knowledge. And faith knowledge is also part of the tacit dimension of knowing. My point thus far has simply been to show that there is another dimension of knowledge – beyond that bordered by reason – that we might call tacit knowledge, and that tacit knowledge cannot be obtained merely by reading a book, watching videotape, or attending a lecture.

So how does one gain tacit knowledge? Well, let me suggest three possibilities – I'm sure there are more. One way that we've already discussed is experience. That's how my twin grandchildren learned to walk and that's how they will one day learn how to ride a bicycle. Another way is *reflection*. Reflection is when you sit back after observing a sunset – or reading a particularly profound paragraph in a book, or experiencing an emotionally moving film, or hearing a unique sermon . . . when you sit back and *contemplate that experience*. A third way of gaining tacit knowledge is by means of a teacher. But in this case it is not the words that the teacher uses, rather it is the nonverbal, non-rational content of such things as body language, tone of voice, inflection, and facial expression that communicate tacit knowledge.

So why is all this important? Well, it's important because the technology of education is developing and changing. With the computer and communication media such as video, film, and, of course, books, we have the capability of teaching and learning in situations where the teacher and learner are separated by great distances and even by time. But when we engage in that kind of separated learning – distance learning, as it's called – we are forced to symbolize the content of the learning in words, numbers, or graphics, in order that it might be transmitted over distance and time. The necessity of that symbolizing means that the kind of knowledge that can be "distance learned" will largely be of the rationally transmittable sort. It is very difficult, or impossible to teach or learn tacit knowledge at a distance. This is unfortunate but tolerable in some cases, for example where the subject matter is highly abstract, as with introductory math or physics. But when the subject matter is less abstract, as with art, music, or creative writing, it tends to be severely truncated by distance learning techniques.

Our Creator has fashioned and called us to live whole, multifaceted lives. Science and technology often requires us to narrow that wholeness by focusing on one or two of the facets. That's OK, as long as we realize what we are doing. In our enthusiasm for technological advancement, however, too often we become blind to the narrowing and truncating character of our artifacts. So the next time you get really enthusiastic about distance learning, consider how you might teach your grandchildren how to ride a bicycle by sending them an e-mail.

CHAPTER 17

HOME SCHOOLING AND CHRISTIAN SCHOOLS

SEPTEMBER 3, 2004

It's late summer and in Northwest Iowa the back-to-school ritual is essentially complete. Dordt College has a new class of freshmen that was welcomed by returning upperclassmen. At Western Christian and Unity Christian high schools the ritual is similar. And Sioux Center Christian School has opened its doors to covenant children from kindergarten through middle school for its 100th year. Christian education is a fact of life in Northwest Iowa. Believing that *all* of life is religion – not just the part having to do with church and Sunday – and that no area of life is neutral or extraneous to the sovereign claims of Christ's kingship or the redemptive grace of his Gospel; a large number of Christian parents see no alternative to Christian education for the children the Lord has given them to raise.

In recent years however, there has been a change in the way some Christian parents are choosing to educate their children. Home schooling – something unheard of around these parts a generation ago – is becoming the choice of a small number of families. Home schooling, of course, is when parents take responsibility not only for seeing to it that their children get a Christian education, but for carrying it out as well. In this essay I would like to look at what I see as some right and some wrong reasons for choosing home schooling, and I want to suggest what I believe ought to be the proper relationship between home schooling and the Christian schools.

Let's look first at what's right about choosing to home school your children. The most obvious reason to do it is when there is no other form of Christian education available. The schools in communities like East Hartford, Connecticut and Iowa City, Iowa – two cities in which I have lived – may be "excellent" by worldly standards, well-funded by a residential community of professionals, but they are not Christian schools, schools that honor Christ as Lord over every area of school life in overt, self-conscious, and distinctive ways. Therefore they are not adequate to the needs of the children God calls us to raise.

That leads to a second valid reason for choosing home schooling. In much of today's society both parents are well-educated and called by God to use their gifts for his service. Moreover, we are increasingly becoming a society where both parents are expected to have full-time work outside the home; for a number of reasons, some more defensible than others. I've got to believe, however, that most Christian parents feel uncomfortable leaving their young children in day-care, or being so tied up with their careers that those young children grow up to be latch-key kids when they reach school age; children who come home from school to an empty house. For families where one of the two parents is trained as a teacher, one way of both employing that parent's talents while providing a safe and loving home environment for all of the children is for that parent who is competent to teach to stay home and do so.

A third valid reason, I'm sad to say, is economic. In society today the gap between the wealthy and those of lesser economic means has widened. And the cost of Christian schooling has steadily increased. While some Christian communities have tried to address this, it remains a significant problem, especially at the grade school and high school level. A large family with only one parent employed outside the home will inevitably have difficulties paying Christian school tuition. If the choices are between going deeply in debt, sending both parents outside the home to work, or having one stay home to home school the children – the one competent to teach, of course – then that last choice may become justified.

One more valid reason is when there is a Christian school in the area but parents come to a conscientious decision that it is inadequate to meet the needs of God's children. A school rooted in a dualistic philosophy, for example, which simply adds a couple of Bible courses to a basically secular curriculum, may pose a worse environment than that of a state school where the faith direction is obviously and self-consciously secular. The point is that it's easier to be a Daniel in Babylon than in Laodicea.

But now let's consider some *invalid* (i.e., wrong) reasons for home schooling your children. The first that comes to mind is a kind of secular elitism. The same mind-set that would encourage a college student to go to an ivy league school, like Harvard or Princeton, instead of a Christian college, like Dordt, Calvin, or Redeemer, might choose home schooling over the Christian school under the dubious assumption that tutorial instruction at home will be more effective than classroom instruction; "effective" being defined by the speed with which the child is accepted into an elite prep school or university. In this case, it's the academic success of the student that is more important to the parents than the holistic development of a child of God; academic success, that is, as defined by the pagan spirits of this age.

A second motivator that too often drives home schooling is the spirit of individualism. We live in a society in which the deepest and most cherished belief is in the freedom and autonomy of the individual person. That's a belief rooted in the Enlightenment, the French Revolution, and humanism; and it is a wretched distortion of the love God evidences for each individual and his calling us to have compassion on persons, to love our neighbor as we love ourselves. When the spirit of individualism thrives, the body of Christ is weakened because we place our own needs and wants above those of the community in which the Lord has placed us. It's a spirit that afflicts every one of us who live in North America in the twenty-first century. And in many instances, home schooling is one of its manifestations. What is particularly insidious about this spirit is that it can piggy-back on the genuine, perfectly valid, and even commendable motivation some Christian parents have to give their children a radically distinctive Christian education – one even more distinctively Christian than they can get in the Christian school.

A third and obviously invalid reason for home schooling is simply to save money so as to live the upscale North American lifestyle. While it's true, as I said earlier, that the cost of Christian education can be unbearable for some families, for most it is simply an impediment to owning the goodies of the American dream, such as a wide-screen TV, SUV, boat at the lake, or dining out multiple times each week. Materialism goes hand in hand with individualism as two of the primary driving spirits of our age.

So there are good reasons and there are bad reasons for parents to home school their children. But what about in the context of Northwest Iowa, where there exists a well-populated and strong community of Christians, numerous church congregations, and well-established Christian schools that, while not immune to the spirits of our culture, conscientiously seek to operate from out of a biblical, reformational worldview? Well, the existence of these schools, and the "fair share" program that keeps tuition costs well below that of Christian schools in other parts of the country, eliminates much of the justification for home schooling. But not all. There will remain a few families for whom home schooling is appropriate and right, based on the legitimate rationale I outlined above.

What then ought to be the relationship between these home schooling families and the Christian school community? Well, given that we are one in our membership in Christ's body, there ought to be a strong and mutually supportive relationship. Home schooling families ought to be supporting the Christian schools, certainly by means of fair share contributions in their churches, but also by other means. We are all part of one local body of Christ that is vitally interested in the Christian education of Christ's children.

Conversely, the Christian schools ought to be supporting these home schooling families as well. Each of the institutions, from grade school to the college level, will need to find ways to encourage and help those who choose to home school. Christian schools are social institutions that unfold and develop over time. In the twenty-first century, in Northwest Iowa, these institutions will need to develop supporting structures for home schooling families. Of course there will need to be some kind of fee structure that is just, in terms of the services provided and the needs of both the families and the schools. But this should not be a problem if we see ourselves first as members of Christ's body.

When parents bring their infant children to be baptized, the whole community of Christians participates. Those members of the congregation present at the service of baptism speak for all of us when they answer, "We do, God helping us," to the question, "Do you, the people of the Lord, promise to receive these children in love, pray for them, help instruct them in the faith, and encourage and sustain them in the fellowship of believers?" Faithful fulfillment of that vow is the basis for Christian education no matter which form it takes.

CHAPTER 18

FOR RACHEL:
THOUGHTS ON CHRISTIAN EDUCATION

MARCH 2, 2005

Last week Rachel – a student at Reformed Bible College in Grand Rapids (now Kuyper College), and my niece – sent an e-mail requesting suggestions for resource material that would help her write a paper on Christian Education. I responded with a list of about six books that I thought worthwhile, and she, in turn, responded, thanking me and sharing with me some of her thoughts as she prepared to write her paper. Rachel has been in Christian schools her whole life, but lately she's been struggling with the idea of Christian education, "wondering if Christian schools separate children too much from society." In the class for which she's writing the paper they have been discussing the all-too-common separation between sacred and secular, making the point that "we should live *all of life* under Christ's lordship without that separation," and using that point as an argument for Christian education. Rachel wonders if the same point might be used as a justification for sending children to the state schools, no doubt seeing Christian education as a kind of separation of the sacred from the secular. She has also been reasoning that an advantage to sending your children to the state school, at least at the high school level, is that their faith will be challenged there. She expresses concern that in Christian schools there is a temptation for adolescents to become mere nominal Christians, where they can comfortably rest in their parent's faith and not have to think about their own faith by having to defend it. Well, I rarely pass up opportunities to engage intelligent college students on important issues such as this. This time, however, I thought I might do so a bit more publicly. So the rest of this *Plumbline* commentary is my response to Rachel regarding her struggling with the idea of Christian education.

Dear Rachel: Thanks for sharing those thoughts on Christian education. The problem that we Christians have of "taking ourselves out of the world," as well as the problem of taking our faith for granted, are very real, and I have to admit, are not uncommon in the Christian school community. There have been many a parent and educator who have wrestled with those same issues – and I certainly include myself among them. Here are some thoughts that have helped me resolve, at least in part, the tensions that arise from those issues.

First, when you consider the dualism of sacred vs. secular, keep in mind that it is a *false* dualism. There really is nothing in this world that is "secular," if by that we mean "neutral." As Abraham Kuyper once wrote, "there is not a square inch in the whole domain of our human existence over which Christ, who is Sovereign over *all*, does not cry: 'Mine!'"* Our task, in a world that has turned its back on Christ, is then to live in such a way that his Lordship is proclaimed, is made manifest for all to see. Of course, we ought to be doing that from the earliest days of our childhood, but education – certainly education during our youth, adolescence, and young adulthood – is preparation for doing that during the rest of our lives. So let's take a look a couple of decades into the future, when a possible child of yours is about to start high school. What will be the difference – ideally, of course – between her going to a state high school or a Christian high school? Well, for one thing her faith will be challenged more in the state school. I certainly agree with you there. But let me come back to that after considering some other differences.

I assume that any child of yours will likely study music in high school. In a Christian school she will learn that music is a part of God's good creation that enables us to respond to him in aesthetic obedience. She will learn that there are creational norms for music, that it's not just a matter of taste, and that music is a way in which she can not only please her Creator, but also a way of serving her fellow humans, and lastly a way of enjoying God's good creation. In the state school, music will be studied as a human-centered activity, the appreciation of which is up to the individual; and she will learn that no one can tell her what is good or bad music – it's whatever she makes of it.

Your child will certainly study history in high school. In a Christian school any history – American history, world history, European history, history of technology – will be seen as fitting in the overall biblical scheme of creation-fall-redemption. History will be the study of humankind's response to God's cultural mandate, his original command to subdue the earth and fill it. And in the light of the Scriptures your child will learn how we humans have responded obediently and disobediently in the past, and thereby gain insight as to how she may live her life in the

---

* "Sphere Sovereignty" [1880] in *Abraham Kuyper: A centennial reader*, James D. Bratt, editor (Grand Rapids: Eerdmans, 1998), 488.

future. In the state school, however, history will likely be taught as simply facts, events that have occurred in time. The focus, without necessarily stating it, will be the development of Western culture, particularly that of the United Sates. The concepts of individual freedom and the market economy, rather than creation-fall-redemption, will color the lenses of the glasses she wears as she reads about past events.

Maybe your child will enjoy natural science and mathematics and take courses in biology, chemistry, algebra, geometry, physics, and trigonometry. But even if she only takes biology and algebra, in the Christian school she will learn to see these subjects as representing part of God's good creation, which he loves and in which he delights. The universe will be seen as multi-faceted, where faith, love, justice, stewardship, and aesthetics are just as real and important as biotic life, matter, energy, and numbers. And she will learn how gaining insight into the natural sciences and mathematics can equip her for service in areas like medicine, government, law, missions, or simply the task of raising her own children. In the state school, however, she will be taught naturalism, either consciously or by default. Naturalism, of course, is the belief that the physical is all there is, that life, emotions, love, and even faith are all the result of complex chemical reactions that obey the laws of chemistry and physics.

Your child will certainly take physical education and health in high school and may very well, like you and your siblings, be involved in sport. In a Christian high school she will be taught to be a good steward of her body and that stewardship involves far more than mere biotic fitness. She will be taught that it involves moral fitness, economic fitness, and aesthetic fitness. Aesthetic fitness, learning how to play before the face of the Lord, will characterize her approach to sport in PE and in interscholastic sport. And she will be sensitized to the distortions that exist in our culture: the worship of sport heroes, the elevation of entertainment – sport or otherwise – to idolatrous proportions. In the state school, in contrast, the focus of PE, health, and sport will be the individual. Competition – whether individual or team – will alone be at the heart of physical activity. And there will be no training in sensitivity to the hedonism and narcissism that characterizes our self-centered, amusement crazed culture.

Well Rachel, I would like to go on and remind you of the differences between the Christian high school and the state high school in areas like English, foreign language, and extracurricular activities. But I'm running out of time. So let me return to that other important topic that you mentioned and tell you why there are times when I would choose to send my child to the state school rather than the Christian school.

You are absolutely right to be concerned about students becoming nominal Christians because they are not challenged to live their faith in the face of possible persecution. But there is something even more deadly than the absence of persecution. It's what you mentioned at the outset of your e-mail: the sacred-secular dualism. If, despite its lofty proclamations, a Christian school acts as if certain areas of high school life are religiously neutral, no different for the Christian school than for the state school – and here I'm thinking of math class, the volleyball team, driver's education, the senior party, and all those other areas that are not directly related to chapel, Bible study, theology, or ethics – then sending your child to the Christian school will be an exercise in hypocrisy and will displease the Lord. But before you make the choice to send your teenager to the state school, remember your prophetic responsibility to the community of Christians to which you belong. Read Matthew 23 a couple of times to gain the strength you need; then rage, rage against the dying of the reformational, biblical light in the Christian high school that has failed you. Call it to repentance and renewal. Work for change before you kick the dust off your feet.

May the Lord bless you and fill you with his Spirit, Rachel, as you write your paper. I look forward to talking with you in the not-too-distant future.

Love,

Uncle "D"

CHAPTER 19

ANIMISM, SECULARISM, AND THEISM

JULY 7, 2006

The prophet Micah is always good to quote when people tend to make light of injustices such as racism, or when Christians temper their outrage at injustice by thoughts of good deeds and "acts of kindness." Here's what the prophet has to say:

> With what shall I come before the LORD and bow down before the exalted God? Shall I come before him with burnt offerings, with calves a year old? Will the LORD be pleased with thousands of rams, with ten thousand rivers of olive oil? Shall I offer my firstborn for my transgression, the fruit of my body for the sin of my soul? He has shown you, O mortal, what is good. And what does the LORD require of you? To act justly and to love mercy and to walk humbly with your God. (Micah 6:6–8)

Those verses went through my mind a number of times last month when I toured with a group of Dordt College staff and friends, places in South Dakota, Minnesota, and Iowa that people don't usually think of as tourist attractions. On three occasions during that week-long excursion I encountered Native Americans, each of whom told their story – the story of Native American people and their encounter with European culture – in very different ways. The first was a young guide who showed us around the Lakota museum in Chamberlain, South Dakota. This young man was painfully bitter about his experience as a Native American. Every encounter between the United States Government and Native Americans – either that he had personally, or that he had heard of in school, or from friends and relatives – was fraught with injustice; often in the form of the government breaking its promises to Native Americans. After his guided tour of the museum a couple of us talked with him and asked him if there was any hope, and in what direction that hope might come. Despite a residual dignity that characterized his demeanor, he could only answer "no." He was hopeless. Western culture and Native American culture were too different from each other, and for this young "brave," the destiny of Native American culture was oblivion.

The second encounter I had was quite different. Our group was exploring the hills and plains surrounding the Missouri River, and one day found us on a reservation where the Missouri meanders in almost a figure eight configuration through some of the most scenic country imaginable. In order to hike in that area, however, we needed an official guide – a kind of forest ranger. This turned out to be a Native American who worked for the Department of the Interior. He was quite a contrast to the museum guide. He was affable, loquacious, informative, and a very positive about the land, his work, and his status as a Native American. On the other hand, being college educated and working for the government, it was clear that he had been assimilated well into North American culture. And that seemed to be his outlook on Native American culture in general: assimilation.

My third encounter was the most interesting. The morning after our visit to the Lakota museum we were gathered together to hear and to speak with Pastor Gabriel Medicine Eagle from the Rosebud Indian Reservation. He gave a fascinating talk that combined elements of personal experience with a history of Native American people over the past two hundred years. As he spoke, I couldn't help think about the nature and meaning of history and how history is so different for Native Americans than it is for those of us with our roots in Western European culture. It struck me that one of the major differences has been literacy. Literacy in Western culture goes way back. One thinks of Homer, the pre-Socratic philosophers, or of the history of Israel. While much of Jewish history wasn't written down until during and after the Babylonian captivity, we know that literacy existed during the time of Moses – just think, for example, of the Ten Commandments. And, of course, prior to the ancient Greeks and Israelites, there was Egypt – Egypt with its hieroglyphic literary culture. As I thought about this it struck me that literacy and cultural development go hand-in-hand. Without literacy there can be no science, no technology, no development in music and the arts, no development in government. And that, in large part, describes the chasm between Native American culture and Western European culture: development, which appears to be dependent upon, at least, literacy.

As an example consider how one would describe the Lakota, Ogallala, or other expressions of Sioux Indian culture from the Great Plains. One thinks immediately of bows and arrows, horses, and buffalo. But wait; horses have only existed on the plains for a few hundred years, introduced by the early European explorers. So the image of the Sioux Indian on horseback, hunting buffalo, is a rather recent one. It would have been foreign to those people prior to the fifteenth century. And we

really have no image of pre-fifteenth century Sioux Indian culture, because that culture – like the culture of Native Americans in general – was
non-literate: it wrote nothing down.

As I listened to Pastor Medicine Eagle, it began to dawn on me
just how different and incompatible the Native American and Western
European cultures really are. And that has helped me to understand the
hopelessness of that young Native American guide in the Lakota museum.

In the question and answer session that followed Pastor Medicine
Eagle's presentation, I asked him about that incompatibility that seems to
foretell the extinction of Native American culture in the face of Western
European culture. I asked whether he thought it might be related to the
issue of "literacy and recorded history." Although he did not directly answer my question, his response provided me with rich food for thought.
He asked me to consider the three mutually antithetical worldviews of
animism, secularism, and theism. What I think he was suggesting is that
animism – the belief that spirits constitute the essence of not only human beings, but nonhuman creatures in nature as well – correlates with
cultures without written traditions, and that it is animism, even more
than the lack of literacy, that begets a stagnant or negative response to
what we as Christians call the cultural mandate. Thus it is that animist
cultures, like those of Native Americans, have no *history* as we know it,
no sense of cultural unfolding. Thus Native American culture, except
for the introduction of the horse by Western Europeans, would not
have changed for thousands of years. And it is that in-born resistance
to cultural change that is the most obvious difference between Western
European and Native American culture. But beneath that difference lie
the Native American worldview of animism and the Western European
worldviews of either secularism or theism, secularism being derived, since
the Enlightenment, from a basically theistic culture.

My excursion on the Great Plains increased my sensitivity to the
injustices that European people and the American government have
wrought upon the Native Americans. But it also taught me that romanti-
cizing Native American culture would be unjust as well. Native American
culture should be remembered; it should be preserved in our history
books – which is something it could not do for itself. But as an animist
culture with no capacity for responding to God's call to be fruitful and
multiply, it is natural and inevitable that it will disappear.

Faithful fulfillment of that vow is the basis for Christian education
no matter which form it takes.

CHAPTER 20

TEACHING CHRISTIANLY

FEBRUARY 2, 2007

One of the major concerns of every Christian teacher is to teach in an authentically and distinctively Christ-centered manner, whether one is teaching theology or calculus. At Dordt College, one way we attempt to measure the success of our efforts is to have our students fill out what are called "Student Evaluation of Instruction" forms, near the end of each semester. As Dean of the Natural Science Division I recently finished reading through evaluations from about fifty courses that were taught this past fall. While I'm pleased with the effort that most faculty make to teach Christianly, I must say that the various written responses by students indicate that we still have room for quite a bit of improvement. There is one question we ask all students in all classes that gets at this question of teaching in a distinctively Christian manner. Here is how it reads.

> As a result of this course and your instructor's efforts, how has your faith grown and your worldview been shaped in a positive Christian direction? And, specifically, how has your understanding/appreciation of Christ's lordship over the particular subject matter of the course been deepened?

The majority of Dordt students make a good effort at trying to answer this question honestly and completely. But the answers given in what we might call the "technical" subjects suggest that our attempts to teach Christianly have had limited results. In courses in the natural sciences, mathematics, engineering, and the like, many students answer by saying that their faith has not really grown as a result of the course, but their appreciation for God's handiwork in creating and upholding a very complex world has deepened. Let's examine that answer just a bit.

Why would students say that their faith has not grown, yet go on to say that their appreciation for God's handiwork has deepened? I suspect it's because they associate the word "faith" with what they identify as "the spiritual realm" of life. Faith, to many Christian students, has to do with Jesus' death and resurrection, with salvation, with the "soul," and with life after death. And despite our harping on the problem of dualism here at Dordt, the worldview of many students remains effectively dualistic: faith belongs to a "spiritual realm" and science and math belong to a "physical" or "natural realm."

Students do, however, appreciate – at least in one sense – the lordship of Christ over the whole of creation. They resonate with the psalmist when he writes that "The heavens declare the glory of God; the skies proclaim the work of his hands" (Psalm 19:1), or that "The earth is the Lord's and everything in it, the world, and all who live in it. . ." (Psalm 24:1). Over and over I've read how students are amazed at the complexity of God's creation and how their appreciation for God's work of creation has therefore been deepened. That's good of course. Although sometimes I wonder if they are only telling me, in a way I want to hear, that this was a challenging course.

Here's the problem: The appreciation – even the awe – that students feel for the marvelous complexity of the created order may be no different from that which might be felt by a deist who believes that an impersonal god created it all in the beginning, but now sits back and simply observes the machine run according to the natural laws that he also created in the beginning. There's really not much difference between that kind of belief and the belief of the atheist naturalist who knows only matter and energy and the laws that govern matter and energy. The difference is that the atheist believes matter and energy to be eternal and self-sufficient, the former replaces that belief with deist belief in a "prime mover," an original being who created the impersonal laws of nature that now govern everything.

The Bible teaches something very different. It tells us that the Creator is also the Sustainer and the Redeemer of all things visible and invisible. It tells us that he was there in the beginning, but that he also took on our flesh and walked among us for period of time. It tells us that the wind and the waves obeyed his command, that he knew the molecules of water so intimately that he could get them to transform into wine – and tasty wine at that. Indirectly it tells us that mathematics, engineering, computer science, biochemistry, and physics are all activities that he calls us to pursue as part of his reconciling work, as a part we play in the history of redemption. It tells us that the whole of creation is waiting in eager expectation for our work to be revealed, for the day when its Creator-Sustainer-Redeemer will return to complete his work of renewing all things. But it also tells us that the Creator-Sustainer-Redeemer is present now: that it is his faithful word that holds all things together, that identifies all things as his servants, and that points all things in the direction of the new heavens and the new earth.

That is why it's important that students in my fluid mechanics class understand that a pump is far more than merely a mechanical device used to impart kinetic energy to a fluid and therefore make it flow. They need to understand the place of that simple device in the story of creation-fall-redemption and consummation. And that's more than seeing the pump as a complex device which we can marvel over and for which we can praise God. It's to understand the pump as a servant, and to understand ourselves as image bearing servants called to be stewards of servants such as the pump.

So we Christian teachers will continue our work to teach our subjects from a distinctively and authentically Christ-centered perspective, particularly those subjects in the natural sciences and mathematics where the challenge is great. Then one day the heavens truly will declare the glory of God, and the pump, the computer, and the Pythagorean Theorem will proclaim the work of his hands.

PART IV
REDIRECTING TECHNOLOGY

Señor, señor, let's disconnect these cables,
Overturn these tables.
This place don't make sense to me no more.
 – Bob Dylan[*]

---

[*] Bob Dylan, "Señor (Tales of Yankee Power)," from *Street Legal*, as referenced in Bob Dylan, *Lyrics 1962-1985* (New York: Knopf, 1992).

Chapter 1

Engineering at Dordt: Redeeming Creation

June 10, 1980

Dordt College is in the process of developing a four-year engineering program. This program, which aims to graduate its first class of engineers by the spring of 1985, will offer bachelor's degrees in mechanical engineering, electrical engineering, and engineering science.

Offering an engineering education to its students makes Dordt somewhat unique among liberal arts colleges, and particularly among Christian liberal arts colleges. There are a number of reasons for this. But today I would like to focus very simply on the question, "Why should *Dordt* offer a program in engineering?"

Dordt sees itself as a biblically directed institution, standing in the tradition of the Reformation. One implication of this is that the sovereignty of God is recognized over all areas of life. There is no area of life, no part of creation that can be called neutral. The Scriptures are very clear on this point. For example, Paul said this about Christ in Colossians 1:15–20:

> The Son is the image of the invisible God, the firstborn over all creation. For in him all things were created: things in heaven and on earth, visible and invisible, whether thrones or powers or rulers or authorities; all things have been created through him and for him. He is before all things, and in him all things hold together. And he is the head of the body, the church; he is the beginning and the firstborn from among the dead, so that in everything he might have the supremacy. For God was pleased to have all his fullness dwell in him, and through him to reconcile to himself all things, whether things on earth or things in heaven, by making peace through his blood, shed on the cross.

In these five verses the Lord Jesus Christ is described in three ways. First, he is the creator of all things, or as John says in the second and third verses of the first chapter of his gospel, "He was with God in the beginning. Through him all things were made." Second, he is not only the Creator of all things, but he also holds all things together. We might say he sustains or upholds the creation and gives it meaning. Finally, Paul teaches us that Christ's death on the cross brings reconciliation to all things. This is most important to understand for we often in our minds tend to limit the atoning work of Christ to the hearts of men, and fail to see, as the Scriptures clearly teach, that the work of redemption involves all of creation. In summary then, Jesus Christ is the creator, sustainer, and redeemer of all things – the whole creation.

Engineering is an activity whereby humans, using their reasoning abilities, attempt to develop the creation. But that development can not take place in a vacuum. It is God's creation that the engineer must work with since all things were created through Christ. And since Christ is the sustainer and the source of meaning for all things, engineering as an activity must have a direction. It is either done in submission and obedience to the will of God, or it is done in ignorance and rebellion against God's will.

The development activity that an engineer engages in can be traced to the original mandate given to Adam to tend the garden. Thus, engineering is a form of stewardship. The engineer has a task of conserving creation; that is, of taking care of our physical environment. He or she also has the task of unfolding creation. By this we mean enabling creation to express itself in ways that would be impossible otherwise. Examples are the development of a synthetic fiber or the harnessing of wind energy by a machine that may convert it to usable electrical energy.

There is a third aspect of the work of an engineer, which is connected with our fall into sin, that is becoming increasingly important today – the work of healing the creation. And for the Christian, it is here where we see concrete evidence of the redemptive work of Christ in all things.

Because we have turned away from God, we have rejected our task as steward. Instead we have become an exploiter of the creation. Our greed and self-centeredness has damaged the creation. Air and water pollution, nuclear waste disposal, the extinction of certain species of wildlife, the shortage of natural resources, the energy crisis: all of these problems are manifestations of humankind's failure to respond obediently to the Word of the Lord in our role as steward over creation.

These problems have caused a great deal of uncertainty and unrest in our society. Many people have become disenchanted with technology, believing that technology itself is the source of our ecological and energy problems. Others have taken just the opposite stance, believing that only increased technological efforts will save us from our current problems. In either case, or in the case of the average citizen who is just baffled by technological developments and problems, there is a recognition that we need answers, that the future will be shaped to a great extent by decisions that are being made now.

Into this situation of turmoil the Christian is called to bring the redemptive healing and the redirection of the Gospel. But to speak to technological issues authoritatively, the Christian must know the Word of God for technology. The only way the Word can be known is by a careful study of creation in the light of the Scriptures. In other words we need Christ-centered engineering schools that will graduate engineers equipped with a high degree of technical competency and directed by the conviction that their task is one of stewardship – conserving, unfolding, and bringing redemptive healing to the creation – in service to their fellow person and to the glory of God.

Chapter 2

Motives for Solar Development
October 9, 1980

About a month ago I had the opportunity to discuss with a group of Christians the possibility of using solar energy to heat our homes here in Northwest Iowa. The occasion was a church retreat that attempted to focus on the problems of obedient Christian living, particularly in terms of our responsibilities as stewards over the resources the Lord has entrusted to us. It was exciting to see the genuine interest for solar energy that existed in just that one group of people. The many and varied questions that were asked indicated to me that the interest was not simply that of a passing fad, but suggested a deeper commitment to change our way of living for the better.

Before dealing with the "nuts and bolts" of solar energy, we considered the general perspective out of which our various motives might grow. We asked why we want to learn more about solar energy and we looked to the Scriptures to provide the context for answering that question.

In the fourth chapter of the book of Hosea, the prophet charges the Israelites with unfaithfulness, lack of love, and a general disregard for the will of God. He laments the cursing, lying, murder, stealing, adultery, and continuing bloodshed that characterized his times. Then in verse three, using words pregnant with meaning for our energy and ecological situation today, he says:

> Because of this the land dries up,
>   and all who live in it waste away;
> the beasts of the field and the birds of the sky
>   and the fish in the sea are swept away.

In other words, the environment suffers because of our sin. Our dwindling natural resources, polluted air and water, and specifically the current energy crisis are not naturally occurring events that must be expected in the evolutionary development of our planet. Rather, they are unnatural events brought about as a result of human greed and our myopic, self-centered vision.

How then do we, as disciples of Christ, deal with these results of our sin? What is the perspective out of which we must view the energy crisis, and the potential of solar energy? The Lord himself has given us a clear and straightforward answer to this question. In the Sermon on the Mount, particularly in those last ten verses of Matthew 6, Jesus asks us to compare ourselves with the birds of the air and the flowers of the field. Commanding us not to be anxious over our material needs such as food, clothing, housing, and the like, he tells us in verses 32 and 33 that:

> . . . the pagans run after all these things, and your heavenly Father knows that you need them. But seek first his kingdom and his righteousness, and all these things will be given to you as well.

So getting back to the question of why we want to learn more about solar energy, we must examine our motives in the light of God's Word, to see if they are "pagan" motives or "Kingdom" motives. I would like to suggest that if we are motivated only by a desire to save money on our fuel bills, or to gain personal or even national energy independence, then we are one with the "pagans" who are anxious only about the material needs of life.

And isn't this the case today in our country as a whole? Consider the current political campaign rhetoric. Isn't it peculiar that the three major candidates for president, all of whom at one time or another have professed to be "born-again" Christians, try to win our votes by appealing to our material self-centeredness when it comes to the issue of energy? Have you ever heard one of the candidates speak of alternative energy schemes for other than the purpose of gaining personal and national, economic and political energy independence?

What then ought to be our motivation, as Christians, for learning about and perhaps developing solar energy in our community? In Matthew 6:33 Jesus said "seek first [God's] kingdom and his righteousness." What can that mean for our current energy situation? I'd like to suggest two answers to that question. First, we ought to concern ourselves with the needs of others. In particular there are many poor people in our country and around the world who will suffer during the winter because of inadequate heat for their homes. We ought to be concerned about their suffering. And since solar energy has such enormous potential for alleviating that suffering, we ought to be busy studying it and planning how we can put it to use for this purpose.

Second, as members of the body of Christ, we are commanded to be the light of the world. Now the "light of the world" must by its very nature be highly visible. What that means for us is that we as a community of Christians must be living very blatantly according to the Lord's norms for stewardship and in the face of a self-indulgent, consumerist society. I have an idea of what that might mean in the not-too-distant future. Some people I've talked to think it's a bit queer. But if you've lived for 25 years within fifty miles of New York City as I have, you can't help but have some queer ideas every once in a while.

During the two major gasoline shortages of the last decade, I lived in the New York metropolitan area, and had the dubious honor of having to get up at 4 o'clock in the morning and wait in line in front of a service station for two and one-half hours, just to get ten gallons of gasoline. I observed many frustrated faces, buried in books and newspapers, trying momentarily to forget their dilemma by reading. The idea that has crossed my mind is this: during the next major fuel shortage, the rich, pampered, self-indulgent consumers of our land, while sitting in gasoline lines or paying enormously high prices for fuel to heat their homes, will again attempt to momentarily forget their frustrations by reading the morning newspaper.

But this time it will not work. For instead of being entertained by Billy Carter's escapades or the latest pronouncement of some Middle East demagogue, the frustrated masses will find on the front page a story of some small midwestern communities, who by seeking the kingdom of God instead of the kingdom of Madison Avenue, have altered their lifestyles; through the cooperation of private groups and local municipalities, these communities have not only adequately met their own energy needs, but are in the forefront of bringing aid to the poor in our country who are most seriously affected by fuel shortages.

The contrast between one individual sitting alone in his big car at four in the morning waiting for gasoline and a community of people working together to meet the needs of the poor and be responsive to the commands of Christ will simply be too intense to be overlooked. When the eyes of the nation begin looking to the "models for responsible living" that are built on Jesus' command to seek first the kingdom, then we will have the satisfaction of knowing that by God's grace we have begun to be who we have been called to be – the light of the world.

While the day that the *New York Times* includes on its front page a story about the utilization of solar energy in Northwest Iowa may yet be a few years off, the time for beginning the work is now. During the next year or two there will be increased opportunity for all of us to learn more about solar energy and begin to plan some meaningful projects. Because of the closeness that we as citizens have to our city governments, the municipally-owned utilities that exist in cities like Sioux Center, and the favorable climatological conditions of our area, the potential for solar energy development is enormous. As we move forward with this development, however, we must constantly test the spirits that motivate us to be sure our vision for the future is a kingdom vision and not a self-centered one.

CHAPTER 3

SERVING WITH SOLAR ENERGY

SEPTEMBER 21, 1984

I have previously described an encounter I had with a salesman who was selling solar energy systems. I tried to make two points: that we ought to be very careful who we purchase solar energy systems from, and we need to develop an organization that will concern itself with the design and construction of solar energy systems, not for the purpose of making a profit, but to serve the genuine energy needs of people. In this commentary I want to describe what I think would be some of the more important characteristics of such an organization.

We read in many places in the Scriptures and we hear every Sunday in church the summary of the Law of God: that we should love him with all our hearts and love our neighbor as ourselves. In the Sermon on the Mount, Jesus said the same thing when, in the context of warning his followers about materialism, he said, "But seek first his kingdom and his righteousness, and all these things will be given to you as well" (Matthew 6:33). The first characteristic of a Christian organization for the design and construction of solar energy systems should be whether it is rooted in this summary of God's law. Its existence must be for the purpose of expressing our love to God and our neighbor. More concretely and specifically, it must exist to serve creation and our fellow man by meeting the genuine energy needs that prevail today. If it cannot be of service to others, then it ought not to exist. If the need that it attempts to address is not a genuine need, then it ought not to exist either.

Let's contrast that with the origins of most of today's solar energy companies. They came into being not because of a heartfelt desire to meet a genuine need in society, but on the conviction that it would be possible to make a lot of money in solar energy. The motivation was financial profit, not service to one's neighbor.

The second characteristic of a Christian solar enterprise, very much related to the first, is that it should be a nonprofit venture. There are many nonprofit institutions that serve societal needs admirably. Dordt College is a good example. Being nonprofit does not mean that the organization would earn no money. It simply means that all of the earnings of the organization would be used to run it. There would be nothing left over to pay bonuses to executives or pay dividends to stockholders. The money earned by the sale of solar collector systems would pay the salaries of the employees, the raw materials out of which the products are made, and the daily expenses of running the business.

A third characteristic of the Christian solar organization should be that the wages paid to the engineers, administrators, or laborers would be based on how much money those people need to accomplish their tasks in the organization and in their homes. The president of the corporation, busy behind a desk or on the telephone doing organizational work, would not necessarily be paid any more than the secretary typing the paperwork or the carpenter constructing the collectors. Also, the exact salary should be determined by how many children there are in a worker's family rather than by how many years he or she has worked for the organization. Compare that with the typical company today, which pays its executives six-figure salaries while paying the minimum wage to those for whom it may not be easy to find another job.

A fourth characteristic of a Christian company should be that its direction would be given by an unpaid board of trustees, elected by the community of those Christians who are interested in using solar energy to serve the needs of society. That community of Christians would constitute a society where membership would be open to anyone agreeing with its purposes and paying a small annual membership fee, say $25. The board would function in much the same way that a Christian school board functions. It would insure that the organization is always directed toward serving society according to biblical norms. Members of the board would be elected by the society and serve for terms of a fixed number of years.

A fifth and somewhat unique characteristic of the organization would be the way it raises capital. Instead of appealing to the greed of those who, having a great deal of money, want to make more by investing it, the society would appeal to its members for donations. If the energy needs are genuine and the cause is just, then we ought to support such an attempt to address those needs with our gifts. Certainly the organization will not be able to exist financially by depending only on contributions, but those contributions will make possible perhaps the most unique characteristic of all.

The sixth and final characteristic has to do with how the organization prices its products and services. The price of a solar collector system designed and constructed by employees of the organization ought to be based on the actual cost of producing it. Raw materials, salaries, transportation, and other costs should be figured in. While the overall price ought to be kept as low as possible, it ought to be fair both to the customer and to the organization. If all costs were figured into the price, then theoretically, so long as sales continued at a steady pace, no outside capital would be needed to keep the organizational books balanced.

However, the overall purpose of the organization is to meet genuine energy needs. And very often those needs are greatest where there is the least ability to pay. Therefore, the organization would develop a system of grants so that those who cannot pay for its services, but who have genuine need of those services, could be helped. Let's imagine that a six-panel, airflow, solar collector system, including storage and controls, can be designed and constructed to meet 75 percent of the domestic hot water and space-heating needs of an 1800 square foot house, for a price of $8000. A family of four with an income of $40,000 per year and no major expenses would be expected to pay the full bill. On the other hand, a family of 7, with an annual salary of only $18,000 might very well qualify for a grant of $6000. In other words, they would only have to pay $2000 for the same system. The $6000 grant would come from donations provided by members of the society.

Is all of this too idealistic? It may be called idealistic. It may even be called utopian. But it is also quite within the realm of possibility. Unlike many schemes that require a great deal of capital to get off the ground, this one requires very little. What it does require is commitment and time. But it can easily be started by two or three people who have a vision for the possibilities of solar energy and are guided by a Kingdom-oriented emphasis on service. Eight to ten hours a week and the use of a garage might very well result in three or four models of low-cost, solar retrofits after a year. And that should be sufficient for a start.

In the year 2000, only sixteen years from now, Lord willing, I hope to pull this essay out of my files and use it as a basis for a new one. Perhaps at that time, something called Sioux Solar Corporation, instead of being an idealistic vision in the head of a dreamer, will be a viable institution and a model of biblically normed energy service, environmental stewardship, and Kingdom-oriented business practice.

CHAPTER 4

WHERE DO WE GO FROM HERE?

OCTOBER 15, 1981

A few weeks ago I received a letter from a former Dordt student who is now working as an electronics engineering technician for a fairly large corporation. He graduated from Dordt before we started our engineering program, and so received his technical training at a state institution.

In his letter to me he raises the question of "the place of engineering in the Christian community." More specifically, he expresses his "disillusionment with" his own work as a technologist within the "secular business environment," asking the disturbing question, "Who are we training Dordt engineers for?" That is, what kind of enterprises will they work for after graduating from Dordt? The question is all the more disturbing because it is so rarely asked, yet it focuses on the very foundation of our efforts to develop an engineering program here at Dordt.

Allow me to quote directly from the letter. After describing his own work with digital and microprocessor electronics and his dissatisfaction with working for a large corporation, he writes the following:

> This brings me to the basic problem I have with the present discussion about the place of technology within the Reformed Christian community. Every article I have read talks in glowing terms about reforming the various fields of technology, but no one says where the graduates are going to get jobs. I have no doubt that jobs are available, but if a person has a vision of the Kingdom of God and a perception that his daily work is to be done to God's glory, then he is not going to be content to work for IBM, Hughes, Hewlett-Packard, or Rockwell International. If this person has a family, he will face even greater struggles over relocating, finding time for his family, and finding a Reformed church and a Christian school. In addition, it has been my experience that an Engineer (or Technician) merely implements decisions that are made in management and cannot be a reforming force. It seems to me, then, that Dordt engineering graduates would be helping a God-denying industry to grow and prosper. I wish this were not so, but if I honestly look at my own work, I can come to no other conclusion.

> I think that the Christian community needs to deal with this question before plunging headlong into educating people for technical careers. At

the very least we need to inform prospective students of the real world situation.

I believe the author of these words is raising a most important question, a question that has not adequately been dealt with in the reformed community. *Are* we educating engineers at Dordt merely to become pawns of some corporate giant?

At Dordt we are trying to develop a distinctively Christian approach to engineering and technology. We are attempting to identify and develop avenues for obediently working out those norms that the Lord has set down for his creation in the area of technology. I think first of norms such as service and stewardship. But suppose we are successful in training young engineers to do their work in ways that genuinely further the cause of the Kingdom, where will they go when they graduate? Can they honestly be the technological light of the world while working for secular industrial corporations? Or will the secular, God-denying, profit-glorifying corporate mentality function as an effective "bushel basket" that will hide their light from a world that so desperately needs it?

Harry Blamires, in his book *The Christian Mind*, does not mince words in answering that question. He says:

> The Christian has believed – and still tries to believe – that he can enter these spheres of activity without yielding anything to the World, that he can enter trailing clouds of spirituality which will magically transform the atmosphere around him, that he can enter without accepting the pragmatic mode of discourse dominating thought and decision in these fields. He has erred. It cannot be done. As a Christian you may enter these spheres determined to be the leaven. But your leavening influence is restricted to the narrow field of personal relations and moral attitudes. You cannot enter these spheres as a *thinking* Christian, for there is no one to communicate with christianly. There is no field of discourse in which your presuppositions can be understood, let alone accepted or discussed. Within these fields you will find yourself inevitably, by acquiescence, subscribing to the furtherance of aims of which you deeply and christianly disapprove. (26)

So, where does that leave us as a Christian academic institution? Ought we to terminate our efforts at developing an engineering program, and counsel our technically talented students to find another vocation – one where their witness will not so easily be compromised and their talents prostituted? I would answer that question with a firm "no." But before I explain that answer, I want to say that there *are* occasions when a question like that deserves an affirmative reply.

I think, for example, of the field of advertising. Advertising today is so totally dominated by the self-centered, materialistic, Madison Avenue ethic that no sensitive Christian could possibly survive as a member of an advertising firm. There has been no effort on the part of Christians to reform advertising, no development of biblical and creational norms for advertising, and all too rarely any prophetic calling of the advertising establishment to repentance. Hence, if my council were sought, I would urge any young Christian not to jump into the field of advertising.

I feel the same way about the military. Assuming for the moment that a Christian approach to military science is not a contradiction in terms, where would a young person go to learn biblical norms for the military? Where would he or she go to practice them? The military is totally and completely controlled by the United States government – which one can hardly deny is anything but an institution for the conscious furtherance of the Kingdom of God on earth. Again, if my advice were asked, I would counsel students away from a military career.

But engineering is different. Granted that modern technology is dominated by anti-Christian spirits; nonetheless, there have been voices calling out in the technological wilderness. Christians have called technology to repentance – I think of Jacques Ellul in France and George Grant in Canada. Christians have undertaken to identify, clarify, and develop normative principles that the Lord has given us for doing our technological work obediently – I think of Hendrik van Riessen and Egbert Schuurman in the Netherlands. And a few Christians have boldly set out to apply those normative principles by establishing enterprises whose primary aims are service and stewardship rather than profit – I think, for example, of the work of E.F. Schumacher in England.

So, while the body of Christ has certainly not been much of an influential force in the history of technology, we have not been left without any witness. And today we can build on that humble but significant foundation.

The question still remains, however, where young Christian engineers will go to apply their talents after graduating from Dordt. My answer is that they should work for small businesses and corporations, which if not already seeking biblical norms for their operation, will be open to the light that young, dedicated, Christ-centered, and service-oriented engineers can bring to them.

Let me conclude by sharing with you part of a vision I have for the 1990's. I see some new corporations being born and growing. I see a firm that designs and manufactures solar energy systems. But because the motivating force of the firm is service and not profit, it is able to produce its systems at a cost considerably below the average market cost for similar products. And because it doesn't concentrate on selling to the rich, but rather on meeting the needs of the poor, it is blessed. It becomes a model to all who come in contact with it, a dynamic demonstration of the basic biblical promise that when you seek first the Kingdom, all the peripheral concerns fall into place naturally. I see another firm that for a few years was the laughing-stock of the motor vehicle establishment because it has the radical notion that obedience to the norms of steward-ship requires the development of "wind-powered electric cars." But soon more and more homes are being equipped with wind generators. And in the garages of such homes there are sets of batteries being continually charged. And also in those garages are small but sturdy, simple but highly reliable, electric cars. They are designed only for trips of 100 miles or less, but the only cost of operation is the replacement, every 5 to 10 years, of the batteries. And the Detroit motor vehicle establishment is no longer laughing.

That's only the beginning of the potential I see for Christian engi-neers in the future. As new technologies develop further, the opportuni-ties for giving biblical direction and true Christian service multiply. The Christian engineer need not sell his soul to the corporate giants in order to use his talents. The Lord will provide exciting alternatives for those who in faith truly seek first his kingdom.

CHAPTER 5

TECHNOLOGY AND THE TECHNOCRATIC FAITH

JUNE 21, 1982

Recently I had the opportunity to address a group of Christians on the topic of technology. The setting was a church retreat. The audience was composed of a high percentage of engineers and technicians, since the church was located in one of our larger cities, home to a sizable electronics corporation.

In our discussion we found it very helpful to distinguish technology from technocracy. The distinction is fairly straightforward: technology is a legitimate human activity that involves unfolding, understanding, and developing the creation. Not only is it legitimate, but it is an important part of the mandate given to humankind by God when he placed Adam and Eve in the Garden of Eden and instructed them to "work it and take care of it" (Genesis 2:15b).

Technocracy, on the other hand, is a perversion of the original mandate. Instead of working and caring for the creation, we make a god out of it and wind up worshipping our knowledge of the creation. This technocratic faith arises when humans allow themselves and their society to be dominated by what we might call the spirit of abstraction.

Abstraction is an important method used to acquire knowledge of the creation. In applying the method of abstraction, we imagine the creation to be broken down into small parts that are thought of as separate from the whole.

Because the parts are small and separated, they are easy to understand. The problem is that we very often stop there, and after acquiring knowledge of the separate parts, we fail to integrate that knowledge back into our picture of the whole.

One result of this abstraction is that scientists and engineers have a tendency to become specialists who know a great deal about one very small and isolated part of creation, but know very little about how that part interrelates with the whole of creation. A second result is the faith in the method of abstraction that begins to dominate society. Technical knowledge, obtained using the process of abstraction, is given primacy over all other forms of knowledge. Technical facts become the arbiter of truth, and humans look to reasoning and calculation to solve all their problems.

The effect on human society is profound and deadly. Men and women become mere functionaries in such a society. They lose such characteristically human qualities as creativity, imagination, reflection, and true religious faith. Humans become estranged from their own spirit, their fellow humans, and the earth itself. In a society dominated by technocracy, everything is approached as a problem to be solved by reasoning and calculation.

An example of the technocratic faith was made blatantly visible in a recent issue of *Newsweek* magazine. One of the larger American industrial corporations took out an advertisement for the purpose of praising technology and strengthening the faith of the masses in the supposed "promise of technology." The ad was entitled "Technology: The lightning rod, Ben Franklin, and the American Revolution," and the introductory note read as follows:

> Much of the credit for the birth and rise of our nation is given to the indomitable pioneer spirit of its people, the vision and wisdom of its leaders, and the successes of its armies. All these were vital to be sure. But mostly, it was the advance of technology that led to the rise of the United States as a world power.

Following this there was a series of paragraphs describing the role of technology in American history. Preceding each of the paragraphs was a statement (one might even call them statements of faith) asserting the importance of technology in the growth of America. Some of the statements read as follows:

> Technology in transportation overcame the problems of territorial size.

> Technology in agriculture turned our vastness to advantage.

> When the Civil War began, the prediction of Frederick the Great [that the U.S. would fail because it was too large] met its ultimate test. And it was technology that saved the Union.

> Technology helped turn the United States from a wilderness into a great nation. Technology will continue being our best hope for the future.

The last statement was particularly revealing. It simply said, "Science and technology can solve many problems. If they don't, what else will?" And at the very bottom of the page, in large bold print, was a statement that summed it all up. It said, "We believe in the promise of technology."

This faith in the saving power of technology amounts to a rejection of God's revelation of salvation through Christ and a worshipping of the creature rather than the Creator. Paul is very clear on this point when he writes in Romans 1:18–25:

> The wrath of God is being revealed from heaven against all the godlessness and wickedness of people, who suppress the truth by their wickedness, since what may be known about God is plain to them, because God has made it plain to them. For since the creation of the world God's invisible qualities – his eternal power and divine nature – have been clearly seen, being understood from what has been made, so that people are without excuse.
>
> For although they knew God, they neither glorified him as God nor gave thanks to him, but their thinking became futile and their foolish hearts were darkened. Although they claimed to be wise, they became fools and exchanged the glory of the immortal God for images made to look like a mortal human being and birds and animals and reptiles.
>
> Therefore God gave them over in the sinful desires of their hearts to sexual impurity for the degrading of their bodies with one another. They exchanged the truth about God for a lie, and worshipped and served created things rather than the Creator. . . .

In the future I will examine some of the modern manifestations of the technocratic faith. For now, let me conclude by saying that technology needs to be redeemed. We ought not to reject it as some evil power that is about to enslave mankind. Technology is not evil in itself. It is most basically a form of service – a way in which we human creatures respond to our Creator. But that response can be in obedience or disobedience to God's Word. And this is where a distinction between technology on the one hand and technocracy on the other is so vital. We ought to promote Christ-centered technology. We ought to become more sensitive to and to reject technocracy, the spirit of abstraction, and faith in technology.

CHAPTER 6

TECHNOCRACY AND DEHUMANIZATION

JULY 26, 1982

In a previous essay I discussed the difference between technology and technocracy – the technocratic faith. I suggested that technology is very simply a way in which humans respond to their Creator in service through the unfolding, developing, and caring of creation. Technocracy, on the other hand, is a perversion of technology. It is the state of affairs that exists when technical knowledge is given primacy over all other forms of knowledge. In biblical terms, it is humankind's worshipping and serving created things rather than the Creator.

The distinction between technology and technocracy is terribly important to us, particularly in today's world, since, as Christians seeking to walk obediently before the face of our Lord, we need to be sensitive to the evil of technocracy while remaining enthusiastic in our promotion of Christ-centered technology. To be the light of a technological world, we cannot hide beneath the "bushel basket" attitude that blindly rejects all technology as "from the devil." To be the salt of the twentieth century earth, we must not lose our savor by blindly accepting the technocratic faith that bridles and enslaves modern technology. As Christ's agents of reconciliation, we must be busy reforming the world in which we find ourselves. This most certainly must include technology.

I would like to consider with you some of the modern manifestations of technocracy. As we do so, let's keep in mind that one of the foremost tenets of the technocratic faith is that everything can be viewed as a "problem" to be solved by reasoning and calculation.

One of the most obvious manifestations of the technocratic faith is the proliferation of technical gadgetry and the accompanying delusion that the "good life" is made up of acquiring as many material possessions as possible. By the efforts of scientists, engineers, and industrialists, our world has come to be populated by such things as telephones, automatic dishwashers, supersonic aircraft, personal computers, cruise control for automobiles, air conditioning, laser surgery, power toothbrushes, stereophonic sound reproduction, automatic lawn sprinklers, electric can openers, automatic garage door openers, instant replay of mass sports events, 120 mile per hour railroads, and imitation bacon bits, just to mention a few.

I don't want to suggest that any of these technological achievements necessarily lead to the corruption of our lives as God's image bearers. In fact, I am certain that while there may be a few technical gadgets whose redeeming value may be very difficult to demonstrate, most of the products of modern technology do have their place and are a very real blessing to particular individuals, families, or groups.

However, along with these gadgets, technocracy brings us the notion that we *all* must have them. We get the impression that no matter who you are, it is better to have air conditioning in your car than not to have it. No matter who you are, it is better to have an automatic dishwasher than to wash your dishes by hand. The phrase "the good life" is pregnant with this crass materialistic notion that acquiring things is the same as acquiring happiness. The result of believing and practicing this faith is not happiness at all, but rather self-estrangement and alienation from our fellow humans, the creation, and the Lord Himself. The cramming of one's life-space with meaningless technological gadgetry results in the inability to function as a genuine human being. We lose our creativity and imagination, we have no time for reflection, we become alienated from our inner self; in other words, we fail to live as the unique creatures we were created to be.

In this vein, perhaps the "ultimate pits" of self-alienating gadgetry today is the computer game. Consider the hypnotic effect on the person who sits in front of the TV set, staring at some two-dimensional ball of light moving from one side to another, remaining that way for hours on end. That person is alienated from society, the surrounding world, and his or her own God-given inner-self.

I realize that some very fine software exists today that can be described as "computer games" and can also be used either to great educational advantage or as a genuine and wholesome form of recreation. I certainly do not wish to discourage anyone from taking advantage of the opportunities that such software may provide. But let me suggest as well that "Pac-Man fever" is most certainly a dehumanizing disease, the epidemic proportions of which are nourished by the technocratic faith.

Another manifestation of technocracy that I will briefly mention is one in which the role of technocracy is most often overlooked. I'm talking about the issue of abortion. In a society given over to the technocratic faith, the problem of an unexpected pregnancy is one, like all other problems, to be solved by reasoning and calculation. The "technical solution" to this "problem" is abortion. In reality, of course, an unexpected pregnancy is a human problem with many dimensions that transcend the merely technical. But the technocratic solution denies the humanity of both the unborn child and the parents. If we Christians are going to be effective in addressing the issue of abortion, we are going to have to see that it is more than a "moral issue" in the narrow sense of the term.

Finally, I would like to consider with you an interesting suggestion made by the French philosopher Gabriel Marcel in his book *Men Against Humanity* (London: Harvill, 1952). He maintains that the spirit of abstraction – which is foundational to technocracy – has an important place among the most dangerous causes of war. If the state, political party, or any other group seeks to gain from someone a commitment to warlike actions against other human beings, then it must first insure that s/he loses all awareness of the individual reality of the human beings who it seeks to destroy. The "enemy" must be transformed into a mere impersonal target, an abstraction; for example THE Communist, THE Imperialist, THE ENEMY. Only by thus reducing our fellow humans to an abstraction as a political object, is it possible to direct human beings, image bearers of God, to transform themselves willingly into button-pushers and trigger-pullers – mass executioners prepared to end the life of countless other image bearers who they see only as abstractions. Does this shed some light on the current foreign policy of the most advanced technological state in the world today – our own country?

Whether it is the Nicaraguans, the Salvadoran rebels, the Russians, or the leftists among our European allies, they are reduced to the abstraction "Communist," and that, of course, is synonymous with "Enemy."

I've only briefly described three of the manifestations of technocracy in our world today. There are many others – some very obvious, like the energy and environmental crises, and some not so obvious but very subtle, and for that reason very sinister. But in closing I want to reassert that my motive is not to dump on technology, but rather to call our attention to the fact that this aspect of God's creation is in very desperate need of redemption and redirection. At Dordt College, our Engineering Department makes a special effort to address this issue. But it must be the concern of God's people wherever they are and whatever they do. We live in an age of technology. We can either reform it for the Lord, or be corrupted by the spirit of technocracy. There's no possibility of simply ignoring the issue.

CHAPTER 7

IMITATING THE JAPANESE
AUGUST 8, 1985

In the business section of its August 5 issue, *Time Magazine* reported that General Motors Corporation has finally decided on a location for its Saturn plant. Saturn is the name of the car that GM intends to build in order to compete with what has been called the Japanese challenge to the American auto industry. Japan has been producing subcompact cars for a cost of about $2000 less than what, to this point, American industry has been able to do. And it is only due to import quotas and tariffs that the American auto industry has had any chance at all of competing with what is generally agreed to be a cheaper and higher quality product.

The reason for Japan's success has been attributed to the difference in the organization and practice of Japanese industry in contrast to American industry. This is most obvious in two areas: first, the greater use of computer and robot technology by the Japanese; and second, the unique relationship of the Japanese worker to the company for which s/he works.

But now an American company, General Motors, is going to attempt to run an automotive assembly plant the way the Japanese do. The car to be produced will be the highly efficient subcompact Saturn, with a fuel economy approaching 60 mpg. The production will involve the latest computer-aided design and manufacturing techniques, and much of the detailed assembly will be accomplished by robots.

But the really radical changes have to do with the workers. According to *Time*:

> Instead of performing a single tedious task like attaching windshield wipers as cars whiz past on a long assembly line, employees would work together in self-directing teams of 6 to 15 people. Each team would be responsible for large sections of the car, and its members would have the latitude to reach consensus on how to divide up and rotate job assignments. Most important, production workers would receive a salary instead of an hourly wage as they do now, and the pay would be directly tied to performance.

The big question is, will it work? Can American industry simply take over many of the practices of Japanese industry that have led to such economic success? To answer that we need a better understanding of the differences between the American and Japanese corporate practices. In a book published in 1981 titled *Theory Z* (New York: Avon Books), William Ouchi has documented those differences. Let's look at just a few.

In Japan, employment is lifetime employment. That is, when you join a company as a laborer, manager, or whatever, you do so with the expectation that you will spend the rest of your working life with that company. On the other hand, in American industry employment is short-term. Sometimes turnover is as much as 50 percent in a year. In fact, the expectation on the part of American workers, especially those in middle and upper management, is that if you stay with a company for more than a few years you are stagnating. Career growth implies changing companies.

In Japan, evaluation and promotion are slow processes, while in American industry they are rapid. This is related to the turnover problem, since rapid turnover necessitates speedy evaluation and promotion. But it also means that workers with little experience in a particular aspect of the industry are expected to take on significant responsibility. Thus, the pace of evaluation must be quickened. This raises the expectation of the workers for promotion. In America, if a worker has not received a substantial promotion in three years, he will inevitably feel that he is a failure.

In Japan, the worker travels what might be described as a non-specialized career path, while his American counterpart travels, for the most part, a very specialized career path. Again, this is related to the long vs. short time employment difference. If you work for a company all of your life, you have time to get to know the company as a whole and can become relatively an expert in all facets of its operation. This requires a general education. On the other hand, if you only work for any given company for a few years, it is impossible to become an expert in all facets of its operation. Therefore you must be an expert in a specialized area that may be common to many companies. Naturally this requires a very specialized education.

In Japan, control of the organization and its members is implicit; that is, much of what in the United States may be called company regulations can be left to an unspoken mutual understanding or to the imagination of the workers. This is possible because of the cooperation and mutual understanding that comes with lifetime employment. In American industry, however, mechanisms of control become explicit and formal, without any of the subtlety and complexity that can exist in cooperative life.

Perhaps one of the major differences is that in Japanese industry there is an emphasis on collective decision-making and responsibility, whereas in American industry the emphasis is on individual decision making and responsibility. This is not to say that Japanese industry is "communistic" while American industry is "capitalistic." Both are expressions of what might be generally called "free enterprise." But both are also expressions of the unique characteristics of their respective societies.

General Motors, as an American corporation, is attempting to build a plant that operates in many ways that are more similar to the Japanese model of industry. Will they be able to succeed? Should they be attempting such a transformation? Are there norms for industrial activity that would allow us to evaluate the Japanese and the American approaches on more than just pragmatic grounds?

Those are questions that we Christians ought to think about. I don't have any easy answers, but in the future, I will suggest a few reasons why I think GM's move is a right one, but why I also think it is doomed to failure.

CHAPTER 8

DISCOVERING NORMS FOR INDUSTRY
SEPTEMBER 11, 1985

In the last essay I raised the question whether General Motors Corporation is doing the right thing as it starts up and plans its new Saturn plant by imitating many Japanese industrial practices. We looked at a few of those practices and how they contrast with the way industry is run in the West. Perhaps the most obvious difference is that employment in Japan is considered lifetime employment – that is, a worker spends his working life with one corporation. In the United States it is quite rare for a worker to spend his entire working life with one employer. In fact, it is quite common for workers to spend only a few years with a given company and then move on to another – usually in an attempt to gain a more prestigious and higher paying position.

Today I would like to raise and suggest answers to three questions. First, how did these differences in corporate practice arise? Second, are there creational norms in these areas of corporate practice? And third, is General Motors doing the right thing by imitating the Japanese industrial model?

In his book, *Theory Z*, William Ouchi sketches two pictures that describe the American and Japanese cultures, making clear the roots of their different industrial forms:

> The *shinkansen* or "bullet train" speeds across the rural areas of Japan giving a quick view of cluster after cluster of farmhouses surrounded by rice paddies. This particular pattern did not develop purely by chance, but as a consequence of the technology peculiar to the growing of rice, the staple of the Japanese diet. The growing of rice requires the construction and maintenance of an irrigation system, something that takes many hands to build. More importantly, the planting and the harvesting of rice can only be done efficiently with the cooperation of twenty or more people. The "bottom line" is that a single family working alone cannot produce enough rice to survive, but a dozen families working together can produce a surplus. Thus the Japanese have had to develop the capacity to work together in harmony, no matter what the forces of disagreement or social disintegration, in order to survive. (54)

On the other hand:

> . . . consider a flight over the United States. Looking out of the window high over the state of Kansas, we see a pattern of a single farmhouses surrounded by fields. In the early 1800's in the state of Kansas there were no automobiles. Your nearest neighbor was perhaps two miles distant; the winters were long, and the snow was deep. Inevitably, the central social values were self-reliance and independence. Those were the realities of that place and age that children had to learn to value. (55)

We might argue that Ouchi's analysis is somewhat superficial in that there have been factors more fundamental than geography alone at work in forming the character of the American and the Japanese cultures. But be that as it may, we cannot deny his conclusion: American culture is one based on individualism, while Japanese culture is quite the opposite − it is a culture that values highly the group and cooperation among members of the group.

This explains some other important distinctions between the two cultures and their respective models of industry. A culture that places a high value on the group and cooperation is one that will foster relationships of trust between people. An individualist society, on the other hand, will not have such a need for trust relationships, but will present a cultural soil fertile for the growth of adversarial relationships. Each individual becomes a competitor in the eyes of each other individual.

The goals of people in these two societies will differ as well. In a society that values group identity, a chief goal will be security within the group. The individual gains satisfaction and a sense of meaning and purpose by being an honored member of the group. In an individualist society, on the other hand, the individual can only find a modest amount of satisfaction in being honored by others, since s/he does not value the opinions of others that highly. Therefore the individual must go beyond relationships to find a worthwhile goal.

The chief goal in American society has become the acquisition of material gain. Members of a society valuing group identity will generally have rather long-term goals and visions. This is because groups are spread out not only in space but also in time. Families and corporations can remain active for many generations, even though individual members are born into and die within those families and corporations. On the other hand, an individualist society will be made of members with generally short-term vision. Twenty years may be a short time in the life of a country, a corporation, or a large extended family, but it is a very long time in the life of an individual person.

So which of the two industrial models, Japanese or American, is more normative and obedient in the eyes of the Lord? I think it should be obvious to all of us that an industrial system based on individualism, with its adversarial relationships, materialism, and self-centered competition, is contrary to the teachings in Scripture. But that does not mean that we necessarily ought to embrace the communal model of the Japanese. Certainly we ought to encourage the development of trust relationships between people and encourage cooperation rather than selfish competition. But group loyalty and finding one's security within the context of the group is only valid when that group is the Body of Christ. And that is certainly not the case with the Japanese cultural model.

Also, the group loyalty motive can be easily taken to an extreme position that makes individualism look good in comparison. Think of the Japanese kamikaze pilot during World War II who was ready, unquestioningly, to lose himself totally for the sake of the group. Or think of modern totalitarian communist states, where the freedom of the individual is crushed beneath the heel of the state.

In Scripture it's clear that each individual is to see himself as having a unique calling from the Lord to be a servant to his neighbors. American individualism and the resulting corporate structure totally loses sight of the nature that each of us has as servants. On the other hand, an absolutist emphasis on community, as is the case with communism, loses sight of the unique calling that each individual has from the Lord.

Another normative principle that can be understood from the Scriptures is that we ought to live, as far as is possible, at peace with those around us. I think we can derive from that an important norm for society: change in society ought, as far as is possible, to be progressive rather than revolutionary. Just as it is absurd and wrong to try tomorrow to bring modern Western technology to a third world agricultural nation, so too it would be wrong to try tomorrow to bring Japanese industrial practices to all American industries. In both cases, in spite of the fact that the changes may ultimately be for the better, we would be involved in facilitating revolutionary rather than progressive change.

So what about General Motors? I think they are doing the right thing in attempting to learn from the Japanese and in trying to implement some of those Japanese industrial practices that have proven to be so successful. By trying it out in one plant, on a limited and modified scale, they are avoiding the error of revolutionary change. Likewise, they are recognizing that the adversarial relationships and the short term vision that are common to the individualist industrial model leave much to be desired. Certainly the Japanese are far ahead of us with their industrial group loyalties, trust relationships, and long-term vision. But I fear that GM's effort is going to fail. For just as we are now learning from the Japanese, the Japanese have been learning from us. Materialism, so deeply ingrained in American business and American labor, is gaining a stronghold in Japanese industry as well. And it is this overriding concern for the acquisition of material gain that will not only prevent us from learning a positive lesson from the Japanese, but is causing a slow change in Japanese industrial society away from the communal model of industry toward a more individualist model.

It is precisely here that Christians need to serve as lights in the industrial world, pointing out what has been right and what has been wrong with both Japanese and American industry in the past, and pointing to the Kingdom of God as the only alternative for the future. I have no hope that GM will succeed in its attempt to develop a better industrial model. But there is hope if Spirit-led Christian businessmen, technologists, and laborers will stop being enchanted by the Lee Iacocca's of this world, and begin to seek together the Kingdom of God in the areas of their callings.

CHAPTER 9

REFORMED CHRISTIANS AND HI-TECH PHOBIA

JANUARY 4, 1988

We live in the age of high tech. The signs are everywhere, from Hollywood to Wall Street, from Washington D.C. to fabled Silicon Valley. Newsstands display stacks of science and computer magazines. Windows of electronics stores are piled high with new products. Foreign policy experts attempt to come to terms with "Star Wars." Engineering stands at center stage.

> Employment of engineers has been increasing at a pace nearly double that of other professions and three times as fast as that of the overall work force. More than 100,000 new students are crowding into American engineering schools each year, double the number of a decade ago. One of every five male college freshmen says he would like to be an engineer.[*]

The quotation above is from the introduction to Samuel Florman's new book, *The Civilized Engineer*. The fact that technology is a major force – indeed, a defining characteristic of our age – is, I'm sure, not news to anyone. But what surprised me was the last sentence: "One of every five male college freshmen says he would like to be an engineer." As an engineering educator here at Dordt College for the last eight years, this simply has not been my experience. At first I wondered whether Florman was exaggerating. But a closer look reveals that he has documented his claim as coming from a report by the Office of Technology Assessment and summarized in an article published in the *New York Times*.

So why the disparity between what I perceive and what apparently is the normal attitude of American college age young people? On the surface the answer is quite simple. My experience is with a small sub-set of American college age young people – those originating from the reformed Christian community. But, of course, I still must ask "why?" Why should the attitude of reformed, Christian young people, particularly those who have attended Christian schools, be different from their public school peers with regard to technology, especially when they share so much in common in other areas of life?

---

[*] Florman, Samuel, *The Civilized Engineer* (New York: St. Martin's Press, 1987).

I would like to suggest three possible answers to that question and argue that although I generally applaud nonconformity and distinctiveness on the part of Christian young people, in this case I think that the difference is in the wrong direction − we Christians ought to be more enthusiastic about technology than the average American, not less.

The first reason I believe that Christian young people do not look as quickly to technology as a vocation has to do with what C.P. Snow called the "two cultures" problem in his book *The Two Cultures*. That problem is the erroneous belief that there exist two cultures in our world, or at least in our academic world: the technical culture and the humanities culture. Traditionally there has been an antagonism between these two cultures. Technically trained people have looked down on the humanities as being of little practical use and, conversely, people trained in the humanities have generally seen technology as abstract and inhuman, and technologists as insensitive, pragmatic, and dehumanized. This false dichotomy has been semi-institutionalized in the existence of liberal arts colleges on the one hand and colleges of science and engineering on the other.

The problem we have as reformed Christians becomes apparent when we consider the nature of Christian colleges. Think of the ones you know: Dordt, Northwestern, Calvin, Trinity, Hope, Redeemer, and even the broader evangelical schools such as Wheaton; all of them identify themselves as liberal arts colleges. These schools are not only the intellectual nurturing communities for a large majority of reformed Christians, but they are also training centers for almost all Christian school teachers. And even though as reformed Christians, because of our holistic biblical view of creation, we know that the two cultures problem is a false one, we nonetheless cannot help but be affected by it and, in turn, help perpetuate it.

A second explanation for the relative disaffection of reformed Christian young people for science and technology has to do with an inconsistency between what we practice and what we preach as our worldview. We may say that Christ is Lord of every area of life, that Christians are called to service in every sphere of creation, and that religion is not some special compartment of life but is our moment-by-moment walk before the face of the Lord; nevertheless, we live as if there exists a spiritual dimension to life, usually identified with morals, ethics, and theology, and a neutral and second-best dimension of life, which includes such things as science and technology. We may say it in words, but in our communal heart we reformed Christians have failed to see that serving the cause of the Kingdom as an engineer, lawyer, or businessman can be just as holy as serving as a missionary, minister, or Christian school teacher. In other words, we are reformed in word only. In deed we are one with our fundamentalist brothers and sisters.

A third reason why Christian young people are not turning the world of technology and science upside-down is somewhat related to the second. Since we see that area of life as neutral, we don't believe that we have anything special to bring to it as Christians. Since we lack a prophetic vision for reforming technology and directing it to service for the Kingdom, we also lack confidence in our ability to make any difference. Technology, like any other calling, requires commitment. But commitment, in order to grow, needs to be fertilized by the confidence that our commitment will make a difference. As Christians we have no communal self-confidence that our work in the areas of science or technology will be any more valuable than that of the non-Christian. Seeing it as neutral, we are deaf to the Word of the Lord for technology. Not only do we not know his will for the areas of energy, electronics, and eugenics, but we don't even believe that it exists.

In summary, three of the reasons why reformed Christian young people are not studying engineering with the same communal enthusiasm as their non-reformed or non-Christian peers are: they lack confidence in their ability to make a difference, they do not see it as in any way related to their Christian faith, and they have been one-sidedly influenced by the humanities pole of the two cultures dichotomy.

It's time reformed Christians begin practicing what we preach. There's a high-tech world out there crying for healing and direction. Only those who hear the Word of the Lord are capable of bringing the gospel of redemption to the world. But to do so we must be filled with the Spirit and have the confidence that our work will not be in vain. We need to begin building that confidence when our children enter kindergarten if we expect them to take up their Kingdom high-tech tasks when they become adults.

CHAPTER 10

WEAPONS OF WAR:
AN HISTORICAL PERSPECTIVE
APRIL 8, 1988

One of my favorite novels is Mark Twain's *A Connecticut Yankee in King Arthur's Court*. I read it long ago, during one of the warm and quiet summers of my early adolescence. I had just finished my freshman year in high school. I was too old to "go out and play" like my younger brothers, too young for a full-time summer job, and so was faced with the prospect of three months of boredom. But the Lord in his providence had different plans. He had in mind for me a trip to Camelot, of all places, accompanied by a Connecticut Yankee who was the product of Mark Twain's imagination.

What fascinated me about the book was the idea of a confrontation of an earlier culture with modern (or relatively modern) technological products. I remember being so impressed with the surprising insights learned by such a confrontation, that the next year in sophomore English class, when assigned the task of writing a one-act play, I did a take-off on Twain's book. I wrote about a twelfth century knight who was somehow transported to our 1962 northern New Jersey high school. The confrontation of the knight with our teenage culture enabled me to point out certain characteristics of that culture that were hard to see otherwise. I also remember that while my English teacher liked the play, I got into a bit of hot water with some of the other teachers and administrators who didn't appreciate the particular light in which the play cast them.

This literary technique of confronting the present with the past in order to gain a different and sometimes clearer perspective on the present is not confined to Mark Twain's *Connecticut Yankee*. Recently it was used very effectively in the film *Star Trek 4: The Voyage Home*, although there it served primarily as the cause of comedy.

The reason I am describing all this is because I wish to use the technique of confronting the present with the past in order to shed some light on the issue of "killing and the quality of life." I agree with the writer of a "Voices" column in the February 29, 1988 issue of the *Banner*, who said that "killing another human being to achieve a better 'quality of life' is a choice that is forbidden to us." I believe that killing another human being always reduces the quality of life for the killer and for the community in which the killing takes place. That's true whether the killing is the abortion of an unborn child, the terrorist bombing of a bus, the leveling of a city by an atomic bomb, or, I would like to suggest, the killing of one soldier by another during a war.

In order to shed a little different light on this point, I want to consider what we take for granted as the "weapons of war" at the end of the twentieth century. Most nations consider the use of poison gas to be immoral. We are generally very nervous about the use of nuclear weapons (although we obviously have not come to any worldwide agreement about banning them). But we take for granted that weapons of defense must be of a sort whereby huge amounts of kinetic energy are given to small particles of metal, the intended result being the death of the human beings who come into contact with those high energy particles.

Now I would like for you to travel back in time with me. Let's go to the city of Dothan in Israel at the time of Elisha the prophet. As we arrive we notice that Elisha has a little problem:

> When the servant of the man of God got up and went out early the next morning, an army with horses and chariots had surrounded the city. "Oh, my lord! What shall we do?" the servant asked.
>
> "Don't be afraid," the prophet answered. "Those who are with us are more than those who are with them."
>
> And Elisha prayed, "Open his eyes, LORD, so that he may see." Then the LORD opened the servant's eyes, and he looked and saw the hills full of horses and chariots of fire all around Elisha.
>
> As the enemy came down toward him, Elisha prayed to the LORD, "Strike this army with blindness." So he struck them with blindness, as Elisha had asked.
>
> Elisha told them, "This is not the road, and this is not the city. Follow me, and I will lead you to the man you are looking for." And he led them to Samaria.
>
> After they had entered the city, Elisha said, "LORD, open the eyes of these men, so they can see." Then the Lord opened their eyes and they looked, and there they were, inside Samaria.
>
> When the king of Israel saw them, he asked Elisha, "Shall I kill them, my father? Shall I kill them?"
>
> "Do not kill them," he answered. "Would you kill those you have captured with your own sword and bow? Set food and water before so that they may eat and drink and then go back to their master." So he prepared a great feast for them, and after they had finished eating and drinking, he

sent them away, and they returned to their master. So the bands from Aram stopped raiding Israel's territory. (2 Kings 6:15–23)

Elisha was someone who obviously knew something about defense. Let's try to solicit from him his views on defense in the twentieth century. In order to do that we are going to have to tell him all about the technological and scientific developments that have occurred between his time and ours. This is going to "blow his mind," to say the least. It will be even more of a contrast than Mark Twain's "Connecticut Yankee" experienced when he tried to tell the Knights of the Round Table about gunpowder.

But Elisha is a man of God. He will understand that our harnessing of chemical and nuclear energy for the production of huge amounts of power is not magic, but is made possible because the Lord put that potential in his creation and has given his image bearers the ability to unfold and develop it. So while our description of modern science and technology may be mindboggling to Elisha, he will not reject it out of hand as either impossible or the work of the devil.

But when we tell him about what we do with the huge amounts of power that we twentieth century people have harnessed for defense purposes, then I think we may get a very strong response. To a man who instructed his king to serve the captured enemy soldiers a feast after he gently rendered them powerless with the help of the Lord, the idea of destroying the lives of soldiers and civilians with the weapons of the twentieth century will seem absolutely barbaric.

But here is where the imaginary confrontation of cultures will get very interesting. You see, Elisha will not simply condemn us as godless barbarians – he will very likely do something worse. He will ask us a question that we have not asked ourselves in 2000 years. He will ask, "How is it that with all your advances in science and technology, you have not progressed beyond violent means of defense? If you are able to develop heart transplants, brain surgery, plastics, and vehicles that could take you to the moon, why can't you develop a way of *gently* turning your enemy's aggression aside?" How would you answer that question?

Technology is not neutral. The products of our technology have the values that our culture adheres to embedded in them. Our culture has never viewed the life of individual human beings with the degree of reverence worthy of those who are created in the image of God. As a result, we have developed a transportation technology that with predictable regularity kills and maims thousands on our highways each year. We have developed abortion as a technological fix for the problem of an unwanted pregnancy, and we have provided the legal props needed to shroud the resulting genocide in a cover of public health respectability. And since the days of Genghis Kahn we have viewed our enemies not as image bearers of God, but as targets that when they become threatening to us are destroyed with the same technological effectiveness and compassion that we give to the bacteria and viruses that threaten our health.

Elisha would be ashamed and outraged to learn that many of us in western culture claim the Judeo-Christian heritage as our own. Perhaps we ought to visit the past more often. In doing so we may learn how we can change the future.

CHAPTER 11

1989 AND TECHNOLOGICAL CHANGE

JANUARY 6, 1989

Is there such a thing as too much progress? That question almost sounds like a contradiction. Isn't progress good? And can you really ever have too much of a good thing?

As this first week of the new year comes to a close I can't help but reflect on all the changes, particularly technological changes, that have occurred during the past year. Most people would call these changes progress, and would see them as good. I might too. But before I join in the chorus of "amens," I want to consider a bit more carefully a little problem I've had with two specific examples of modern technology, a problem that has arisen because of the rapid rate of technological advance during the past years.

Because I teach engineering students, I have a great deal of interaction with computers. Not only do I use them to solve engineering problems, but the computer has also become a tool that greatly aids much of the administrative work associated with teaching. The word processor has almost completely replaced my typewriter. The database has greatly simplified my record keeping and filing. And the spreadsheet has made budgeting almost enjoyable. About eight years ago I purchased an Apple computer for my home use and have never regretted the considerable investment it required at that time. It has served my family well.

When I went to college in the late 1960's we never used computers. I only got to see one in the last semester of my senior year. During my four undergraduate years I used a slide rule. Most of today's students, on the other hand, have never seen a slide rule. They have used calculators and personal computers during virtually all of their education.

It's interesting to plot the recent history of the computer in education. During the '60s it was for the most part nonexistent. There were, of course, computer science students as well as graduate science and engineering students who had access to large mainframe computers. But that was a very small percentage of the nation's high school and college population. In the '70s, the handheld calculator replaced the slide rule. By the end of the '70s, most science and engineering students were working with large computers, and virtually all high school and college students involved in any kind of computation were using calculators. But that still left a large majority of students who never touched a computer of any kind.

During the '80s, the personal computer burst upon the scene. Now everyone uses a computer, not just the science and technology students. Today English majors do their writing on word processors, business majors have to become totally familiar with the database and spreadsheet, and there is not an office to be found where the computer has not replaced the typewriter.

But what is going to happen in the '90s? The problem I sense is one of shrinking periods of technological change. For example, whereas it took a decade for the calculator to mature and replace altogether the slide rule, it took only five or six years for the personal computer to come into its own and replace the typewriter. In fact, I would suggest we are now already well into the third period of personal computer development. The first is perhaps best exemplified by the Apple II. The second involved such computers as the Apple IIe, the IBM PC, and the Apple Macintosh. And today we talk about "286" and "386" computers, numbers which identify the main "chip," as it's called, which is the heart of the computer.

These periods of development are becoming shorter and shorter. The result is that if you purchase an up-to-date computer and then spend six months learning how to best use it, it is out of date by the time you begin putting it to real use. The problem is one of stewardship. How do I make best use of my time and monetary resources in an environment of rapid technological change? Or, to go back to my original question: ought technological change occur so rapidly that a culture is unable to keep up with it? Is this the way the Lord intended things to go, or are we doing something wrong? And if we are, what is it?

Let me give a second example. I have enjoyed music ever since I was in sixth grade. When in high school I used to read Lafayette Radio catalogs and imagine what it would be like to listen to a really good stereo system. In 1967 I bought my first system. It was only an average one for the time, and the cost was modest. But it was a quantum leap better than my portable Sears Silvertone record player. Three years ago I purchased a new system, including a compact disc player: another quantum leap. To listen to Brahms' *Requiem* or Bach's *Magnificat* or to some good folk music by Bob Dylan or Joan Baez, without any background hiss, with crystal clarity, and with 90 decibels of dynamic range, is an experience that can't help but make you think of Psalm 19. Now it isn't only the heavens that are declaring the glory of God, but the compact disc as well.

Unfortunately, the problem I described with respect to the computer is beginning to arise in the area of audio and video reproduction. Each time period of technological advancement brings a new breakthrough in audio and video. And the time periods are getting smaller. We barely have time to fully appreciate the experience of listening to Handel's *Messiah* on compact disc when suddenly something like concert-hall processing, or surround-sound becomes available.

The last thing I wish is for this essay to be construed as a promotion of high tech consumerism. Perhaps the easiest solution to the problem I've posed is to simply ignore it. After all, we can, and certainly have in the past, lived very nicely without compact disc players or 386 computers. The comforts of a healthy family, sufficient food and shelter, and perhaps a comfortable seat and a few good books ought to make any of us feel that our cup is running over. And given the knowledge that millions of God's image bearers in this world have not even the barest necessities to sustain basic biotic existence, we ought to be very careful before we become one with the consuming masses of our materialistic culture.

But the Lord did create us in his image. And part of that means being called to develop his creation. Computers and sound reproduction systems are the partial results of that legitimate process of unfolding and developing. When I seriously consider Colossians 1 and am impressed with the fact that Christ is creator, sustainer, and redeemer of the whole of creation, I realize that these technological marvels exist because God cared enough about his whole world to send his Son to redeem it. That lends real meaning to the appreciation of my compact disc player.

However, the problem remains. Is our technology advancing too fast for us to really appreciate it? I don't have any easy answers to that question. But I do have two warnings. First, don't get caught up in consumerism. Just because a technology has been surpassed by a new advancement does not necessarily mean that it is not useful for you. Coveting in a technological age may mean discontent with the blessings you have not even learned to appreciate yet because an advanced model has attracted your attention. And second, don't turn your back on all the products of modern technology. That would be like closing your eyes to the clear night sky or to a morning sunrise. Keep in mind that Psalm 19 was written for us as well as for people of ancient Israel.

CHAPTER 12

GUNS AND TECHNOLOGICAL DISOBEDIENCE

MARCH 1, 1989

> On a weekday last month, shortly after children spilled onto the Cleveland Elementary School playground in Stockton, California, for midday recess, a man dressed in combat fatigues stepped out of his 1977 station wagon, parked nearby. Patrick Purdy, 24, left a Molotov cocktail behind to blow up the car and walked over to the schoolyard where he played as a child. Hoisting to his shoulder a Chinese-made AK-47 assault rifle he bought in an Oregon gun store for $350, Purdy opened fire on 450 students.

Reading stories such as this one [five children were killed and thirty others were wounded], which was printed in the January 30 issue of *Newsweek*, reminds us that we live in a fallen world that suffers under the curse brought about by the sin of mankind. It wasn't just the sin of one man that brought so much misery to Stockton, California, on that dreadful day last month.

As an engineer and an American, I can't help but feel some communal responsibility for what took place on that school playground. While madmen can no doubt do considerable violence with their hands alone, we multiply the potential for that violence by placing a technological tool like an assault rifle at their disposal. AK-47 assault rifles don't grow on trees. They are designed by engineers, manufactured by industrialists, sold by businessmen, and controlled (or left uncontrolled) by the governing authorities.

Unfortunately there are a number of people who believe that technological tools are inherently neutral. In other words, it is only the activity in which the tool is utilized, not the tool itself, that can be judged as right or wrong, obedient or disobedient to the will of our Creator. I would refute that. I believe that all products of technology have values designed and built into them. Those values are distinct characteristics of the tool that can be judged as positive or negative.

For example, an automobile has many values built into it. If it is costly to purchase, and if it guzzles gas, it can be said to be disobedient to the norm of stewardship. On the other hand, if it is so reliable that it almost never needs repair, then it is obedient to the norm of trust. If it is beautiful to look at and is very satisfying to drive, it has a positive aesthetic value built in. But if, like a Cadillac or Lincoln Continental, it is designed to be a symbol of wealth, of having "made it" in the materialistic sense of the word, then that negative value, that characteristic of being a status symbol, cannot be separated from the car itself any more than the other values can. In this example, reliability, beauty, and comfort are positive values that are consciously built into the car. Extravagance, self-indulgence, and greed are negative values that have been consciously built-in – and, I might add, are usually exploited in the advertising that promotes the car.

Consider another example, the hypodermic syringe. The last time I was at the dentist I remember very well the shot of Novocain I received prior to drilling. I remember it because, unlike most such shots, I didn't even feel it. I remember thinking to myself, "three cheers for the engineer who designed, and the manufacturer who produced a syringe so sharp and effective that I can't even feel it!" And I'm sure that particular brand of syringe is appreciated even more by people, like diabetics, who must use it regularly. Some might argue, however, that the hypodermic syringe is a poor design, that it contains certain characteristics that contribute to its misuse such as in drug addiction. In addition to becoming a tool for drug use, it becomes one of the means by which diseases such as AIDS are transmitted.

But I would argue that those characteristics of the hypodermic syringe that make it useful to drug addicts are not negative values that have been either consciously or unconsciously designed into it. It's more like a baseball bat that is designed for playing a harmless game, but is used by a violent person to kill someone. On the other hand, if the role of the hypodermic syringe became such that it was aiding and abetting drug abuse in a significant way, it would be the responsibility of the designers to consider that potential misuse and to redesign it so that it would serve its positive purpose without contributing to abuse. Meanwhile, it would be the responsibility of government to control the distribution of syringes.

Now let's return to the case of assault rifles. I would argue that these guns have a negative value designed into them – they are created specifically to kill people. Like neutron bombs, napalm, and anti-personnel grenades, their characteristics are based on the assumption that at least some human life ought to be destroyed. And like cocaine, assault rifles are being used throughout our country to threaten, if not to destroy, image bearers of God. There are two differences with cocaine, however. The first is that cocaine can be used, under a doctor's careful prescription, to relieve pain and help people to live. Assault rifles can only kill. The second difference is that you need a prescription to legally buy cocaine. To buy an assault rifle, all you need is $350.

Obedient technology requires that we design products that promote Kingdom values. Death is not a Kingdom value and engineers should not be designing perversities the likes of assault rifles. Societal justice requires that we promote Kingdom values. That suggests to me that a society must create laws to protect its citizenry from the threat of disobedient technology. That means either the complete outlawing of all guns – or at least the very strict control of those that may have a legitimate purpose – and the outlawing of the rest.

The existence of such technological products like the one used to ravage the Cleveland elementary school playground is a mark of a sick society – a society that values individualistic freedom above human life and forces you to get a prescription to get life-prolonging medicine, but places no restraints on purchasing instruments of death.

As an engineer, and as a representative of the Lord Jesus Christ, I can tell you with some authority that there will be no assault weapons in the New Jerusalem. Let us, with the apostle John, pray that the Lord does indeed come quickly.

CHAPTER 13

A JOHN DEERE PENCIL SHARPENER?

MAY 4, 1990

A couple of weeks ago I was invited to speak to a class of sixth-grade students about solar energy. I was delighted to find that, despite the apathy these days regarding energy stewardship, these young people were intently interested in learning how we might use the sun to meet energy needs. They asked a lot of good questions: some were kind of technical and specific to solar energy, others more general, having to do with environmental conservation and stewardship of resources. The discussion got really interesting when we talked about the topic of appropriate technology, or more specifically, appropriate energy end-use.

To get the point of appropriate energy end-use across, I used some analogies. We considered four different energy use situations, each involving energy transformation processes familiar to us all.

First consider what it would be like to be a diligent sixth-grade student, who does a lot of writing with a pencil, and who has to get up to sharpen that pencil at least six times each day. Now the classic Boston pencil sharpener that hangs on the wall and requires you to turn the handle a dozen or so times to sharpen each pencil works rather well. But wouldn't it be easier and faster if somehow we could have a power pencil sharpener? What if we parked our father's John Deere tractor just outside the classroom, ran it at idle during the school day, and hooked up the power take-off to the pencil sharpener so that anytime we needed to sharpen our pencil all we would have to do is insert it for an instant in the sharpener? What would be wrong with such a scheme?

Or consider the difficulty many of us have in slicing off a pat of butter to put on our pancakes in the morning. Unless you wait around for the butter to soften, slicing that pat from a stick of butter recently taken from the refrigerator can be a hassle. What if we ordered an extra-fine blade for our chain saw and kept it handy in the kitchen. Then whenever we needed that pat of butter we could immediately and with great ease slice it using our chain saw. Is there a problem with this idea?

Here's a third situation: What if we ran out of bread and needed to go to the store to buy a loaf? The store is four blocks from our house and it would take us 20 minutes to walk there, or almost 8 minutes to ride our bicycle. And sometimes the weather isn't very nice. Suppose, instead of walking or riding a bike, we started up the 4,000 pound steel automobile in the garage, accelerated it to 30 miles per hour, decelerated it to a stop at the store, bought our loaf of bread, then started up the car again, accelerated and decelerated it, and finally parked it back in the garage. We'd have been able to transport that 16-ounce loaf of bread from the store to our house and would have cut the travel time down to under three minutes each way. Anything wrong with that?

Finally, let's consider the problem we have in keeping our homes warm during the winter. In this case warm means about 70 degrees Fahrenheit. Our ancestors kept warm by using a form of stored solar energy – wood. I say solar energy because it's the sun that is the energy source for the photosynthetic process by which trees convert water from the ground and carbon dioxide from the air into what is eventually wood. But heating with wood has its problems. For one thing it's very inconvenient. You have to cut and dry the wood, tend the fire regularly, and, of course, see to it that you have a source of wood for the future. That means setting aside a large plot of land that you devote to the growing of trees. Then every time you cut down a tree you plant another so that your energy source is renewable, so that it will always be there. Today that's not practical for another reason. There are so many of us living on just a fixed amount of land that using wood to heat our homes would mean requiring an unreasonable percentage of that land to grow trees. Worse than that, the smoke from all those wood fires would pollute the air so badly that we probably wouldn't be able to grow the trees anyhow.

But today we have another way of using the sun's energy to heat our homes. We can design our homes so that they utilize the solar energy immediately available during the day, store it for use during the night and during overcast days, and lose very little of it to the cold winter environment. To do that, however, means investing in insulation, caulking and weather-stripping, active or passive solar collectors, and perhaps putting up with the occasional discomfort associated with not having your house toasty warm every time you get up in the morning.

But, alas, using direct solar energy and fixing our houses up so that they are truly energy efficient costs money and effort. And right now there is a way that is a little less expensive (in the short run), and a great deal easier. We can bring into our homes certain chemicals that are cheap (in terms of money) and contain so much concentrated energy that they could easily raise the temperature of our house to over 5000 degrees Fahrenheit if we wanted them to. These chemicals represent a form of solar energy too. The only problem is is that it takes the sun millions of years to produce them. So there's no hope of making them renewable. Once they are used up they are gone forever. And one other factor needs to be considered. These chemicals (I'm talking about natural gas, oil, and coal, of course) are also exceptionally useful for creating very high temperatures, for making fertilizers, pharmaceuticals, pesticides, plastics, and herbicides – the kind of things that we need to feed the hungry, to heal the sick, and to synthesize new and useful materials for all sorts of applications.

The point I'm making is that burning gas, oil, or coal to produce a mere 70 degree temperature for our homes is an inappropriate end-use of an energy resource. It's being unstewardly. It's like carrying 4,000 pounds of steel with you in order to quickly transport one pound of bread four blocks. It's like cutting butter with a chain saw. It's like using a John Deere tractor to power your pencil sharpener.

Isn't it time we repented of our energy selfishness and began asking ourselves what the Lord requires of us as stewards of the various sources of energy he has created? It's good to know that at least one class of sixth-graders is concerned about these questions.

CHAPTER 14

STEWARDING THE LORD'S ENERGY

APRIL 2, 1991

Every spring semester I have the enviable task of teaching two related courses. One, *Technology and Society*, deals with the question of norms for technology, or in other words, what is the Word of the Lord for how we carry out our task of developing his world? The other course, *Solar Energy Engineering*, deals specifically with one way in which engineers can respond to that Word in the area of energy stewardship. This spring an event occurred that gave me the opportunity to put into practice, in a small way, some of the things I teach in both of those courses: my water heater broke.

Sioux County is notorious for the short life-expectancy of its water heaters. Most barely outlive the five-year warranty that the majority of them carry. So it wasn't a great surprise to me that my eight-year-old heater finally began leaking. Being in the midst of teaching my students about norms for technology, and specifically norms for energy utilization, it was natural that I inquire about the energy efficiency of the models available as replacements. I learned that a standard hot water heater is about 75 percent efficient and costs $300. A high-efficiency model, on the other hand, would be closer to 85 percent efficient and would cost $500. That sounded a bit expensive – a $200 premium to get only a 10 percent increase in efficiency. But there was one other economic benefit to the high efficiency model. Its warranty period was ten years instead of the usual five. That was sufficient incentive for me to put my money where my professorial mouth had been and spring for the more expensive, high-efficiency water heater. I'd pay an extra $200, but over the next ten years I would certainly save at least that much by reducing both my natural gas and water heater consumption.

This little incident illustrates a number of points. First, it's obvious that there exist ways of being more stewardly than we currently are with regard to energy resources. Second, it's also obvious that currently, if you want to be more stewardly, you have to do at least two things that the average person usually seeks to avoid: you have to initially spend more money for a stewardly alternative, and you have to spend time searching for that alternative. Not only did I have to pay a $200 premium for my energy-efficient water heater, but since my plumber only carried standard models in stock, I had to wait a few days, nursing my leaking heater so as to provide hot water without flooding the basement, until the high efficiency unit could be shipped.

But why should we bother about what I'm calling "energy steward-ship"? Most people asking the question "What's the advantage?" conclude that the little bit of money saved in the long run doesn't outweigh the initial expense and the bother. As Christians, however, we need to reject that self-centered criterion by which our culture determines its actions, and instead of asking "What's in it for me?" we ought to ask, "What does the Lord require of us?" If we could answer that question properly – biblically – then we might approach our energy stewardship tasks with the same sense of responsibility that mom and dad approach their task of parenting, that deacons approach their task of ministering to the needy, and that school board members approach their task of assuring the biblical direction of their Christian school.

In the remainder of this essay I want to discuss four things that I believe the Lord requires of us as we utilize the energy resources he has provided. We might call these four "norms for energy utilization." First, the way we use energy ought to be appropriate to our cultural situation. Second, it ought to reflect the fact that we are caretakers who the Lord has charged with the task of managing a complex and fragile system. Third, our use of energy ought to reflect God's more general norm of justice. And fourth, it ought to exhibit love: love for God, love for our fellow image-bearers, and love for the creation. There are other norms that should govern our energy use, but perhaps these are the ones that are most critical as we approach the end of the twentieth century.

By saying that our energy use ought to be appropriate to our cultural situation I mean that we need to consider the social and technological context in which we find ourselves. For example, it would be culturally inappropriate for us to return to the horse and buggy as our primary means of transportation even though it may reduce our dependence on fossil fuels. Likewise, while the wood-burning stove may be appropriate for people living in sparsely populated and heavily wooded areas, it would create enormous problems if everyone tried to heat their homes with wood. On the other hand it's clear, given both the political and environmental problems associated with petrochemical sources of energy, that we need to find alternative resources to meet our transportation and space-heating energy needs. Two appropriate resources that we have not adequately pursued in this country are solar energy and our ability to adjust our lifestyles.

The norm associated with our being caretakers implies that we treat the creation in a conserving and efficient manner. A glaring form of disobedience to this norm is our continual use of fossil fuels to provide low level heat for our homes. Using natural gas or oil to heat our homes is like raising a 2000 pound cow and then slaughtering it only for its liver, allowing the rest of the cow to be wasted. Or perhaps a tragically more realistic example is the slaughter of near extinct elephants for the purpose of collecting only their ivory tusks. This norm requires that we understand economics in more than just a monetary sense. We need to fully appreciate the value of all God's creatures, whether that value can be expressed in dollars and cents or not.

Doing justice in our use of energy means that we seek to give every person and every creature the opportunity to be the person or creature that God calls them to be. Our current fossil fuel gluttony prevents many peoples of the world from living fruitful lives. The turmoil in the Middle East is only the most obvious example of this. Likewise, that gluttony wreaks havoc on the creation in terms of air pollution, oil spills, and the like. Thus the air, the birds, and other kinds of creatures too numerous to mention are prevented from being what the Lord intends for them to be.

Finally, the norm of love means that in our energy use we have a genuine and heartfelt concern for our fellow image bearers and for the creation. We ought to be outraged and saddened when we see the land stripped barren for the sake of mining coal, when the sunset is obscured by the haze of air pollution, or when the nightly news brings oil-drenched waterfowl into our living rooms during their last torturous moments of life. And our love for our fellow human creatures ought to move us to compassion when we consider the plight of many in this world, some even here in the United States, who do not have available or cannot afford the energy needed to cook a meal or drive away the winter chill.

The United States government, and particularly the Bush Administration, believes in the free-market economy. That faith, along with a fear that promoting efficiency would violate free market principles, has led to an anti-normative energy strategy emphasizing increased fossil fuel production and ignoring conservation. The legacy of such a strategy can only be: greater human death and destruction as more wars are fought over oil; environmental pollution and degradation; and a world where our grandchildren will have to struggle for energy resources that we in our ignorance throw away today. The voice of the Christian church needs to be heard in the midst of such chaos, first prophetically proclaiming the will of the Lord for our lives as energy stewards, and then living in ways that model that will, erecting signposts of the coming Kingdom where God's will for energy stewardship will be honored by all.

CHAPTER 15

WHAT'S A LEGITIMATE NEED?

JULY 2, 1991

During the past two weeks the Engineering Department at Dordt College has been conducting its *Young Scholars in Engineering* program. With the help of the National Science Foundation, the program brings thirty eighth-grade students to campus in order to acquaint them with and hopefully get them excited about engineering as a possible career choice. The students learn of such things as computers, mechanics, electricity, trigonometry, energy use, electric motors, and even engineering ethics. Most of these topics are covered during a two to three hour lab session or field trip. Another engineering topic we try to acquaint them with is design. This we accomplish by having the students work in teams of three on a particular design project of their choice. But this is an exercise that cannot be completed in a two to three hour lab session; so we introduce the topic early in the first week, and the program culminates in a public session where each group presents the results of their design work.

During the first session on design we ask the question, "how does an engineer begin working on a given topic?" and introduce the idea of identifying a legitimate need. It's easy to see that there are many products in the technological market place for which there is no compelling need. After all, who really needs an automatic popcorn attachment for their VCR? Who really needs buffalo fur upholstery or a plastic, dashboard hula dancer for their car? And these are only the most obvious. If we take a minute to look around us, most of us will agree that we a drowning in a sea of high tech gadgetry for which we have little or no need. But the idea that a product ought to meet some legitimate need before it is designed and built is unfortunately new to most eighth-grade students. In fact, if you assign a given group of grade school or high school students the task of thinking up an invention that they might like to design, produce, and market, you will very likely receive a list of highly imaginative candidates all of which have two characteristics in common: first they are high tech (usually very energy intensive), and second there exists no genuine need for them.

So the first thing we do with our young scholars is to force them to begin thinking about needs. We tell them that *the purpose of any engineering design is to either meet some kind of real need or to meet that need better than it is currently being met.* To determine whether or not a need is legitimate we first have the students ask themselves the following: "If the need has to do with people, will meeting it help them be better people?" Then we discuss a set of twelve criteria by which they might judge whether or not these imaginary people are better because we met that particular need. The criteria go something like this.

First, will people be more energetic because you met this need, or will they become lazy?

Second, will meeting this need enable people to be physically healthier, or will it be harmful to their bodies. We probably ought to ask this question about every new candy, fast food, or labor-saving device that is introduced.

Third, will people be emotionally healthier, or will meeting this need destroy their psyche and their senses? I wonder whether this question was asked before the "Walkman" was invented. Or consider cigarettes. A "Walkman" can lead to hearing loss. Cigarettes lead not only to the loss of your sense of taste and smell, but also to the psychological disability of becoming dependent on the nicotine (not to mention the emotional strain of being avoided by people who can't stand your stench).

Fourth, will meeting this need enable people to reason more clearly, or will it lead them to become intellectual vegetables? This question brings to my mind many of the computer games that children have become addicted to. They may learn eye-hand coordination but at what intellectual price?

Fifth, will meeting this need enable people to become better at designing and building other useful things and being good stewards of the rest of creation, or will it lead them to become destructive?

Sixth, will people be able to communicate better because you met this need? Or will it simply increase confusion?

Seventh, will meeting this need improve social interaction, or will it promote alienation? This criterion makes me think of certain computerized teaching tools that relegate young children to a little cubby-hole where they sit and work with a computer program all day long. It may save money for a school system but the price to the social development of the children is far too costly.

Eighth, will meeting this need enable people to make proper use of their resources (for example, time, money, and talents), or will it lead them to become wasteful, exploitative, or miserly.

Ninth, will people's abilities to play be enhanced? And by play I'm thinking of such activities as music, art, sports, acting, dance, or simply telling a good joke. Or will meeting that need lead them to become un-imaginative, dull, stuffy, and boring?

Tenth, will it help them act more justly to their neighbor, or will it turn them into rip-off artists?

Eleventh, will it help them act more lovingly to their neighbor, or will it encourage them to "look out for number one." This conjures up images in my mind of a driver speeding down the highway at 80 miles per hour, secure in the knowledge that his new radar detector will pro-tect him from being caught by the police and ticketed. His self-centered disregard for the safety of other drivers is only amplified by the radar detector.

Finally, we need to ask if meeting a given need will help people put their trust in the right places. Will it strengthen their faith in God, or will it lead them to gullibly trust empty things (horoscopes, or large bank accounts) or un-trustworthy people?

Each of these criteria deals with needs that people may have. But what about needs that nonhuman creatures have, needs that are not di-rectly related to people? Here our guiding principle is much the same. We must ask whether meeting such an imagined need will improve them. Or will it create pollution, destruction, extinction, and death.

There is a sense in which concerning yourself with the legitimacy of a need can stifle your enthusiasm and creativity – especially if you are a thirteen-year-old boy. But I'm pleased to report that such was not the case with our thirty young scholars this year. And the list of final projects indicated that they had begun to understand the silliness of VCR pop-corn popper attachments and rocket-powered skateboards. For the needs they sought to meet with their designs were for the most part genuine: the need to keep poisonous substances away from little children, the need to keep sidewalks clear of ice so that especially the elderly would be able to walk safely in the winter. These are only two of ten excellent design projects that put to shame the marketing strategies of Madison Avenue. Perhaps there is hope for the future after all.

CHAPTER 16

THE INTERNET AND ECONOMIC NORMS

OCTOBER 6, 1995

A few days ago a colleague of mine made a comment that caused me to pause and reflect on some things that we ordinarily take very much for granted. I had just been telling him about an upcoming educational conference that I knew he would be interested in. What was unique about the conference was the fact that the registration was free and attending it would not require us to travel any distance. It was an Internet conference. That is to say, the conference was going to be conducted over the Internet, the conferees communicating with each other by computer. A two-week period was identified, during which prospective conferees could sign-up, sending their e-mail address to the conference organizers. Then, on one specified date, the four papers to be presented would be e-mailed to each of the conferees. For the following two weeks the conference would take place over the Internet as two to three days of e-mail discussion would be devoted in turn to each of the papers. It sounded like a great idea. My colleague's response, the comment that made me pause and reflect, went something like this, "Isn't it amazing how so many people do so much work for free in order to keep the Internet going?" You see, currently, the Internet is supported by people who do their work *pro bono*, that is, without pay.

Well, in one sense it certainly is amazing. It seems that many people involved in technology today are simply out to make money on it. We've even raised to the status of folk heroes people like Steve Jobs and Bill Gates, merely because they were successful in making millions of dollars in the computer business. But in another sense, maybe it isn't so amazing that people would work for free in order to keep the Internet going. After all, as Christians we believe that the Lord calls us into existence in order to work, that by our work we have an opportunity to bring Christ's redemptive healing to a broken world, and that the ultimate reward for our work is not money, but rather to hear the Lord say, "Well done good and faithful servant! You have been faithful with a few things; I will put you in charge of many things. Come and share your master's happiness!"

(Matthew 25:21).

But wait! This is America, the land of the entrepreneur. This is where the twin virtues of free enterprise and competition saved us from sharing the fate of the Soviet Union. To work without concern that you are being paid "what the market will bear," well, that's downright un-American! If free enterprise is a virtue, then working without concern for pay must be . . . a vice.

Is it? Or is all this pro bono technical work in support of the Internet instead a glimmer of light that by God's grace is breaking through the darkness of our self-centered and materialistic culture? Doesn't this kind of thing happen elsewhere? How about when the hobbyist woodworker spends hours in the workshop crafting a piece of furniture, only to give it away. There is no monetary remuneration, just the pleasure of knowing that the gift is appreciated. Or what about when grandmother works for days purchasing food, carefully preparing it, cooking it, and serving it on a day when the extended family is gathered together just to visit with one another? She neither gets nor looks for money as payment for services rendered. But again, there is the pleasure of knowing that her work is appreciated.

There are of course many more examples that one might give of work – hard work – that is done without thought for monetary remuneration. Some of them, of course, might be classified as work that is done in connection with an act of mercy or an act of love. But many, like that of the hobbyist woodworker, have no such connection. The work is performed without thought of monetary remuneration. Although, again, there is usually great pleasure either in the doing of the work or in the knowledge that it is appreciated for its quality. Are these all un-American, violations of those twin virtues of free enterprise and competition? In other words, vices? I should hope not.

On the other hand, consider the following. During the 1960s the Engineering Council for Professional Development had written in its code of ethics the following statement: "Engineers shall not undertake nor agree to perform any engineering service on a free basis."[*] Many other engineering codes likewise insisted that it is the obligation of engineers to require "adequate" compensation, in other words, monetary remuneration at the going fee scale (335). It seems it is considered a vice to work for free because doing so destroys competition. Or, more importantly, it eliminates a source of income that might slake the avaricious thirst of

---

[*]   Martin, M.W. and Schinzinger, R., *Ethics in Engineering*, 2nd ed. (New York: McGraw Hill, 1989), 334.

fellow engineers. If that sounds strange, consider this second example: I recently talked with a government official in a small municipality regarding a possible energy efficiency project that could serve both the residents of the municipality as well as the cause of good stewardship. But I was immediately rebuffed because the project would involve the municipality in making a product available to the residents at a price lower than what they would pay if they bought it from a local vendor. Notwithstanding the fact that the product was an insignificant part of the local vendor's business, the mere threat to competition – more hypothetical than real – was enough to quash any further consideration of the project.

My point in all this is simply that the free enterprise system as we know it is seriously flawed. The success of the Internet, supported as it is by people who work without concern for monetary remuneration, is one small demonstration of that flaw. Sadly, that very success is about to be the undoing of the Internet. Right now it is possible to send text, graphics, and even pre-recorded sound messages over the Internet. In a very short time it will be possible to send not only pre-recorded sound messages, but real time sound messages as well. If that were to occur, anyone with access to the Internet would be able to communicate with someone else, anywhere in the world, in a manner that would be indistinguishable from that of the telephone. But the big telephone companies will never let that happen. One way or another the Internet will be turned into a vehicle for making money. Then, of course, you will have to subscribe to use it. And you will have to put up with the equivalent of "commercials" and "advertisements."

Some day, after the Lord returns, we will no longer work for money. We will work, motivated by the satisfaction that our efforts serve and are appreciated by others. Economics will then mean stewardship. And there will probably be something like the Internet. But it will not be amazing that people work without monetary pay to support it.

CHAPTER 17

IDEALISM AND THE ORDER OF CREATION

JUNE 3, 1996

In the spring of each year I teach a course to senior engineering and computer science students entitled *Technology and Society*. It's a kind of capstone course that attempts to tie together everything the students have learned during their four years by focusing on the Christian technologist's responsibilities with respect to the direction of technological work. One of the things that I have the students do is to design, on paper, a small business. The students work in groups of five or six and each group is assigned to a particular aspect of the business. For example, one group works on the internal structure of the business, that is, how the various offices relate to one another. Another group works on how the business functions economically: its relation to its customers, its profit or not-for-profit status, and the price it will charge for its services and/or products. One group tries to develop a salary package for all those employed by the business, another group works on developing guidelines for hiring employees, and yet another group works on how the business will relate to the surrounding community and the impact of the business on the environment. The overall purpose of the exercise is to give the students an opportunity to apply those fundamental principles of justice, stewardship, and service in a practical context. This past year the business was a recycling company located here in Sioux County.

The greatest barrier to successfully completing the project is the perception on the part of students that applying the biblical principles of justice, stewardship, and service to a North American business enterprise is idealistic and impractical. For example, it is deeply ingrained in the worldview of many Christian students, especially engineering and computer science students, that the fundamental purpose of a business is to make money. If the business provides a genuine and meaningful service to society, well, that is viewed as a desirable bonus. I would argue that a biblical perspective rejects that, insisting that the fundamental purpose of a business is service and that any monetary profit earned from the business ought to be seen as a tool for improving that service.

So a few days ago I was musing over the written student evaluations of *Technology and Society*, and lamenting the fact that a handful of students again made the comment that the approach taken to the small business project is, as they say, too idealistic. At about the same time I was studying the Old Testament and reading a book by Professor David Holwerda of Calvin Seminary entitled *Jesus & Israel: One covenant or Two?* As you might guess, since it is the basis for this essay, these various elements came together to provide me with one of those delightful little moments of insight whereby I learned something I had always wondered about.

On reading the book of Exodus, I had paused briefly in Chapter 23 to contemplate that strange prohibition that the Lord gave the Israelites regarding cooking a young goat in its mother's milk. Exodus 23:19 reads as follows: "Bring the best of the firstfruits of your soil to the house of the LORD your God. Do not cook a young goat in its mother's milk." I have always wondered why the Lord proscribed such an action. After all, what difference does it make to the young goat whether you boil it in water, ordinary goat milk, or the milk of its mother? In any case it gets boiled and, very likely, eaten. Sure, to us humans, it may seem somewhat cruel if we imagine ourselves in the place of the young goat or the young goat's mother. But if we make that imaginative leap, then just killing and eating the young goat seems cruel. There must be a better explanation.

I found that explanation when reading Holwerda's book. In the context of a discussion of righteousness as understood in the Gospel of Matthew, particularly Christ's teaching (6:33) that we ought not seek first after material wealth but strive instead for God's kingdom and righteousness, Holwerda refers back to that strange passage in Exodus. He writes,

> The prohibition against boiling a kid goat in its mother's milk was a warning not to practice rites by which Israel's neighbors sought to gain fertility from the deities. Israel's God was not a nature deity capable of being manipulated by magic or fertility rites. Israel's God is a God of the covenant who sends blessings, including fertility, on those who keep his covenant.[*]

What God requires of us, first and foremost, is faith. We demonstrate that faith by trusting that he will provide for all our needs. That faith, that trust, gives us the freedom to live obediently – some might even say, idealistically – by being the loving servants that he calls us to be. In terms of business, this means that we seek first to be of service to our fellow image-bearers, to be good stewards to the rest of the creation, and to

---

[*] Holwerda, David E., *Jesus & Israel: One covenant or two?* (Grand Rapids: Eerdmans, 1995), 143.

subject all of our business activities to the norm of justice, trusting God that the profits necessary to run the business will flow naturally. Doing so is keeping covenant with the Lord who has ordered his creation in such a way that righteousness is the very foundation of economic success. I quote once more from Holwerda:

> Israel was warned not to follow other avenues in search for economic success, such as worshipping the fertility deities and boiling a kid in its mother's milk (Exodus 23:19). Instead, Israel was called to keep the covenant and its righteousness and thereby discover the truth that it is God who gives grain, wine, and oil. As a matter of fact, the earth as God's creation is ordered by righteousness, justice, steadfast love, and mercy (Hosea 2:19), and the earth as God's agent showers its abundance on those who seek God and observe the order that he has established. To follow other norms is to seek another God, and the result will be judgment, not only on Israel but on the nations, even if it takes seventy years to appear. Thus the righteousness articulated in the Sermon on the Mount is foundational because it is rooted in God's created order. (143–144)

So it is not really idealism when we say that a business needs to seek first to be of service and the profits will follow naturally. It is righteousness. And we can be confident that such a business enterprise will be blessed because that righteousness is foundational to the very order of the creation.

CHAPTER 18

STEWARDSHIP FOR PROFLIGATE TIMES
APRIL 5, 1996

Sometimes it's understandable why various peoples in the world view all of reality in terms of cycles. Certainly the daily, monthly, and annual changes in the natural environment suggest the reality of cycles. But increasingly, the older one gets, events in the social environment tend to exhibit a cyclical character as well. I was reminded of this recently when my wife and I accompanied our youngest son and his wife for an afternoon of "open houses," as they began shopping for a home in Sioux Falls.

No doubt my work as an engineering educator at Dordt College has kept me sensitive to energy stewardship concerns, but I think that merely living through the seventies and eighties has been enough to imprint indelibly on my mind a concern for energy conservation. During those two decades I was involved in the purchase and maintenance of three different houses and four different automobiles. And during those two decades we, as a culture, experienced what we called at the time, "the energy crisis." The "crisis" was at its worst during the mid-seventies, when crude oil shortages forced the price of gasoline to suddenly jump from under fifty cents to well over a dollar per gallon. During the same time period, electricity and natural gas prices rose steadily, and by 1980 one's monthly utility bill comprised a significant portion of the cost of owning a home. Thus it was, when looking at houses in Sioux Falls with my son, I naturally raised questions regarding the insulation quality of the walls, the efficiency of the windows, and the overall monthly utility bill. To my surprise, however, these concerns were low on the list of those selling points promoted by the army of realtors we encountered. I was puzzled to find that many moderately priced houses were fitted with windows of questionable energy efficiency. And one realtor, in response to my question about the insulation quality of the walls, told me that no one was building with six inch walls anymore, everyone is back to using 2x4 construction. Her explanation was that the quality of the insulation in the walls had improved greatly. I knew, of course, that explanation was either an outright lie, or a betrayal of the Realtor's ignorance, because the heat conducting value of fiberglass insulation has not changed significantly in thirty years. So why was there no concern for energy efficiency among the people selling, and assumedly among the people buying houses in Sioux Falls? The answer became obvious when I looked more closely at the brochure handed to me at one of the houses. Listed were all the individual costs that added to the overall monthly price of owning the house. The heating budget was listed as $50 per month – out of a total budget of $1200 per month. Seventeen years ago, on moving to Sioux Center, I remember trying to find a house with a heating budget under $50 per month, but out of a total budget that I was trying to keep under $400 per month. So in those seventeen years it seems that the accepted cost of owning a home has tripled, but the cost of heating it has stayed relatively constant.

The same kind of phenomenon has occurred with respect to owning and operating an automobile. Seventeen years ago you could purchase a very nice, new car for around $6000. Today a similar model will cost almost $20,000. Yet the price of gasoline has actually decreased slightly.

I conclude from all this that the energy crisis is over. Not only that, but we are now living in an energy-economic context much like that of the late fifties and early sixties. Energy costs are very low and there is little concern for energy conservation on the part of the average home owner or automobile operator. The recent raising of speed limits on the nation's highways is further evidence confirming this point.

"So what?" you may ask. Perhaps we ought simply to be thankful for low energy prices. Well, perhaps we ought to be. The problem is that we are not responding like the grateful stewards of the creation that we are called to be. We are returning to our self-centered and profligate patterns of consumption. It was, in part, our gluttonous consumption of fossil fuels that triggered the energy crisis in the seventies. Back then, however, that consumption was attributable to both greed and ignorance. Today ignorance is no excuse.

My point in this essay is not to warn you of the next energy crisis. I'll save that for opportunities I may have to address groups of people who are not necessarily disciples of Jesus Christ. For readers of this essay, I want to remind you that the death and resurrection of our Lord was to atone for our sin, for the sake of the whole creation. To ignore the energy lessons of the past is to reject our calling as stewards of God's good creation. It is sin.

Proverbs 26:11 is the most appropriate commentary on this kind of sin: "As a dog returns to its vomit, so fools repeat their folly." On the other hand, those who repent and follow Jesus – the Creator, Sustainer, and Redeemer of all creation, petrochemicals included – are freed from the curse of cyclical existence.

CHAPTER 19

BUYING A CAR

JULY 8, 1996

A few years ago my wife and I began to realize that our car was getting up there in terms of both miles and years and that in the near future we would have to consider replacing it with a newer one. So we have been planning and saving and expect that in the fall we will make that rather major purchase. But what kind of car ought we to buy? What ought to guide us as we shop around for a machine that will meet our transportation need for perhaps the next six to ten years? As the time for us to buy gets closer, those questions get more serious, and the need for some guiding principles becomes more acute. In this essay I want to consider some possible guiding principles for purchasing a car, and also some powerful forces that tend to make us insensitive to those principles.

Purchasing a car in 1996 in North America is a religious act. It is a response to God, who calls us to live obediently as stewards of our energy and monetary resources, and commands us to glorify him and be a witness to his truth in every corner of our lives. Thus, the details of our car purchase demonstrate to the Lord how serious we are when on Sunday morning, for example, we recite the Apostle's Creed. They also stand as a witness to our neighbors of both the Word of the Lord and our faithfulness to that Word.

Let's consider just a few norms for purchasing a car. I will briefly discuss economic, aesthetic, juridical, ethical, and faith norms.

Economic norms are the easiest to understand but not always the easiest to follow. Since the Lord calls us to responsible stewardship of his creation, we ought to choose a transportation device that does not waste, but preserves those resources. Well, that may mean that we ought to first see if we can live *without* this device we call the automobile, since one has to admit that it is a very unstewardly example of modern technology. Walking, using a bicycle, or mass transportation are all far more stewardly ways of traveling than driving a car. But if you live in northwest Iowa, traveling by foot, bicycle, bus, train, or airplane severely limits where you can go. A car becomes a virtual necessity, or, a necessary evil, one might say. But what car we buy ought to be as stewardly as possible. So we will naturally want to purchase one that minimizes the waste of resources. That says to me that I ought to look for a car that gets good gas mileage, that, if not recyclable, will at least last a long time, and that will not require an inordinate amount of monetary resources. In other words, it ought to get high mileage and be reliable and inexpensive. Nothing controversial or earth shattering about that.

When we say that a car ought to be aesthetically normative we are saying that it ought to be delightful to use and that it be playfully allusive. In other words, it should be user-friendly and it ought to suggest certain positive things about the people who own and operate it. User-friendliness is a quality that most people look for in an automobile. But how can a car be playfully allusive in a way that pleases the Lord? Well, consider that the parts of a car, or the car as a whole, may suggest things about itself and its owner. For example, the tail fins on a 1959 Cadillac suggest rocket plumes. The spoilers on many of today's cars are meant to suggest such power and speed that an airfoil is needed to keep the car from flying off the road. As a mechanical engineer, let me assure you that they *only* suggest – they are not functional in any way. Make-believe rocket plumes and spoilers are at worst, silly. But there are some other kinds of allusions that certain cars may embody. Traveling in Washington, DC, recently, I saw many a Mercedes Benz, Lexus, and BMW. Such cars suggest wealth, power, and fame – the gods of our culture – and for that reason we ought to be very careful about our relation to such icons. I would imagine that Daniel, since he refused to bow down to Nebuchadnezzar's Babylonian idol, probably also disassociated himself with those Babylonian artifacts that suggested the gods of that culture. I'm not saying you ought not drive a new Cadillac. But you ought to be very aware of the kind of allusions that owning such a car entails – particularly if you are called as a model to young impressionable Christians, as are those of us in the teaching profession.

For a car to be juridically normative it must be legal to own and operate. In addition, it must embody justice, both in its manufacture and its use. Thus, if you know that a particular high mileage and low cost import is made by slave laborers in a repressive, dictatorial country, you probably ought to refrain from purchasing it. Likewise, if owning it prevents you and other members of your family from being the whole people that the Lord has called you to be, you also ought to refrain from purchasing it.

For a car to be ethically normative it ought to help you love your neighbor as well as the nonhuman creation. Thus you want to find a car that is safe and non-polluting. Likewise you would want to avoid spending so much money on a car that you were left unable to adequately donate to the charities and other good causes that you normally, out of love for your neighbor, support.

Finally, a car is normative with respect to faith norms in at least two ways. First, a faith norm for any technological artifact is reliability. You need to be able to trust that the car will do what it is supposed to do – especially when driving at highway speeds. Secondly (and reflecting what I've already said regarding aesthetic normativity), purchasing, owning, and operating a car ought to demonstrate your trust and love for God. That means, for one thing, that you ought not to be swayed by those commercials that try to tell you how wonderful you are and how you *deserve* a particular kind of car. Those commercials work on the fact that many Americans like to consider themselves rugged and autonomous individuals who trust only in themselves.

As I said earlier, purchasing a car in 1996 in North America is a religious act. Think about that the next time you are in the market for one. I will be thinking about it during the next few months.

CHAPTER 20

CLONING, DETERMINISM, AND APRIL FOOLS

APRIL 1, 1997

Today is what is traditionally referred to as "April Fools' Day." And perhaps some of the faithful *Plumbline* listeners, to whom this essay was originally addressed, arrived at their radios expecting to hear an entertaining, tall-tale of sorts, followed by that well-used punch line, "April fools!" Well, maybe I'm just not imaginatively up for such a recitation . . . or perhaps it's simply that I have been too struck by another kind of foolishness as I observe the remaining number of twentieth century April Fools' Days wind down to less than the number of prongs on a dessert fork.

I'm thinking about all the silly babble that is going on these days regarding what is called "cloning." Oh, I realize that most of it is media hype, concocted to sell magazines and to keep people glued to their TV sets. But it nonetheless surprises me to think that, at the close of the twentieth century, so many adult human beings can fall for such foolishness.

You see, we have gone through all this before. More than a century ago. Back then it was called mechanical determinism. We've updated the adjective "mechanical," so that now we are talking about "genetic determinism," but, after all, determinism is determinism.

Let me explain. Back in the eighteenth and nineteenth centuries, as a result of the work of scientists and mathematicians like Isaac Newton and René Descartes, foolish people developed the silly notion that we human beings are composed of nothing more than the complicated but strictly lawful motion and interaction of little particles called atoms. Thus, while we have the illusion of a free will, all our thoughts and actions are really determined by the physical laws that those atoms obey. Listen to one influential thinker of 111 years ago, who drew a perfectly logical conclusion from this hypothesis of physical determinism. His name was Pierre Laplace, and in 1886 he wrote:

> An intellect which at a given instant knew all the forces acting in nature, and the position of all things of which the world consists – supposing the said intellect were vast enough to subject these data to analysis – would

embrace in the same formula the motion of the greatest bodies in the universe and those of the slightest atoms; nothing would be uncertain for it, and the future, like the past would be present to its eyes.*

What Laplace is telling us is that once we have sufficient memory in our computers we will be able to analyze the motion of the present atoms, and then calculate what those atoms must have been doing in the past. In other words, we will be able to determine what happened one minute, one day, or one century ago, based on what is happening today. Likewise, we will be able to determine what will happen at any time in the future. It's all a matter of calculation based on the state of the atoms that presently constitute the universe. That's called physical determinism because all past and future events are *determined* by the present physical state of the universe.

Today, the whole popular debate about cloning rests on almost the same kind of deterministic underpinnings. Only now instead of physical determinism we have genetic determinism. Somehow the popular culture has gotten the absurd notion that human beings are derived of nothing more than the complicated but strictly lawful arrangement of genetic material, and that the clone of an individual person would be an exact duplicate of that person. Well, that is silly, of course. Everyone knows that identical twins have exactly the same genetic structure. And while identical twins may look very much alike, they are always two different persons with different personalities, identities, responsibilities, and ways of responding to their Creator.

In 1864 the Lord blessed his servant, Fyodor Dostoyevsky, with the wisdom to respond appropriately to the physical determinism of his day. That response strikes me as singularly appropriate today, as we contemplate the present, silly discussions regarding cloning. I will close this essay by quoting from it.

> . . . you say, science itself will teach man (though to my mind it's a superfluous luxury) that he never has really had any caprice or will of his own, and that he himself is something of the nature of a piano-key or the stop of an organ, and that there are, besides, things called the laws of nature; so that everything he does is not done by his willing it, but is done of itself, by the laws of nature. Consequently we have only to discover these laws of nature, and man will no longer have to answer for his actions and life will become exceedingly easy for him. All human actions will then, of course, be tabulated according to these laws, mathematically, like tables of loga-

---

* Laplace, P., 1886, *Introduction à la théorie analytique des probabilités, Œuvres Complètes*, as cited in Çapek, M., *The Philosophical Impact of Contemporary Physics* (New York: Van Nostrand Reinhold, 1961), 122.

rithms up to 108,000, and entered in an index; or, better still, there would be published certain edifying works of the nature of encyclopedic lexicons, in which everything will be so clearly calculated and explained that there will be no more incidents and adventures in the world. . . .

Man is stupid, you know, phenomenally stupid; or rather he is not at all stupid, but he is so ungrateful that you could not find another like him in all creation. I, for instance, would not be in the least surprised if all of a sudden, *à propos* of nothing, in the midst of general prosperity a gentleman with an ignoble, or rather with a reactionary and ironical, countenance were to arise and, putting his arms akimbo, say to us all: "I say, gentlemen, hadn't we better kick over the whole show and scatter rationalism to the winds, simply to send these logarithms to the devil, and to enable us to live once more at our own sweet foolish will!"**

** Dostoyevsky, F.M., *Notes from the Underground*, translated by Constance Garnett (New York: Macmillan, 1864).

CHAPTER 21

ENERGY STEWARDSHIP AND "THE BOTTOM LINE"

MAY 5, 1997

One of the assignments for students in my course on solar energy is to perform an energy audit on a home in Sioux Center. The focus of the audit is space-heating energy, and so they go through a very detailed analysis of such things as the amount of insulation in the walls and ceiling of the house, the condition and number of the windows, the spatial layout of the house, and they even perform a blower door-test, that is, a test to determine the infiltration losses of the house – how much heat is lost due to cold air leaking in and warm air leaking out. After collecting all the necessary data, they subject it to a computer analysis that predicts how much energy it should take to heat the house during a typical winter. Then the students compare the results of that analysis with the actual utility bills paid by the homeowner and the climatological records that describe the severity of the winters covered by those bills. If the students have done their analysis well, the predicted energy use and the actual energy use should be the same. This year, each energy audit group in my class of engineering students was successful in getting very close agreement between their predictions and the actual energy use of seven different homes in the local area.

One purpose of the assignment is to hone the analytical skills of students who will be doing this kind of work in the future. Another purpose, however, is to develop in the student an awareness of energy stewardship norms and a fervor for working toward the realization of those norms in the situations of daily life. One way of realizing those norms is to educate the public about the possibilities that exist for conserving energy. The assignment, therefore, culminates in a report to the homeowner that details how the house is losing energy and how the homeowner can make changes to conserve energy in the future. Often those changes are modest, such as caulking around windows and doors so as to reduce infiltration losses, or insulating basement walls to reduce heat conduction losses.

There is, however, one ambitious recommendation that the students must make to the homeowner. That is the addition of an active solar energy system. The students are required to perform a solar energy system design for that specific house on which they performed the energy audit. I describe this as an "ambitious" recommendation because, unlike caulking around windows or increasing the insulation a bit, adding a solar energy system is rather costly. In fact, whereas most of the energy saving measures recommended by the students range from a few dollars to a few hundred dollars in cost, the addition of an active solar energy system typically costs in the ballpark of $5,000. And while most of the energy savings measures pay for themselves in reduced utility bills within a few months to a few years, the payback period for a solar energy system is usually greater than ten years.

So, while the students are required to recommend a solar energy system to the homeowner, it is not expected that any homeowner will act on that recommendation. This raises the question, "Why bother?" And that leads into a discussion of the cost of energy, the limited character of our present energy resources, the next energy crisis, and the motivation for conserving energy.

Right now, energy in the United States is priced artificially low – as if it were a renewable rather than a finite resource. Sometime in the not-too-distant future that will change. And when that change occurs, we will have an energy crisis that will make the energy crises of the seventies pale in comparison. Assuming that you will still be able to purchase traditional forms of energy, you can get a feel for the coming crisis by imagining that by September of this year the price of gasoline would rise to $10 per gallon and that your monthly utility bills would increase about five-fold, let's say from an average of $100 per month to $500 per month. Imagine the consequences for society that such changes would make. We will adjust, of course. But that adjustment will involve quite a bit of suffering, particularly for those whose family incomes fall below the national median.

The suffering that we will encounter will be a direct consequence of the motivation (or lack thereof) that we have today for conserving energy. As some of my more pragmatic students often point out to me, if people are not going to realize monetary savings by energy conservation measures, they simply will not pursue them. And while, for the most part, I have to admit that that is true, it seems contradictory to me that it should be true for Christians. After all, we do not realize any monetary savings when we purchase furniture for our homes, when we go out to a restaurant for dinner, when we take our family for a week of vacation, or when we donate to the building fund of our local congregation. And that is as it should be. For we are called to be stewards of our families and churches, and good stewardship means far more than economics. But we are also called to be stewards of God's good creation. Why is it, then, that we will only exercise those creational stewardship responsibilities when we can reap a monetary payback from doing so?

In this regard I am reminded of two verses from the eleventh chapter of the book of Proverbs:

> Those who trust in their riches will fall, but the righteous will thrive like a green leaf. Whoever brings ruin on their family will inherit only the wind, and the fool will be servant to the wise. (11:28–29)

Perhaps we ought to understand "one's family" as referring more broadly to one's household, including the energy stewardship responsibilities that we have for that household.

CHAPTER 22

PLAYING GAMES WITH "DEEP BLUE"

JUNE 2, 1997

One of the front page stories in the 12 May 1997 issue of the *Sioux City Journal* ran under a headline that read, "Deep Blue demolishes champion in final game." The first line of that story ran as follows:

> In a dazzling, hour-long game Sunday, the Deep Blue IBM computer demolished world chess champion Garry Kasparov and won the six-game chess match between man and machine.

The essence of the story is that humankind has been beaten by a machine in an area that has, to this point in time, been considered the exclusive dominion of flesh-and-blood people. Consider the following two lines from that newspaper story. First, Garry Kasparov, the chess champion who lost to the computer, is quoted as saying, "I feel confident that the machine hasn't proved anything yet." Then C.J. Tan, the scientist who was in charge of the IBM Deep Blue computer states, "We on the IBM Deep Blue team are indeed very proud that we've played a role in this historic event."

Apparently there are a lot of people who are interested in this "showdown" between human and computer, otherwise the newspaper would not have printed the story and given it such a prominent place on the front page. Let's take a moment to consider why that might be so.

During the track season, one year when I was in high school, I ran a 25-meter sprint against a 1959 Volkswagen Beetle. It was a lark, a whim, the kind of thing that high school students do. But no one took it seriously. A little knowledge of physics enables one to realize that, if the distance is short enough, virtually any human can beat any automobile in a sprint race. Likewise, beyond a certain distance, even Michael Johnson is going to lose to the most modest of cars. It is simply a matter of acceleration and top speed ability. There is no genuine interest in races between humans and automobiles because there is no confusion between humans and automobiles. An automobile is a machine designed to enable humans to travel longer and faster than they can if they depend on their bodies alone.

But it appears that there may be some confusion between humans and computers. In fact, there are people who genuinely believe that, some day in the not-to-distant future, a computer will be designed that equals or exceeds all the capabilities of the human brain. That belief is based upon the assumption that human beings are essentially very complex machines.

From a Christian perspective, of course, this is nonsense. We know from God's Word (Genesis 1:26–27) that he created us in his image. We read in Psalm 8 that he made human beings "a little lower than the angels and crowned them with glory and honor" and "made them rulers over the works of [God's] hands." Thus this fascination with chess-playing computers must be fallacious and we ought to be able to show those caught up in that fascination where the fallacies are. It is sort of like that gullible high school friend of mine, the one who owned the Volkswagen. Not knowing much about physics, he simply believed that any car could defeat any human in a race. After demonstrating that a person could outrun a car over a 25-meter distance when starting from rest, I had no difficulty in explaining to him that his fallacious belief rested on an ignorance of the laws of physics.

Yes, the laws of physics can help us expose the fallacies in the fascination of some with chess-playing computers. But while these realities are considerably more complex than the concepts of velocity and acceleration, they *are* rooted in the way God has structured the creation. I am thinking particularly about the aesthetic aspect of creation and the normative structure for what we call "play" and "games."

A chief characteristic of play is surprise. And surprise is based on uncertainty, or unpredictability. What makes a game of baseball enjoyable, as a game, is the unpredictable character of the events that make up the game. Certainly there are rules that provide a boundary of predictability. But within those rule boundaries, it is uncertainty and surprise that gives baseball the nature of a game.

To appreciate this better, consider the "game" of tic-tac-toe. When we were little children it may have been fun for us to play tic-tac-toe. But it wasn't long before we figured out how to prevent our opponent from winning. From then on tic-tac-toe ceased to be fun. It ceased to be a game. The reason is that we figured out that there are a finite number of possible moves, and we knew how to counter any one of them. Thus tic-tac-toe lost its unpredictable nature for us and ceased to be a game.

What about chess? Well, chess is like tic-tac-toe, only much more complicated. There are many more rules and so many possible arrangements of the chess pieces on the board that no human mind can contain all of them. Thus, when two people play against each other, there always exists uncertainty. The game involves not only the spatial arrangement of pieces on the board, but also logical skill taken to its limit, the exercise of psychological abilities, planning for future moves, stewarding the limited resources of time and number of moves, and, of course, playfulness.

When a computer is used to compete in a game of chess, the game is reduced to the spatial arrangements of the pieces on the board and the rules that govern their movement. All the logic has been done in advance by a programmer (a human). Thus, when "playing the game" the computer does not engage in aesthetic, creative, psychological, or even logical activity. It is simply a complex series of physical processes, the movement of electrons through a microprocessor, in response to inputs that represent the spatial arrangement of the chess pieces on the board and the way the software has been put together by the programmer. If the program has been written to take advantage of the computer's capacity and speed, then, in theory, every possible chess move can be accounted for in advance and the game becomes one like tic-tac-toe. There are no possible surprises. When that happens the game ceases to be a game and playing chess against a computer is reduced in playfulness to the level of running a race against a car.

CHAPTER 23

DISTANCE LEARNING:
CULTURAL FORMATION OR DEFORMATION?

MARCH 7, 1997

A few weeks ago I attended a teleconference dealing with energy stewardship. I was in a classroom in the Maurice-Orange City High School, a
classroom that is equipped with modern telecommunications equipment
and connected to the Iowa Communications Network – the ICN. The
conference originated in Washington, DC, and the Orange City site was
just one of many across the country composing the conference network.
During the ninety-minute conference, a moderator conducted a discussion with four panelists who represented a variety of viewpoints on energy conservation and management. One of the panelists was an official
from the federal Department of Energy. Another was Amory Lovins, the
well-known advocate of "soft-energy paths," that is, energy production,
distribution, and utilization measures that are renewable and environ-
mentally benign.

This essay, however, is not about energy technology – it is about
communication technology. For as I sat there, participating in the conference, I was thinking as much about the new mode of communication
that I was experiencing as I was about energy.

Recent changes in communication technology are just expressions
of the more comprehensive digital computer revolution that started in the
late 1940s. When I was in college in the sixties, it was still possible to get
an undergraduate engineering degree without encountering a computer.
For most calculations, engineers relied on their trusty slide rules. By the
mid-seventies however, the hand-held calculator had displaced the slide
rule, the integrated circuit was revolutionizing products such as radios
and televisions, and the typewriter was preparing to follow the slide-rule
into extinction. That happened in the eighties, the decade of the personal
computer, when the word processor and the spreadsheet changed forever
the way ordinary people wrote and calculated. Now, in the last decade
of the twentieth century, the personal computer is well established and
communication technology has taken the stage and been given the spot-

light. The Internet, the World-Wide-Web, teleconferencing via the ICN, voice mail, and cellular telephones are changing the way we interact with each other. If we consider the next decade, that is, the first decade of the twenty-first century, we may be able to foresee – somewhat fuzzily and in broad outline, of course – some even bigger changes that will result from the advancement of digital computer technology. But rather than tantalize you with what is not yet, I would have you consider a development in communication technology that is happening right now. It is called "distance learning," and, as the name might suggest, it is the application of computer and communication technology to the task of education in order to overcome the traditional barriers of space or location.

The idea behind distance learning is not new. When, on a Monday morning in September of 1964, a high school kid walked into his first college class at Rensselaer Polytechnic Institute – a chemistry lecture session – he walked into a huge auditorium with 900 other freshmen. The professor, on the stage in the front of the auditorium, could hardly be seen. But bolted to side walls of the auditorium – four on each side – were television monitors. In addition, a PA system insured that everyone could hear the professor. Thus every student in the auditorium had at least some opportunity to see and hear what was going on. The professor lectured and sketched notes using an overhead projector, the image of which was transferred to each of the monitors in the auditorium.

Today "distance learning" is much improved. Instead of simply overcoming the distance in a large lecture hall, modern communication technology makes it possible to overcome distances like that between Washington, DC, and Orange City, Iowa. The monitors are now at least thirty-two inches instead of twenty-one, the pictures are relatively sharp, and they are in color. The rooms in which the distance learners sit are usually small and comfortable, handling no more than thirty or forty people. And most importantly, the potential exists for back-and-forth communication. That is, there are facilities in the rooms so that the image and voice of the distance learner can be transmitted back to the central site, enabling the distance learners to interact with the "professor" by asking questions and getting an immediate response.

So now we arrive at the heart of this essay. I will put it in the form of a question. *Is distance learning an appropriate use of technology that advances education by enabling students to more effectively fulfill the tasks that the Lord calls them to?* Instead of offering a definitive answer to that question, I want to take the remainder of this essay to suggest an approach to *how* we might answer it.

At the outset, it is important to recognize that education is a human activity and as such it is multifaceted. That is to say, it is not merely a matter of information transfer, of socialization, or of culture formation, although those are three important aspects of education that must be given serious consideration. Of course there are some really basic aspects to the educational process that almost "go without saying" – there must be a physical space *in* which, and a time *during* which the learning can take place. That space must not only be conducive to human life, it needs to have at least a minimum of comfort and be distinguishable as a learning space so that the mind of the learner can remain focused on the task of learning. Beyond those basics, however, I suggest that the following questions need to be adequately answered.

First, is the mode of learning culturally appropriate? In other words, is there a kind of historical continuity with previous modes of learning so that the student is enabled to make a smooth transition from previous learning experiences to the new learning experience? For example, attempting to use textbooks with a preliterate culture is clearly inappropriate. Requiring that all writing be done in longhand cursive for a class of students that has grown up using the word processor would also be culturally inappropriate.

Second, does the communication technology engage the learner in a delightful and harmonious way. This is a question of user-friendliness, and it builds upon the question of cultural appropriateness and the physical space foundation previously mentioned. Another way of asking this question might be, "Does the communications technology result in a delightfully harmonious interaction, at the human-technical interface, whereby the technological artifacts dissolve into an extension of the user?"

Third, does the mode of learning facilitate open and unambiguous lines of communication. In other words, are all the human avenues for the conveyance of meaning open and unobstructed?

Fourth, does the mode of learning promote social interaction and development both for the individual learner and for the society to which that learner belongs? For purely individualized modes of learning the answer to this question would have to be, "largely no."

Fifth, does the mode of learning promote good stewardship of resources such as time, energy, and, needless to say, money?

Sixth, does the mode of learning promote justice? In other words, does it balance the opportunities that all people have to be the persons that the Lord calls them to be? A mode of learning that, for example, required a great deal of familiarity and ease with the latest technological artifacts would favor technophiles at the expense of, shall we say, the technologically challenged. Likewise, given our understanding of the role of individual learning styles, it may be easy for a particular mode of learning to favor persons comfortable with one learning style while alienating those comfortable with a different learning style.

Seventh, does the mode of learning promote love of one's neighbor, care for the creation, and love of God? It may be argued, for example, that a mode of learning that eliminates all direct contact between the learner and the natural creation will fail to promote the kind of care for the created order that the Lord requires of us.

Finally, is the mode of learning trustworthy and does it promote in the learner ultimate trust in God rather than in some created thing. This question covers a lot of ground. For example, it requires that the mode of learning be technically reliable, not subject to interruptions, outages, and down-time. It also requires, for example, that the mode of learning point to dependence of the learner on God rather than on her own rational abilities. In this sense the Cartesian methodology of radical doubt – ending in *I think, therefore I am* – is a failure.

I have only raised some questions in this essay, I have not really given any answers. I hope that those of you who are contemplating the adoption of distance learning will carefully consider it in light of these, and perhaps other, questions.

CHAPTER 24

Y2K, TECHNICISM, AND "THE CROWD"

JANUARY 5, 1999

Well, the year 2000 is now less than twelve months away and as it approaches, the prophets of doom are becoming louder and more strident. This is nothing new, of course. At the turn of each century there have always been those who predicted catastrophe and social disruption, if not Armageddon itself. What is unique this time around is the way in which the characteristics of our age have permeated the very vocabulary of millennial prophecy. Consider the infamous "Y2K." "K" is the symbol used in the metric system of units to indicate a multiple of 1000. Thus "Y2K" stands for "year 2000" – innocuous enough. But of course, it stands for more than just the year 2000. It stands for a particular computer problem that is alleged will appear on January 1 of the year 2000.

It's not my purpose in this essay to discuss all the details of the "Y2K" bug. Frankly, I don't expect that it will create major problems. But it *is* symptomatic of a more serious problem.

Y2K is a potential crisis for a society that has become dependent on the computer but has not had the foresight to look ahead a few years as it writes programs for its computers. As most people know, the Y2K bug arises when a computer reads dates as only two digit numbers – 68 stands for 1968, 97 stands for 1997 and so on. The problem is that in the year 2000, those two digits will be 00, the same as they would have been had we been using computers in the year 1900. Computers naturally recognize 00 as a lower number than 99, and thus translate 00 to mean 1900 unless they have been specifically programmed to read it as 2000. But most have *not* been programmed that way, and thus there has been a frantic race to *reprogram* them before January 1, 2000 arrives. The majority will be successfully reprogrammed before that happens.

But I said that Y2K is symptomatic of a more serious problem. That more serious problem is *analogous* to Y2K. If, in the description of the problem, you replace "computers" with "people" and replace "programming" with "nurturing," you may begin to grasp the analogy. In the past 50 years we have been programming computers without giving much thought to their complexity, their potential importance, and the ways they will have to function in the future. Likewise, for the past 50 years we have been nurturing people without giving sufficient thought to their complexity and importance as image bearers of God, and to the ways in which they will be called to live in the future. In fact, we have allowed what might be called the computer spirit of the age to dictate the nature of our nurture.

Let me describe just two ways in which we have yielded to this computer spirit of the age. The first has to do with the nature of our thinking and has been well described by Neil Postman in his book *Amusing Ourselves to Death: Public discourse in the age of show business* (New York: Penguin, 1985). During the last 2.5 millennia, we have moved from orality to literacy. That is, we once communicated ideas and maintained traditions by word of mouth, by telling stories. Early in our cultural history we invented and began using the alphabet, and so began making the transition from storytelling to story writing. With the invention of the printing press in the fifteenth century that transition was complete. Now we are a *literate* culture. We communicate ideas and maintain traditions by using words – words that are written down as well as spoken. That "written-down-ness" of words gives them a kind of permanence that is absolutely necessary to abstract thought. And the ability to think abstractly – which made possible Greek philosophy, the Bible, science, the arts as we know them, and technology – is fundamental to what we know as our human creatureliness. I will go as far as to say that – at least for the last two thousand years – to be human is to be literate. To be illiterate is to be an impaired human being.

Today, however, as a result of the computer spirit of the age, we are becoming less literate. Television, film, video, the personal computer, and the Internet are moving us from literacy to iconography. That is, we are moving away from communication by words and sentences – which promote abstract thought – to communication by pictures, either moving or still, which as a steady diet, I would argue, inhibit abstract thought. Therefore, if to be literate is to be human, we are being dehumanized by the computer spirit of the age.

But I want to describe one other way in which we are yielding to the computer spirit of the age. It has to do with homogeneity (i.e., "sameness"), individual responsibility, and what the philosopher Søren Kierkegaard called *the crowd*. Computers work by categorizing and classifying. They necessarily ignore the fine distinctions that exist in the world so that they can organize data according to a predetermined scheme.

Likewise, the computer spirit of the age leads people to categorize and classify themselves and others. Thus the individual gets lost in the crowd of homogeneity. Conformity becomes the order of the day. The word of the crowd becomes the word of law. Leaders are qualified not by justice and righteousness, but by crowd-appeal.

Listen to what the Christian philosopher Søren Kierkegaard had to say about the crowd back in the nineteenth century – a time when, although there were no computers, the dehumanization characteristic of the computer spirit was already manifesting itself. Kierkegaard writes:

> . . . a crowd in its very concept is the untruth, by reason of the fact that it renders the individual completely impenitent and irresponsible, or at least weakens his sense of responsibility by reducing it to a fraction.
>
> The crowd is untruth. Hence none has more contempt for what it is to be a man than they who make it their profession to lead the crowd.... For it is not too great a risk to win the crowd. All that is needed is some talent, a certain dose of falsehood, and a little acquaintance with human passions.
>
> . . . to honor every man, absolutely every man, is the truth, and this is what it is to fear God and love one's "neighbour." But . . . to recognize the "crowd" as the court of last resort is to deny God, and it cannot...mean to love the "neighbor." . . . But the thing is simple enough: this thing of loving one's neighbour is self-denial; that of loving the crowd, or of pretending to love it, of making it the authority in matters of truth, is the way to material power, the way to temporal and earthly advantage of all sorts – at the same time it is the untruth, for the crowd is the untruth.[*]

If we are to live as sincere Christians, we must resist the computer spirit of our age. That means being counter-cultural: resisting the homogenizing forces of conformity, developing a healthy contempt for the crowd-manipulating techniques of modern advertising, and openly castigating the so-called leaders of the day who seduce the mindless masses with amusement. And most importantly, it means we must strive to preserve literacy by doing – yes, the unthinkable – we must cultivate reading . . . and restrain the ubiquitous proliferation of television, film, video, and the Internet that has already impaired so much of twentieth century humanity.

---

[*] Kierkegaard, Søren, 1859, *The Point of View for my Work as an Author*, (New York: Harper Torchbook, 1962), 112–119.

CHAPTER 25

A NEW KIND OF DEPRAVITY

SEPTEMBER 2, 1999

Human beings and digital computers are different kinds of creatures. That sounds like I'm stating the obvious, doesn't it? Of course no one really confuses a computer with a person today – treating a PC (personal computer) as if it had sensitivities to injustice, aesthetic nuance, or was capable of love. But what about the other way around? How does your computer treat you? In his book *The Invisible Computer: Why good products can fail, the personal computer is so complex, and information appliances are the solution* (Cambridge, MA: MIT Press, 1999), Donald A. Norman argues that the PC is intrusive and overbearing, precisely because the way that humans and computers interact today forces people to conform to the basic characteristics of the computer. In other words, the personal computer is a device that treats people as if *they* were computers.

Norman argues that people seek to be beings that get at the *meaning* of things and for whom the *details* do not always matter very much. Humans have characteristics that computers can only approximate. For example, even if I stammer, use colloquialisms, or even misuse a word here or there, you will still get the gist of what I am saying. That's because in listening to me speak, you focus on the overall meaning, not on the details of what I say. You understand context. And if I misspeak, you quickly disregard my error, subconsciously saying to yourself, "Oh, he said *X*, but he really meant to say *Y*." A computer cannot do that. It is a slave to accuracy and precision. From a human-centered point of view, people are compliant while computers are rigid. We are attentive to change while computers are insensitive to change. People are creative and resourceful while computers are unoriginal and unimaginative. But from a computer-centered point of view, people are vague while computers are precise. We are disorganized and distractible, while computers are orderly and undistractible. And people are emotional and illogical, while computers are unemotional and logical.

Over the course of the last few centuries, we in the West have come to idolize those attributes and have bestowed them upon the computer. Now don't get me wrong. There is nothing wrong with encouraging people to try to be precise, orderly, and logical when the situation calls for it. That's a part of God's good creation that he has called us to appreciate. But it is possible to absolutize a part of creation – to raise it to a governing position above everything else. When we do that we fall prey to what Paul describes as exchanging the truth about God for a lie, and worshiping and serving created things rather than the Creator (see Romans 1:25). The result is a distortion of our very humanity, a form of depravity. I fear that the role of the personal computer in our society, specifically the ways that we have created to interact with it, is evidence of distortion and depravity.

Let me give you two examples that I have gleaned from Donald Norman's book that show how we have duped ourselves with the computer-centered point of view. First, compared to the difficulty many of us – shall we say, more experienced people – have with the computer, consider the ease with which young people, those who have grown up with the computer, quickly adapt to working with it. In his book, Norman writes the following:

> To me, the maddening point about those who have grown up with the technology is that they don't realize that there might be a better way. When I have problems, I fret and fume and suggest a dozen better solutions. When they have those very same problems, they shrug their shoulders. They have grown up believing that it is natural and correct to spend a large portion of every day redoing one's work, restarting systems, inventing "work-arounds." What a horrible heritage we have passed down to them."

A second example is the conventional wisdom that we are currently living in truly revolutionary times and that the technological changes that are occurring as we move from the twentieth to the twenty-first century will have an even greater impact on civilization than the invention of the printing press. Countering that assertion, Norman writes, "Today, the 'revolution' in which we live consists mainly of improvements in what has already existed." In making this point, he cites the far-reaching changes that occurred during the latter part of the nineteenth century. Rapid travel, for example, did not exist in the nineteenth century. The end of the nineteenth and start of the twentieth century gave us first ocean liners and the railroad, followed by automobiles and airplanes. Likewise, talking to someone over long-distance was not possible for most of the nineteenth century. In the twentieth century we began regularly using telegraph, telephones, and television. Instant recordings of sights and

---

** Norman, p. 90.

sounds (photographs and the phonograph) were developments that oc- curred over the turn of the century. As we move from the twentieth to the twenty-first century, these are all being improved – but not radically changed. Norman might be faulted here for not giving sufficient recogni- tion to the revolution *beneath* the surface – the analog to digital revolu- tion. But going beneath the surface is to adopt the computer-centered point of view. The analog to digital revolution is revolutionary primarily for engineers and scientists, *in the context of their work*. For most citi- zens of the twenty-first century, the changes are those of "improvement" rather than revolution.

Once upon a time people bought personal electric motors. Seriously! They also bought peripheral devices to go with their personal motors; devices like fans, mixers, vibrators, and drills. If you think I'm exaggerat- ing, check out a copy of a 1918 Sears Catalog. You will find that Sears sold personal motors, along with an array of peripheral devices to go with them. Nobody purchases personal motors today. Electric motors are embedded in the appliances that serve specific and well defined purposes. Those appliances include fans, refrigerators, power tools, and copy ma- chines. The electric motor has disappeared, embedded unobtrusively into the interior workings of specific appliances. It's time for the computer to perform the same disappearing act. Then we might begin to free ourselves from a peculiarly late-twentieth century form of depravity.

CHAPTER 26

HYPERLEARNING AND THE
NON-DISCIPLINE OF CONSUMPTION

OCTOBER 1, 1999

It has been said that the dawn of the third millennium represents a new age of technology that will change forever the way people work, play, travel, learn, and even think. That's quite a claim. In this essay I wish to consider that aspect of the claim having to do with how people learn – particularly with respect to college level education.

Which technological artifact best represents the character of our age? "The personal computer," you may suggest. Well, yes, the PC is certainly a representative of our age. But there is a better one, one more expressive of who we are. It is the credit card. You see, the credit card not only suggests the technological infrastructure that supports modern commerce as well as our infatuation with technological gadgetry, it suggests something even more basic about the character of our communal personhood. It expresses better than anything else does our commitment to *autonomous consumption*. By "autonomous consumption" I mean our absolute and unfettered freedom to acquire material possessions. And by "unfettered" I mean our not being burdened by the constraints of time, place, and the decisions of other people.

So what has this to do with college level education? Quite a bit, actually. As the new age of technology dawns on the college classroom we find subtle but significant changes taking place in methods of instruction. As the chalk dust settles for one last time, we find that even the relatively new whiteboards with their colorful dry-markers are giving way to presentations prepared in advance with software like Microsoft *PowerPoint* and presented with the help of computers and powerful video projectors. Textbooks increasingly contain CD ROMs. And often there are websites that students can visit, offering information that the textbook by itself cannot contain. Professors increasingly use the Internet not only to communicate with their students, but also to post information about their courses: syllabi, supplementary materials, past classroom presentations, solutions to past homework problems, and so on. The Internet environment becomes like a shopping mall in which, ideally, the students can select those materials that best enable them to learn the subject matter. And not only is there choice of material, there is choice of time. If a professor places her class notes on the Internet, then the students have less compulsion to attend class at the particular time those notes are being presented. Their freedom extends not only to what materials they may learn from, but also to when they may choose to do the learning.

The destination of this move toward increased freedom for the student by locating the apparatus of education on the Internet is what some have called hyperlearning, distributed learning, or, more commonly, distance learning. There, in the freedom of cyberspace, the shopping mall model of higher education is realized. Students are disburdened from the constraints of time, place, and the decisions of other people. In the spirit of autonomous consumption they are free to choose what they want to learn, when to learn it, and what tools to use in the process. The pleasures of consumption have penetrated the dense, arduous walls of the academy. But wait. Is it possible that the shopping mall model of learning has overlooked something essential about the learning process? Consider these words of wisdom from Albert Borgmann, a philosopher of technology from the University of Montana. He writes:

> The pleasures of consumption require no effort and hence no discipline. Few proponents of course would claim that distance learning will be effortless. But they fail to see that the discipline needed to sustain effort in turn needs the support of the timing, spacing, and socializing that have been part of human nature. . . . [*]

The "timing, spacing, and socializing" that Borgmann is referring to occurs only when persons other than the learner first provide it. It is not something that we can provide for ourselves until we have learned from others how to provide it. In other words, before we can attain the self-discipline needed for responsible scholarship, we need to place ourselves under the discipline of others. Driven by the spirit of autonomous consumption, the proponents of distance learning overlook this interdependency quality of our humanity. And this itself is a characteristic of our age: we disregard, and at times despise the wisdom of past ages. It would thus be well to listen to the opening words of the Book of Proverbs. There the writer tries to explain the purpose of the book. The proverbs of Solomon are:

---

[*] Borgmann, Albert, *Holding On to Reality: The nature of information at the turn of the millennium* (Chicago: University of Chicago Press, 1999), 207.

. . . for attaining wisdom and instruction; for understanding words of insight; for receiving instruction in prudent behavior, doing what is right and just and fair; for giving prudence to those who are simple, knowledge and discretion to the young – let the wise listen and add to their learning, and let the discerning get guidance – for understanding proverbs and parables, the sayings and riddles of the wise. The fear of the LORD is the beginning of knowledge, but fools despise wisdom and instruction. (Proverbs 1:2–7)

This academic year I have the privilege of teaching a class of 40 freshman students who are aspiring to be engineers. They come to my class with an incredibly wide range of high school preparation backgrounds. But the majority of them have one trait in common. They lack the self-discipline and the time management skills needed simply to read their textbooks. Over the next four years, by interacting with their environment in ways that we in the Engineering Department have structured and prescribed, they will achieve a basic and common level of understanding of mathematics, science, and their particular fields of engineering. And they will gain the maturity to learn on their own so that one day they will not need our structured prescriptions. But right now they are perhaps the perfect witnesses to call if one wishes to make the case that the widespread use of distance learning for educating college students will be about as efficacious as the widespread use of flowers for bringing about peace in the Middle East.

CHAPTER 27

SAVING YOUR MIND:
DURABLE STORAGE MEDIA AND THE SELF

NOVEMBER 1, 1999

Almost thirty years ago I stood before my first class of students and – with the zeal of a recent convert – tried to get them to wrestle with what I saw as the reformational issues of the day. One of those issues involved realizing that many Christians over the centuries have adopted a Platonic attitude regarding the body. Most evangelicals grow up believing that we are souls, imprisoned in bodies until the time of our death. Then, "putting off this earthly dwelling," our bodies return to the dirt from which they were originally made and our souls – our essential selves – spend eternity with God in heaven. In the late sixties I learned that this view is not biblical. I learned that it had origins in pagan Greek thought, that it became an institutionalized part of Christian thinking during the Middle Ages when Greek philosophy was synthesized with Christian theology, and that it was reinforced by the rationalism of the Enlightenment when many Christians accepted the notion that our humanity is evidenced exclusively by our ability to reason. I remember reading with enthusiasm Abraham Kuyper's *Lectures On Calvinism*, particularly the lecture on "Calvinism and Science," where Kuyper decries the "dualistic conception of regeneration" that neglects "to give due attention to the world of God's creation" especially the body, because it cares "too exclusively for the soul."* Thus, over the last three decades, I've generally thought that I had a pretty good sensitivity to the problem of "soul-body" or "mind-body" dualism, seeing it as a kind of Christian heresy. I've recently learned, however, that I have not been as sensitive as I originally thought.

I'm currently reading a book entitled *How We Became Posthuman*** by Katherine Hayles. It's a rather ponderous book that one could classify as philosophy of technology. It is not written from a specifically

---

* Kuyper, Abraham, *Lectures on Calvinism*, Grand Rapids: Eerdmans, 1931), 118.

** Hayles, N. Katherine, *How We Became Posthuman: Virtual bodies in cybernetics, literature, and informatics* (Chicago: University of Chicago Press, 1999).

Christian perspective. But it is making very clear to me that this problem of "mind-body" dualism is far more complicated and pervasive than I once thought it to be. Most striking are the apparent convergence of biotechnology and information technology, the assumptions about our humanity that are driving that convergence, and the image of our humanity that such convergence places before the eyes of the general population.

Since the late 1940s, researchers in the areas of information theory and cybernetics have been working at developing "artificial intelligence." For many, the theory has been that one day we will be able to build computers that are as powerful as or more powerful than the human brain. The assumption behind that theory is that the human brain is a very complicated computer. And the belief behind that assumption is that all of reality is made up of the same basic elementary particles, governed by the same natural laws, and can – in theory – be explained by our knowledge of those particles and laws. This belief, or faith – for that is what it truly amounts to – is known as *naturalism*. Lately however, a subtle shift has taken place in naturalism, particularly among those involved in information technology. A new form of naturalism – which Hayles, in her book, refers to somewhat pejoratively as *posthuman* – gives a privileged place to information *patterns* as over against the material in which those patterns are manifest. Thus reality, including human thought and action, is reduced even further than it is in classical naturalism. It is no longer the lawful behavior of subatomic particles that is allegedly ultimate. Rather, it is the information pattern structuring the particles that is ultimate. For human beings, "embodiment in a biological substrate is seen as an accident of history rather than an inevitability of life" (2). Hayles argues that in this posthuman view, consciousness, which has been regarded as the "seat of human identity" in the West since before the Enlightenment, is seen as just another product of evolution – and one that is not even necessarily tied to biological systems (3).

Now, consider, if you will, that while these developments have been occurring in information technology, a similar pattern of developments – although starting much earlier – has been occurring in biotechnology. It was perhaps in the Middle Ages when, with the invention of eyeglasses, humankind first realized that it could improve the body. Soon after that, developments in chemistry and biology led to a blossoming in the field of medicine. A generation ago medical research provided the drugs and procedures for fighting, first, disease and then for curing some congenital defects. Today, heart and other organ transplants are common and genetic engineering is able to treat our bodies even before we are born.

Taken together, these developments in information technology and biotechnology lead to the conclusion that the human body is simply the "original prosthesis," which we have learned to manipulate. "Extending or replacing the body with other prostheses becomes a continuation of the process that began before we were born" (3). In other words, the recent convergence between information technology and biotechnology has affirmed with a vigor the old Greek belief and Christian heresy that we are essences – or *souls*, if you will – temporarily occupying mortal, flesh and blood bodies. Our bodies take on the character of *fashion accessories* (5), and temporary ones at that. It is thus easy to become seduced by fantasies of disembodied immortality, whereby our minds or souls or essential selves – assumed to be nothing but information patterns – can be transferred to more durable storage media when our "earthly dwelling" begins to wear out. Perhaps we will inhabit the dynamic RAM memory of a large computer for a while. Or maybe we will be stored in the read-only-memory of a CD or DVD, waiting the time when we can be downloaded to the hard drive of the newest model robot – one that we have previously selected, as we might an automobile from the new-car showroom. But wait, do you hear the contradiction in these words about minds and the "storage media" that they might inhabit? Do you see that we cannot talk even about the simplest information patterns without mentioning the media that manifest those patterns? Doesn't that suggest something about the inseparability of the storage medium and information pattern, about the unity of body and soul? It should. For here we have an example of God's good creation resisting, on the one hand, our attempts to reduce what is irreducible, and foiling, on the other hand, our attempts to deny the unity in creation as we attempt to absolutize those fallible and temporary categories of soul, body, mind, and matter.

So as we witness the convergence of information technology and biotechnology, it is all the more important that Christians have a good understanding of the unity in creation and of the problem of soul-body dualism. Behind the words of Abraham Kuyper and the reformed thinkers of our day we ought to hear those words of Job when, sitting on his ash heap on the edge of despair, he said,

> I know that my redeemer lives, and that in the end he will stand upon the earth. And after my skin has been destroyed, yet in my flesh I will see

God; I myself will see him with my own eyes – I, and not another. (Job 19:25–27a)

As I stand before my students these days I realize that some things do not change very much. Thirty years later I'm still trying to get them to wrestle with the reformational issues of the day. And some of those issues have not disappeared. Like the issue of mind-body dualism, the form has changed along with changes in technology. They still require us to faithfully exercise our Christian worldview.

CHAPTER 28

GADGETS, SOFTWARE UPGRADES, AND CHASING AFTER THE WIND

FEBRUARY 1, 2000

Last week I received an advertisement in the mail inviting me to upgrade my Bible software to the latest version. I must confess that I was tempted. The program on my computer that enables me to quickly find any verse in the Bible, either by selecting the verse by name or by searching for key words, is the same one that I've been using for the last five years. I'm pleased with the way it works and thus have resisted spending seventy dollars to upgrade it to something with a little more glitz and pizzazz, but what I believe would be no significant increase in functionality. And so last week I resisted again. But I *was* tempted. And while many of you may consider it harmless – perhaps even humorous – to be *tempted* to upgrade one's *Bible software*, it provided me with an occasion to reflect on one of the chief distortions of our modern society: the frantic quest for the newest technological improvement. Then, just the other day, a student* reminded me of a point made by Professor Bob Goudzwaard when he visited Dordt College a few months ago. Goudzwaard had argued that our society has become too "goal oriented."

> He said it is like we are living in a long tunnel. We cannot look back. We cannot take any detours. We can only look to the future and keep moving ahead. (1)

What Goudzwaard is referring to, and what the Bible software update advertisement evidences, is a characteristic of the spirit of *technicism*. It is a form of dissatisfaction that results in an urge to change one's present condition for what is perceived to be a better one. It is a chief driving force behind the seemingly infinite developments of technological gadgetry that characterize our present age. Computers are only the most obvious example. Buy a brand-new computer today and in less than three years you will be bombarded with advertisements seducing you to discard it and buy a new model. The same spirit, however, drives the

---

*   Matt Roozeboom, *EGR390 Journal*, January 26, 2000.

design, production, and sale of everything from televisions, sound systems, and camcorders, to power tools, automobiles, and kitchen sinks. And sad to say, it is even infecting areas of our lives that we do not easily associate with technology. Consider the exponential increase in divorce and remarriage during the last few decades. Sure, the primary causes are self-centeredness and unfaithfulness. But this spirit of technicism, which engenders dissatisfaction with that which is not "new," nourishes that self-centeredness and that unfaithfulness.

Like all perversions, this extreme goal-orientation finds its origin in a part of God's good creation. As his image bearers and servants, we are called to be busy "tending the garden," so to speak. And that kind of service always has an aspect of goal-orientation associated with it. The trouble occurs when we sinners turn our focus from God to ourselves. We then turn that goal-orientation into an absolute. *It* becomes the god that we effectively serve. Paradoxically, that absolutized goal-orientation turns upon itself and *becomes the goal.* Like a dog chasing its tail, we get ensnared in the vicious circle of never being satisfied and always wanting more. The Scriptures speak directly to this perversion. Consider those words of wisdom from the book of Ecclesiastes:

> Whoever loves money never has money enough; whoever loves wealth is never satisfied with their income. This too is meaningless. As goods increase, so do those who consume them. And what benefit are they to the owners except to feast their eyes on them? (Ecclesiastes 5:10–11)

The posture of obedience, which stands in stark opposition to this paradoxically recursive materialism, can be found in the New Testament. Consider Paul's letter to the Hebrews:

> Keep your lives free from the love of money and be content with what you have, because God has said, "Never will I leave you; never will I forsake you." (Hebrews 13:5)

Even more to the point are Paul's words to Timothy:

> But godliness with contentment is great gain. For we brought nothing into the world, and we can take nothing out of it. But if we have food and clothing, we will be content with that. Those who want to get rich fall into temptation and a trap and into many foolish and harmful desires that plunge people into ruin and destruction. (1 Timothy 6:6–9)

The Scriptures focus on the pursuit of wealth and we generally read that to mean money. But that word "wealth" applies with equal validity to technological gadgetry, such as computers, camcorders, and software. Our society knows nothing of Christ's command that we deny ourselves, take up our cross, and follow him as obedient servants seeking to meet the needs of our neighbors before our own needs. In fact the whole economic system of the west is predicated upon the assumption that we are each out to serve ourselves first with the best we can afford. The individual pursuit of happiness, the good life, that's the American way. But it's the precise opposite of *the Way, the Truth*, and *the Light*. And so we arrive at the hilariously ironic point in history where Bible software is hawked to consumers by appealing to their self-centered pursuit of the latest and greatest.

If Jesus called the Pharisees of his day a "brood of vipers," what words do you think he might choose to describe the *techno*-Pharisees of today?

CHAPTER 29

"READING" AUDIO BOOKS

AUGUST 1, 2000

A few weeks ago, I had an interesting experience traveling from Sioux Center to Grand Rapids, Michigan. I was driving by myself and one of the methods I chose to overcome the boredom of the 12-hour trip was to select an audiotape – or perhaps one might call it an audio book – to play in my car cassette player. It was a special 2-tape set by Mars Hill Audio on the life and thought of Michael Polanyi, a chemist and a philosopher. I also brought with me a book that I was in the process of reading: Neil Postman's new book, *Building a Bridge to the Eighteenth Century*. I planned to read that after I arrived at a motel in Michigan City, Indiana, and probably finish reading it the next morning, during which I had a couple of hours of free time.

The tape set and the book were, in themselves, very interesting. In this essay, however, I want to describe what I learned simply by reflecting upon the experiences of "listening to a tape while driving," on the one hand, and "reading a book while seated in a comfortable chair in a quiet room," on the other.

I've been listening to Mars Hill Audio tapes while driving in my car for about two years. I find the regular bimonthly journal tapes to be interesting and informative. And, they almost completely eliminate the tedium of long drives. Each tape is a kind of audio magazine, with Ken Meyers, the producer of the tapes, interviewing about eight different persons on a variety of topics of interest to thinking Christians. The tape I chose to listen to for this particular trip, however, is a little different. First of all, it is two tapes – making it a total of three hours in length. And more importantly, it deals with only one topic: the life and thought of Michael Polanyi. Polanyi was a chemist who became a philosopher. He lived from 1891 until his death in 1976. His greatest contribution was breaking with conventional views on scientific knowledge and developing a theory of knowledge that emphasized what he called *tacit knowing*, a phrase that describes the kind of knowledge of the world that cannot be articulated in words or propositions. A good example is the knowledge that an eight-year-old child has of riding a bicycle. It's obvious that he knows how to ride the bicycle because we see him doing it. But he cannot express in propositional or verbal form how he does it. The tapes are divided fairly equally between first describing the life of Michael Polanyi and then discussing his theory of tacit knowing. As you might infer, it is the kind of audio program that takes some concentration in order to appreciate.

The book that I read, *Building a Bridge to the Eighteenth Century* by Neil Postman, attempts to analyze modern trends in technology and education in terms of historic, Enlightenment thought. It presents a reasoned and common sense critique of both modernity and postmodernity by reconsidering the thought of historical figures like Voltaire, Rousseau, Jefferson, and Thomas Paine. Although it is what one would consider a "popular book," it, like the audiotape about Polanyi, takes some concentration to appreciate.

What I would like you to consider as you read this essay is the difference between listening to the tape while driving a car and reading the book while seated in a comfortable chair in a quiet motel room. There are *obvious* differences, of course. One listens with one's ears and reads with one's eyes. And the environment of a car traveling on Route 80 in Iowa at 10:00 a.m. in the morning is different from that of a Michigan City motel room easy chair at 7:00 a.m. the next morning. Again, that's somewhat obvious, the car environment being potentially more distracting than the hotel room environment. But we begin to get a hint of something interesting now, especially if I consider what is the *same* about the two activities. In both the listening and reading activities I am attempting to learn something. I am trying to understand a set of propositions – in the first case by listening to a recorded voice and in the second case by observing the words on the page. In both cases I am trying to understand something – and that takes concentration.

Now I've already said that the automobile environment is potentially more distracting than the motel room environment. But the advantage of the tape is that I can keep my eyes on the road while concentrating most of my mind's efforts on listening to and understanding the sentences being spoken. In other words, the audiotape technology has a property that resists certain kinds of distractions better than does that of the book. Perhaps that is obvious to you too, and you too have enjoyed music, stories, or other kinds of entertaining programming on a tape player while driving in a car. But take one more step with me in the direction of considering the properties of audiotapes and books with respect to resisting distraction and facilitating understanding. While driving on Route 80 and listening to the audiotape, I was occasionally frustrated; losing my concentration as I battled an occasional 18-wheeler. As a result, I missed a sentence or two, possibly even an important point. Oh, I could have rewound the tape. But the rewind button is unfriendly – it probably wouldn't have taken me to exactly the place I wanted to go. In addition, I would have had to take my eyes off the road in order to operate it. So I never rewound the tape. As a result, my comprehension of the tape's message was limited by those occasional distractions and resulting losses in concentration.

Now compare that with reading the book in the quiet motel room the next morning. Despite the quiet, there were some distractions. A couple of times, for example, I got up to get myself a cup of coffee. When I came back, my bookmark told me where I had been. But I often re-read the last paragraph in order to enhance my reading continuity. And like everyone else, I assume, my mind would simply wander at times as I was reading – even without any outside distractions. When I found that happening, however, it was a simple matter to reposition my eyes on the page and read again what I had read previously without comprehension.

My point is this. The Lord has created us in such a way that print technology has properties that favor concentration and comprehension in situations where in-depth understanding is critical, that is, where continuity in a sequence of propositional statements is required. Audiotape technology favors comprehension where understanding can withstand occasional discontinuities in the message being presented and where the concentration must be shared by another activity – driving a car, for example.

So if you want to hear music or listen to a story, audiotape technology can work wonders, particularly if you need to be driving a car at the same time. But if you wish to plumb the depths of epistemology or chemical catalysis, you better stick with print media. It seems the book, in one form or another, will be with us for some time to come. It's well designed for in-depth learning and understanding. And that's a good reason to encourage our children and grandchildren to read and to develop their reading skills.

CHAPTER 30

IMMORAL MACHINES

DECEMBER 1, 2000

During the summer following my freshman year in college, I worked for a company that braided asbestos insulation on thermocouple wire. But this essay is not about asbestos – we didn't know it as a health hazard back then. Rather, I want to talk about the nature of technological artifacts.

In the factory where I worked that summer I had about five different jobs. What job I had during a given week depended on which regular worker had gone on vacation. My favorite job was overseeing the braiding machines as they braided together wire from about twenty individual spools and produced one cable. The nice thing about that job was the long periods of time during which I had nothing to do except listen to the hum of the machines – or read. Naturally I chose to do the latter, and during that summer I read about three books as the braiding machines hummed away. Occasionally one would stop because the wires on its spools had run out. Then I would simply replace the spools with fresh ones and set the machine to running again. On very rare occasions a wire would break. Then I would have to get out the welding kit and fix the break. But 75 percent of the time I spent simply reading.

The job that I really hated was the one where I had to spin the wire onto those little spools that went into the braiding machines. There I had one huge wheel of wire – a supply that could last a week – and my job was to produce little spools, one at a time. You had to place the little wooden spool on the spooling machine, thread the wire onto the spool, turn on the machine, wait about two minutes for it to fill up, shut off the machine, cut the wire, place the filled spool in a bin, grab an empty spool, and start all over again. The whole cycle took only three minutes. That meant that in one hour I could load twenty spools with wire. It also meant that I went through the same repetitive motions twenty times each hour – no variations, no decisions for me to make, just the same repetitive motions, twenty times each hour. It was, in a word, *dehumanizing*. Whereas with the braiding machines I had responsibility for keeping them running, as well as fixing things when they went wrong, the spooling machine, in a sense, had *responsibility for keeping me moving* in the same repetitive way. Some of the older workers who regularly had to do that job told me that they overcame boredom by losing themselves in daydreams. Their minds wandered away to more pleasant thoughts while their bodies remained slaves of the spooling machine. I was never able to master that particular technique.

I was reminded of this experience of slavery to a machine when I was traveling this past summer. One morning, after spending the night in a motel room, I walked to the motel lobby in search of a cup of coffee. This particular motel had a technologically sophisticated coffeemaker and dispenser. I noticed that the dispenser had directions written on it. "Push button," the directions said. So I grabbed a cup, held it under the spout and pushed the button. Now, the directions did not say whether I was supposed to hold the button in or simply push and release it. So I held it in, thinking – somewhat absentmindedly – that I should control how much coffee would be dispensed. But I was wrong. The dispenser stopped after giving me about two-thirds of a cup of coffee. Again, somewhat absentmindedly, I pushed the button. You know what happened, right? The coffee dispenser dispensed another two-thirds of a cup. And two-thirds plus two-thirds makes four-thirds. Which means that a third of a cup overflowed and was wasted. I stood there imagining that all the other people in that hotel lobby were staring at my overflowing cup thinking, "how selfish and wasteful! Why can't he just follow the directions?" After overcoming my embarrassment and assuring myself that no one in the lobby really cared what I was doing, I thought to myself, "What a stupid machine!" Then I realized that the coffee dispenser was a technological artifact that had been designed by some engineer somewhere. That engineer designed into it the assumption that persons using the dispenser ought to have minimum responsibility for getting their cup of coffee – the machine would do it all, including deciding how much coffee to dispense. The person is reduced to the role of a button pusher.

If you have followed my two examples you now have a good idea of what "dehumanizing technology" means. It is any artifact or procedure that either removes responsibility from human beings, treats human beings as if they were machines, or, as is usually the case, both. And if you are a sensitive Christian, you realize that removing responsibility from human beings or treating them as if they were machines is both unloving and unjust. It is unloving in that it does not evidence genuine care for persons. It is unjust in that it prevents persons from being the full, responsible selves, before the face of the Lord – which they are created and called to be.

Some people think that technological artifacts are morally neutral – that it's only in how we use them that questions of ethics and justice arise. I hope that from my two examples – the spooler machine and the coffee dispenser – it's clear that, far from being neutral, technological artifacts have moral properties designed into them. If so, we will also recognize that the more important properties an engineer might design into a machine are not efficiency and reliability – but love and justice. And when one recognizes that, one is well on the way to understanding what *obedient engineering* is all about.

CHAPTER 31

INTERNING IN BABYLON (OR, "DEVELOPING DORDT'S DILBERTS INTO DANIELS")

MARCH 2, 2001

A few weeks ago my engineering students asked me to help them reflect on what it might mean for a Christian engineer to work in a large, secular corporation. Since what I told them applies to any conscientious Christian who finds him or herself in American industry, I thought I would include it in this little book.

So, you are a Dordt engineering student! Good! You've been blessed with technical talents and interests and you find that when you put your hands and mind to the task, creatures that intimidate other people – like computers, diesel engines, and Laplace transforms – are subdued, and behave as they ought! Excellent! And now you are eager to take your place in the twenty-first century, to turn this confused technological world upside down, and to bring the light, the healing, and the renewal of Christ's Kingdom into the darkness and brokenness of the cyber-age! Wonderful! That's why you were born when you were and blessed with those talents that you have!

But you've got a problem. As you look forward to your new job after graduation, or to your summer internship if you are not yet a senior, you don't see "space, the final frontier," opening up before your eyes. And you don't see yourself – chief engineer on the real Starship Enterprise – setting off to "boldly go where no man" or woman has gone before. No, when you turn your realist eyes upon the future, what you see is . . . a cubicle! And when you look into the mirror, you see . . . well . . . you see Dilbert!

You're not the first to raise such concerns. Even before Dilbert was created, there were voices – Christian voices – raising similar concerns. Back in 1981, before anyone had graduated from Dordt's engineering program, while we were still in the process of starting it up, a Dordt alumnus wrote to me expressing his concerns. This fellow had majored in one of the sciences while at Dordt and had gone on to work as a technologist in the newly emerging digital electronics industry. But he was frustrated in his work. When he read the announcements about Dordt's new engineering program, he wrote to me the following:

> Every article I have read talks in glowing terms about reforming the various fields of technology, but no one says where the graduates are going to get jobs. I have no doubt that jobs are available, but if a person has a vision of the Kingdom of God and a perception that his daily work is to be done to God's glory, then he is not going to be content to work for IBM, Hughes, Hewlett-Packard, or Rockwell International. If this person has a family, he will face even greater struggles over relocating, finding time for his family, and finding a Reformed Church and a Christian school. In addition, it has been my experience that an Engineer . . . merely implements decisions which are made in management and cannot be a reforming force. It seems to me then, that Dordt engineering graduates would be helping a God-denying industry to grow and prosper. I wish this were not so, but if I honestly look at my own work, I can come to no other conclusion.

So, has it come to this? After all your hard work studying calculus, physics, chemistry, and those innumerable engineering courses, is this some dirty little secret that we have managed to keep from you? Is all you have to look forward to a fat paycheck to keep you comfortably numb while spending 50 hours per week for the next 45 years in your "drab, boorish, and perhaps dehumanizing" Dilbert cubicle?

Well, I won't lie to you. That's one possible scenario. But it's not the one your engineering professors envision for you. To view another we need to look at western civilization through the lens of history. Technology has always played a significant role in the development of cultures. From the Egyptian pyramids to the roads and aqueducts of ancient Rome, from the cathedrals of medieval Europe to the factories of the industrial revolution, for good or for bad, engineering design has been one force directing the history of the West. But it wasn't necessarily the major force. In the early middle ages the fledgling Church was of profound importance. And so Augustine was blessed with the talents needed – and called to that time to serve. In the sixteenth century the Church waned in influence, society became more differentiated, the state grew in importance, and the Lord called and equipped John Calvin to the task of giving direction. In the late nineteenth and early twentieth centuries when the failures of the Enlightenment were becoming increasingly apparent, the Lord equipped and called Abraham Kuyper to reform scholarship, statesmanship, and the Church. We are now at the very beginning of the twenty-first century and it is no mystery to anyone that the dominant cultural force is and will be technology. The major problems that the world will face in the next 50 years will be those brought about by misdirected technology. And you? Well, you *really were* born when you were and given the talents that you have in order to take your place in line with Augustine, Calvin, and Kuyper, and to address those twenty-first century technological problems that loom on history's horizon.

But wait a moment. Lest you triumphalistically run off to find yourself a white horse on which you can go tilting at cyber-windmills, or lest you despair at ever rising to the challenge and crawl off in resignation to your Dilbert cubicle, consider three other of the Lord's servants in history. The first was an arrogant teenager who got his eleven brothers so peeved at him that they sold him into slavery. He spent years in Egypt growing, learning, and maturing before he was able to fulfill the culture-forming tasks to which he was called. The second barely survived infancy. He spent almost his whole life in training in Egypt. It really wasn't until he was an older man that he even recognized and accepted his calling to lead God's people. The third was carried off into slavery in Babylon when he was a boy. But he remained faithful to the Lord and made the best of a bad situation, learning from the Babylonians so that he was fully prepared to serve when the Lord called him to influence the culture of his day. What I'm trying to say is that like Joseph and Moses in Egypt, and like Daniel in Babylon, you may very well need to spend some time in the Babylon of the twenty-first century – not yielding to it, but learning from it, and maybe even influencing it. But maybe not. Maybe you won't influence very much the first few corporations you work for. But keep at it. There will come a time, either in one of those big corporations or in a small organization run by yourself and like-minded fellow-servants, where you will see that the Lord has used you, that your life has been one of fruitful service, and that the world is different because you have been here. But it will never occur because you do it on your own. We always serve as members of Christ's body. So don't lose touch with the fellow servants you are privileged now to know. And your name may never be as well-known as that of Augustine, Calvin, or Kuyper. And that's OK, because the Lord prefers work that is done quietly and humbly to that which makes a splash. But be assured at those times when you sit in your cubicle in near despair: there's a world of difference between the Dilberts of this world and the Daniels. And you are a twenty-first century Daniel.

CHAPTER 32

NATURALISM AND THE DECLARATIONS
OF THE STARRY HOST

MAY 1, 2001

"By the word of the LORD the heavens were made, their starry host by the breath of his mouth." This sixth verse of Psalm 33, and many others from the Scriptures of the Old and New Testaments, asserts unequivocally that what we call "the universe," or "reality," is not self-existing, but is the creation of a self-existing person, the sovereign God of Abraham, Isaac, and Jacob, and the father of our Lord Jesus Christ.

In addition, the Scriptures make clear that the Word of God structures reality so that its essential meaning is service to its Creator. This is the point of Psalm 119: 89, 91, where we read: "Your word, LORD, is eternal; it stands firm in the heavens. . . . Your laws endure to this day, for all things serve you." In other words, creatures – including humans, stars, DNA molecules, lions, dandelions, and the dreams of little children as sleep brings rest to their bodies – are servants.

Further, Scripture reveals that God not only calls all things into being for service, but by his Word he upholds and directs all things. Speaking metaphorically rather than biologically, one might say that the Word of the Lord is the *very life* of all things. And that is exactly what the apostle John writes in the first chapter of his gospel, where we read, "In the beginning was the Word, and the Word was with God, and the Word was God. He was with God in the beginning. Through him all things were made; without him nothing was made that has been made. In him was life, and that life was the light of all mankind." This is even more explicit in some other parts of the Scripture. For example, Paul writes in Colossians (1:17) of Christ, the incarnate Word of God, "He is before all things, and in him all things hold together." And in his epistle to the Hebrews (1:3) he writes that Christ is "sustaining all things by his powerful word."

Thus we know from God's revelation in the Scriptures that he created all things, that he upholds and sustains all things by his Word, and that the very meaning of all things is service to the Creator. Moreover,

Scripture makes it clear that some of this is obvious even without God's special revelation. In the poetically beautiful words of the psalmist we read "The heavens declare the glory of God; the skies proclaim the work of his hands. Day after day they pour forth speech; night after night they reveal knowledge. They have no speech, they use no words; no sound is heard from them. Yet their voice goes out into all the earth, their words to the ends of the world" (Psalm 19:1–4). And, if you prefer expository prose to poetry, you can turn to Paul's epistle to the Romans (1:20) where we read, "For since the creation of the world God's invisible qualities – his eternal power and divine nature – have been clearly seen, being understood from what has been made, so that people are without excuse."

"Without excuse?" Without excuse for what? Well, for being blind. For the inability to see the obvious: that reality is dependent for its very being – both in terms of its coming into being and the sustenance of that being – upon God. That kind of blindness must have been a problem in Paul's day or he wouldn't have written about it. But it is even more of a problem in these post-Enlightenment times when the average person's worldview is clouded by an over-zealous faith in reason and scientific knowledge. That faith, which I will refer to as *naturalism*, believes that there exist natural laws – the most important of which are the laws of logic and rational thought – and that the universe is governed completely by these natural laws. Moreover, *naturalism* believes that we can come to understand these laws and thereby come to see how the universe works, and even how it came into being. In other words – the words again of Paul in his epistle to the Romans (1:25) –people have "exchanged the truth about God for a lie, and worshiped and served created things rather than the Creator."

In recent years many Christians have become enthused by a movement known as "intelligent design." That enthusiasm stems from the impression that intelligent design may provide a rebuttal to naturalism and an answer to the old creation-evolution debate – an answer that defeats evolutionism, the view that the universe and everything in it has come into its present form through its own natural law processes and without a creator. What intelligent design attempts to do is to point to irreducibly complex structures in the universe – structures like that of the eye in humans and animals – and use the mathematics of probability to argue that these structures could not have evolved by chance, governed only by natural laws. Such an argument shoots holes in evolutionary theory and suggests that, at least for some elements of the universe, there must have been the intervention of an outside agent, an agent who intentionally designed those irreducibly complex elements. For Christians, of course, that "outside agent" is God.

Now I respect the Christian commitment as well as the scholarly effort of those involved in the intelligent design movement. But I think the movement suffers from at least one serious flaw. That flaw is its perceived attempt to find confirmation of God's handiwork only in those parts of creation that indicate intelligent design. In order to find a common starting point for debate, it appears to yield the other parts of creation to the scientific explanations of the naturalists. But why should the lawfulness that we find in the creation, even the lawfulness involving stochastic − or, what we call "chance" − processes, declare the glory of God any less than irreducibly complex processes? It's almost as if the laws of mathematical probability are not creatures of God as much as is the human eye. I think it yields too much to the naturalists − all for the sake of argument.

Imagine that a friend walks up to you accompanied by a third person and attempts to introduce you to that third person. What if that third person, instead of greeting you, looks right past you and says to your mutual friend, "Wait just a minute! I see this supposed person you are trying to introduce to me. I can even hear him. But how do I know he's real? After all, sight and hearing are only sense impressions and can easily be deceived. Unless you prove to me with sound logic that he exists, I won't shake his hand or take any part in this introduction." Suppose then that your friend looks sadly at you, turns to the third person and says, "OK, I see your point. Wait here while I try to gather some irrefutable evidence for his existence − perhaps a birth certificate, a high school diploma, and the written testimony of some people whose existence you do acknowledge − then we can continue with this introduction." Wouldn't you feel a little "put out?" After all, you are standing right there. What greater proof could there be?

It's like that with intelligent design arguments. The whole creation is shouting of God's glory, from the Milky Way Galaxy to the electron, from the laws governing cell division to the laws of probability, from rulers of nations to babies waiting in their mothers' wombs to be born. But we are willing to concede to the naturalists that these things say nothing convincing about God, and we run off to find only that which will meet their rationally grounded tests for intentional, intelligent design.

It's all a matter of faith, of course. The naturalists have put their faith in the sovereignty of reason and what they call natural law. And thus they are deaf to the heavens as they declare the glory of God. But if your faith is in the Creator-Sustainer-Redeemer of all things, then even those things that, with our finite minds, we say happen by chance – I dare say, even chance itself! – will proclaim the wisdom of our Creator.

And lest you imagine that my example of your introduction to a person who doubts your existence is a bit far-fetched, let me remind you of one occasion when something not unlike that really happened. You may recall that after our Lord's death and resurrection, Thomas refused to believe that he arose unless he could see and touch the wounds from the nails and spear. Jesus invited him to do just that, and Thomas believed. But consider Jesus' final words to Thomas. He said, "Because you have seen me, you have believed; blessed are those who have not seen and yet have believed" (John 20:29).

CHAPTER 33

DISCERNING GOD'S WORD FOR
ENERGY USE IN 2001 AMERICA

JUNE 1, 2001

Well, here we are in the twenty-first century – 30 years later – and it's all happening again. Those of you who were around in the 1970s – particularly if you were driving a car back then – know what I'm referring to. It's the issue of energy use, of course. Back in the '70s we called it an "energy crisis." It occurred then in two waves, once in 1974 and again in 1979. We experienced gasoline rationing – red or green flags outside gasoline stations told you if they had any gas to sell. There were shortages of natural gas causing some industries to temporarily shut down. There were "brown-outs," where supplied electrical voltage was reduced by a small percentage – unnoticeable in most situations but devastating to electrical equipment that had to draw more current to make up for that reduced voltage. And, of course, there were increased energy prices. In 1974 gasoline went from 35 to 56 cents per gallon – a 60 percent increase. In 1979 it jumped another 60 percent to about 90 cents per gallon. Today, as history seems to repeat itself, the new Bush administration believes it's all a question of supply and demand – capitalist economics.

What should Christians believe? More importantly, how should Christians – who are called to be the light of the world – how should Christians live their lives when it comes to making decisions regarding energy resources?

To answer those questions, let's start by going back to the basics. *First*, consider that all things are God's creatures. Whether persons, animals, or plants; whether petrochemicals, rivers, or technological artifacts – all things owe their existence ultimately to God their Creator. *Second*, consider that to be a creature is to be called into being for service. Thus all things are servants. *Third*, consider that God's call to his servants is that they serve him *according to their kind*. Thus the sun serves in accordance with the nature of thermonuclear reactions. The winds and the rivers serve in accordance with how the Lord created them. Petrochemicals serve, or ought to serve, in accordance with the Word of the Lord for petrochemicals. People serve in many different ways, but all of them characterized by what it means to be made in the image of God. Technological artifacts like machines and factories and societal structures like economies and governments ought to serve according to their diverse, God-given natures.

If you have followed me this far, you are probably about to say, "Okay, but how do we know how such things as societal structures and technological artifacts are supposed to serve?" And that's a very good question, because it is we, God's image bearers, who are responsible for that service. But that gets us to my *fourth* basic point: we humans *are* responsible; it is *we* who are called to the task of stewardship with respect to the rest of creation.

So now let's get back to that critical question about how we can know *how* God calls different creatures to serve him. It's a critical question because it is the fundamental question of stewardship. It's a critical question because you can't simply go to the Bible and find the chapter and verse that speaks about how we should be good stewards of electricity or petrochemicals. It's a critical question because before you can properly answer it you must be gripped by a heartfelt concern for God's good creation and by an understanding of justice that includes nonhuman as well as human creatures.

So, assuming that we understand and embrace the four basic points I mentioned earlier, and assuming that we care about all God's creatures and share his outrage when they are prevented from being who he calls them to be, then we can move to the more specific question of how to gain insight into God's will for how they should serve him. Let's use energy resources as a specific example.

We gain insight into God's will for how energy resources should serve him by studying them in the light of Scripture. That's what Christ-centered natural science and Christ-centered engineering is all about. When engineering students at Dordt study thermodynamics, fluid mechanics, heat transfer, and solar energy, e.g., they learn a variety of concepts that enable them to be good stewards of thermal-fluid systems and the resources that enable those systems to function. One of the more obvious things that they learn is how and why we ought not to waste energy. Energy efficiency is one way in which we function as good stewards of energy resources. Obviously. But did you know that to be good energy stewards we have to go beyond simple efficiency and attempt to *match the nature of the energy resource with the nature of the energy application?* For example, powering a transportation vehicle requires concentrated energy and thus concentrated energy sources, like petrochemicals or storage batteries, must be used. But heating a house, or heating hot water, requires only diffuse, low-temperature energy. We ought therefore to find a diffuse, low temperature source for those kinds of applications. It doesn't make sense to heat your house to 70 degrees or your water to 140 degrees by burning a petrochemical at 2,000 degrees. It makes far more sense to use a diffuse source of energy like solar, which you can collect at 150 degrees or less using a fairly simple solar collection system.

As I reflect on how the impending energy problems of our day merely repeat the energy crisis of the 1970s, I cannot help but think of how the prophets of the Old Testament brought the Word of the Lord to Israel – and yet over and over again Israel went back to its wicked, selfish ways of living. So in closing this essay, here is one piece of concrete advice I hope will be taken seriously. The next time you have to make a decision involving the purchase of an energy system – a car, a hot water heater, an air conditioner, or other appliance, or the energy systems for your new house – don't let your decision hinge on whether the initial high price you pay for energy efficiency is made up by the low operational cost that results from that high efficiency. It's anathema to this generation, but it's none-the-less true: the Lord wants you to care more about his creation than about whether you save money.

CHAPTER 34

TECHNOLOGICAL FREEDOM

AUGUST 2, 2001

A few weeks ago my wife and I traveled to Colorado Springs to attend the annual convention of Christian Schools International. We drove our car 750 miles to get there, 750 miles to return, and in doing some sight-seeing put a couple hundred miles on the car while we were there. In addition to the Psalm 104 kind of pleasure that accompanies driving in the mountains, I was impressed – positively and negatively – with other aspects of the trip. I was surprised, for example, how the 75 miles per hour speed limit on long stretches of Interstates 80 and 76 allowed me get there much quicker than I had calculated using my rule-of-thumb average speed of 60 mph. I was frustrated with the bumper-to-bumper traffic in the cities of Denver and Colorado Springs – day *and night*: a function of the population density and the inevitable summer roadwork. Overall, though, I had more of a sense of freedom than frustration. The beautiful mountains, the good behavior of my middle-age but well-tuned car, and the wealth of new places to explore gave me a sense of harmony with the creation and a sense of freedom to be the person the Lord is calling me to be.

That sense of freedom, however, did contrast with what I read in the newspaper early last month when the recent decision of the New York State Legislature to outlaw cellular telephone use while driving was announced. Many people think that such laws curtail our freedom. Thus you could read a few columnists and a good number of letters to the editor decrying the New York State Legislature's decision. But I think they are wrong. In my judgment – my judgment as an engineer as well as a private citizen – laws such as the one outlawing cell phone use while driving are needed in order to guarantee our freedom. Let me explain.

We live in an age that is dominated by technological change. That technological change originates in the design work of scientists and engineers, and its impact is felt by the everyday citizen. One of central characteristics of this technological change is complexity. No individual engineer or scientist can initiate such change without the cooperation of other members of society. And no individual citizen can assess adequately or be completely prepared to respond properly to such change – it takes a community of citizens if there is to be any hope that the response to such change will be beneficial. Take the automobile and cellular phones as two separate examples for a moment. It took more than Henry Ford to bring what we today know as the automobile into existence. Without civil engineers to design the highways, statesmen to legislate appropriate traffic laws, medical and ergonomic experts to deal with the implications that mass travel by auto has for the human body, and business persons to make not only the automobile, but also gasoline and other supporting products available on a mass scale and at a reasonable price, "cars" would have remained a mere technological oddity like the personal submarine or, sadly, like vehicles that operate completely on renewable energy. And I've only mentioned a few of the numerous aspects of an automobile that make it complex and require the cooperation of different kinds of experts in society.

The cell phone is similar, only it's at an earlier stage in its technological history. Still, without the social, economic, and legal infrastructure to support its refinement and use, the cell phone would have remained at the stage it was 50 years ago: a toy functioning more in the imagination and play of children; or it would be associated with Dick Tracy and other comic strips, where it would function completely in the imagination.

Most of us realize the importance of a legal infrastructure and that laws licensing persons of appropriate age to drive, laws prohibiting drunk driving, and legislation requiring that cars meet minimum safety standards are valuable and actually promote the freedom of citizens whether or not they drive a car. Without the assurance that such laws will exist and be enforced, no conscientious engineer would consider designing such a dangerous artifact as the modern automobile. And any sensible citizen would avoid putting his life in jeopardy by driving one. But that's because the automobile has been around for a long time and the average citizen has a good idea of its possibilities and limitations. The cell phone as a mass marketed phenomenon is relatively new. Most people are ignorant regarding the nuances of how its properties will cause it to interact with other technological artifacts. Even the engineers who design cell phones are learning that the hard way – by social experiment. Enough evidence is available, however, for experts to assess the danger of using a cell phone while driving. That assessment is being communicated to governing bodies responsible for insuring freedom, and at least one of those governing bodies has now created legislation to restrict the unsafe use of cell phones.

True freedom is the opportunity to be who the Lord calls you to be. That opportunity can be curtailed for an engineer if there is no legal protection of the people he serves from the misuse of his designs. That opportunity is threatened for the average citizen when he has no legal protection against unsafe technology. Thus laws like the recent New York State law prohibiting simultaneous driving and cell phone use are obedient responses, on the part of government, to the Word of the Lord for society. The negative reaction to such laws is too often a disobedient response to God's Word that is rooted in individualism and self-centeredness.

Chapter 35

The Devaluing Potential of E-mail

May 2, 2002

A basic rule of economics is that resources are valued in proportion to scarcity. In the present age, where the god of economism reigns, this rule has powerful implications for all areas of life, not only those that are justifiably qualified as essentially economic in nature. We are all aware that as crude oil becomes increasingly scarce, its value increases, and thus the price of gasoline goes up. Likewise, because the skill to play basketball at a level of professional competency is scarce, professional teams seek after the best college players with lures of absurdly high salaries. And so it is that almost anything that is scarce will have a high monetary value attached to it – even such trivia as comic books and baseball cards.

The inverse of this rule holds true as well. The easy availability or ubiquity of any resource devalues that resource in economic terms. Even though fresh water, for example, is vital to our lives, its relative easy availability – at least at this point in history for most of North America – has rendered it cheap, both in terms of its monetary cost and the wasteful habits of North Americans who run it down the drain without the slightest qualm.

This basic economic rule has analogies in other areas of life. One's aesthetic appreciation of a single flower, for example, is diluted when, like a florist, one is surrounded by flowers, day in and day out. And although it is awful to admit it, our caring for our neighbors is devalued when our neighbors are transformed from a single family in a nearby town whose house has burned down to the millions of families on the other side of the world whose lives are threatened by war or famine.

Devaluing also occurs with modes of communication. When in 1964 I was a freshman in college, away from home for the first time, I greatly valued the letters that I received, faithfully written each week by my parents. It was the scarcity of letters – or of any kind of mail, for that matter – that helped significantly to elevate their importance. Had my parents been in the habit of writing a letter every day, or possibly ten, or fifty letters each day, I surely would not have valued them as much.

Well, you know where I'm going, don't you. That's right, I want you to consider with me for a moment the characteristics of e-mail. Tomorrow morning I will go into my office and find about ten new e-mail messages. If I were to wait until 5:00 p.m. to check my e-mail – and if tomorrow were a typical day – I would find about 100 e-mail messages waiting for me. That's over 500 e-mails in a week – over 25,000 e-mails in a year. They have become ubiquitous, ever-present, and profligate. They have become legion. Like the Sorcerer's Apprentice, I beat back the e-mail waves with the delete key or the archive command . . . only to find myself drowning as my head sinks beneath the tide of e-mail-conveyed information.

E-mail, like one's voice, like pen and paper, and like the telephone, is a medium of communication. And like every other medium it has particular properties that characterize it and influence what it conveys. I want to suggest that one property of e-mail is that it facilitates the proliferation of messages – and thereby cheapens the content of those messages. The very same message, constructed from the same words in the same order, will generally be more valued by the reader if they are conveyed by paper and ink through ordinary mail channels, than if they are sent via e-mail. That's hard for most people to understand because in our post-Enlightenment culture we associate messages with abstract, rational propositions that we believe have meaning "in themselves," un-affected by their context. But that is an untruth. When I receive a message from another person it arrives in a visual and sonic context. It has shape and texture about it and it has a greater or lesser association with other messages that I receive around that same time. My eyes are cast at a particular angle and move in a specific way as I read it. And when I'm done reading it, I must do something with it because it always comes in some physical form: glowing phosphors on a cathode-ray tube, semi-conductor switches in the memory chips of my computer, or a piece of paper overlaid with typed or handwritten symbols. And that should help you see a second property of e-mail that influences the way we value the messages it conveys. The medium of e-mail is transient, ephemeral. If you don't make a concerted effort to save the message to a trustworthy electronic storage medium, or print it out and save it in your file cabinet, it will disappear. And that property of impermanence that characterizes the medium cannot help but influence the way the message is perceived by the one receiving it.

In summary, the sheer volume of e-mail renders any one e-mail mes-sage of less value than, perhaps, it ought to be worth. And e-mail's transient or ephemeral character reinforces that devaluation. Like water running carelessly from an open faucet and down the drain while we brush our teeth, so the flood of e-mail messages runs through the cyberspace of our lives, flushing with it communications that, at another time, in another form, may have held greater meaning for us.

For *Plumbline*,* I'm . . . well, wait a moment: just one last thought. If by some chance your technophilic sensibilities have been offended by my arguments. If you think that my reasoning is wrong-headed. Let me know. Send me an e-mail message.

---

* Well, after all, these were originally *Plumbline* radio addresses. And although I've gone through and edited the word *Plumbline* to "essay" for most of them, I decided to leave this alone. You see, the proper way to end a *Plumbline* is to say, "For *Plumbline*, I'm D. Livid Vander Krowd" (or something like that).

CHAPTER 36

BETWEEN TECHNICSM AND BIO-ROMANTICISM

NOVEMBER 1, 2002

In the book of Genesis we read that the Lord creates the heavens and the earth, fills them with life, and then installs his image-bearing servants to oversee and manage it all as it develops throughout time. "Develops throughout time" – I use that phrase deliberately, because the nonhuman creation has a dynamic, unfolding character to it – much like the way a child does, developing from infancy through childhood and adolescence, with all that is potential in the child being slowly and progressively actualized during different stages of maturation. And like the developing child who is in need of parents, the creation is in need of a steward for it to unfold and develop properly, according to the Word of the Lord. That is why the Lord charged his image-bearing servant with, what Reformed Christians call today, *the cultural mandate*.

But there is a problem, one that deeply affects the cultural mandate. God's image-bearer – humankind – rebelled, sought to live independent of God, independent of his fellow servants, and even independent of God's good creation – though such an attempt is absurd, since humans are finite and temporal creatures, and very much an integral part of creation. The result of this rebellion, this fall into sin, has been a curse upon the relationship between humankind and the rest of creation. Creation resists humankind's attempts to steward it. And humankind, in turn, has attempted to subjugate the creation to its own selfish ends.

But the Lord, in his mercy, has provided a means for healing the brokenness that has resulted from this rebellion. In Christ, not only are we humans restored to our right relationship with God, but also to our right relationship with the rest of creation. That's why Paul writes in Romans (8:19–21) that

> . . . the creation waits in eager expectation for the children of God to be re-
> vealed. For the creation was subjected to frustration, not by its own choice,
> but by the will of the one who subjected it, in hope that the creation itself
> will be liberated from its bondage to decay and brought into the freedom
> and glory of the children of God.

That liberation, however – that healing of creation and the rift between the human and nonhuman parts of creation – will not be completely accomplished until the Lord returns. Until that time Christians are called to be stewards, earthkeepers, concerned for being both agents of healing and reconciliation on the one hand, and faithful stewards and culture-formers on the other hand. In other words, our task is both to help liberate creation from its bondage so that it can be what the Lord calls it to be, and to unfold and develop creation according to the original cultural mandate so that creation can become what the Lord calls it to become. In simple everyday terms, we have the complimentary tasks of conservation and development.

But here lies a problem and the central point of this essay. Due to our finitude and our sin, we Christians have developed a tension between these two aspects of our stewardship role. Some of us are motivated primarily (and perhaps triumphalistically) by a vision for the unfolding and developing of creation as we imagine how it may have occurred had the fall into sin not occurred. Claiming every square inch of creation for the King of creation, we seek to boldly develop it in whatever direction seems possible. Unconsciously influenced by the unbiblical spirits of technicism and economism – spirits that lie to us by saying that whatever is technologically possible ought to be pursued, and that the accumulation of wealth is the only road to happiness – we overlook our calling to heal and to conserve. On the other hand, some of us are motivated primarily by our awareness of the history of abuse that has characterized humanity's relationship to the nonhuman creation. Unconsciously influenced by the unbiblical spirits of naturalism and romanticism – spirits that lie to us by saying that nature is good "in itself," independent from humankind, seeking to convince us that all development is an inherently evil, arrogant attempt to impose our selfish wills on something that would be much better off without us – we then deny our mandate as culture formers.

The only solution to this problem starts by listening to God's Word. And, of course, that means studying his whole Word, from Genesis to Revelation, and studying the creation in light of that Word. In concluding this essay, however, let me refer to just those two key verses in Genesis where our role with respect to the nonhuman creation is first described. In Genesis 1:28 we read, regarding humankind:

> God blessed them and said to them, "Be fruitful and increase in number; fill the earth and subdue it. Rule over the fish of the sea and the birds of the air and over every living creature that moves on the ground."

But what is meant by "Be fruitful," "fill the earth and subdue it," or "Rule over?" We gain insight into what those phrases mean when we read the creation account in Genesis 2:15:

> The Lord God took the man and put him in the Garden of Eden to work it and take care of it.

To "work it and take care of it": the meaning here is clearly one that will not support our ravaging the creation for our own selfish wants. Neither, however, will it support our placing the nonhuman creation on a pedestal where it is best left untouched by human hands – or footsteps, as some are inclined to say. The nonhuman creation is given its proper due when, like an abused human child, it is treated with the tender care that brings healing and renewal, and when it is enabled to flower and develop before the face of the Lord – and according to his Word.

CHAPTER 37

HOW MOUNTAINS AND HILLS PROSPER

JANUARY 1, 2003

If you ask a child what he might like for his birthday, you will usually get a fairly detailed answer. If you ask a college student what they hope to do in the decades after graduation, you may encounter a bit more uncertainty, but you will eventually get some kind of answer. And if you ask a middle aged man or woman what it might mean for them to prosper in the time left to them, you likely will get a very detailed answer, based on a wealth of past experience. But what if you wanted to know how a mountain might prosper and flourish. What if you were interested in the future of a particular valley? Mountains and valleys do not speak in the language of words. To know whether the land is prospering or how it might prosper in the future, we humans need wisdom. We need to gain understanding of the land and its relationships to the myriad creatures – including humans – that it supports. More importantly, we Christians need to listen to the Word of the Lord as he instructs us in our role as caretakers of his creation.

But listening to the Word of the Lord is not always easy. We have a tendency to be deafened by the noise of pagan ideologies, like technicism and economism on the one hand, or the worship of the nonhuman environment on the other. That deafening noise tends to distort our hearing so that we either dismiss the land as having only human-centered, utilitarian value, or we idolize it, imagining that it would be far better if humans never set foot upon it. Both of these ideological extremes prevent the land from prospering and deny to humans the caretaking role assigned to them by their Creator.

In this essay I would like for us to listen carefully to the Word of the Lord and to think about what it might mean for our relationship to the land today. First consider the Old Testament prophet Hosea who was called to prophesy against the unfaithfulness of Israel during the reign of Jeroboam. Almost immediately, in the second verse of the first chapter of the book of Hosea, we read of a strange command whereby the Lord says to Hosea: "Go, marry a promiscuous woman and have children with her, for like an adulterous wife this land is guilty of unfaithfulness to the LORD" (Hosea 1:2). Now granted we all find it strange that the Lord commands Hosea to marry an promiscuous woman. But isn't it also rather strange that he accuses *the land* of being unfaithful to the LORD – unless, of course the Lord is using *the Land* as a metaphor for his people Israel. But we soon find out that that is not necessarily, or at least not exclusively, the case. Listen to these accusatory words from the first three verses of chapter 4:

> Hear the word of the LORD, you Israelites, because the LORD has a charge to bring against you who live in the land: "There is no faithfulness, no love, no acknowledgment of God in the land. There is only cursing, lying and murder, stealing and adultery; they break all bounds, and bloodshed follows bloodshed. Because of this the land dries up, and all who live in it waste away; the beasts of the field, the birds in the sky and the fish in the sea are swept away. (Hosea 4:1–3)

Clearly there is a relationship between humans and the land that goes far beyond what a typical industrialist or environmentalist might imagine. It seems that to air and water pollution – conventional ways of poisoning the land in the twenty-first century – we ought to add moral pollution. Somehow the land is affected as much by our abuse of the creatures it supports – like our fellow human beings – as it is by our directly abusing it.

Now consider the Old Testament prophet Ezekiel who prophesied around the time of Israel's exile. In the first section of the book of the Bible bearing his name he brings the Word of judgment to Israel and Judah, foretelling the destruction of Jerusalem and the exile of the Jews. But Ezekiel went along with the Jews into exile and, as is recorded in a later section of the book, was given the task of bringing God's Word of comfort and reconciliation to his people Israel *and to the land*. Consider the following seven verses from Chapter 36, where Ezekiel is told by the Lord to actually speak to the land. Here the Lord says to Ezekiel:

> . . . prophesy concerning the land of Israel and say to the mountains and hills, to the ravines and valleys: "This is what the Sovereign LORD says: I speak in my jealous wrath because you have suffered the scorn of the nations. Therefore this is what the Sovereign LORD says: I swear with uplifted hand that the nations around you will also suffer scorn. But you, mountains of Israel, will produce branches and fruit for my people Israel, for they will soon come home. I am concerned for you and will look on you with favor; you will be plowed and sown, and I will cause many people to live on you – yes, all of Israel. The towns will be inhabited and the ruins rebuilt. I will increase the number of people and animals living on you, and they will be fruitful and become numerous. I will settle people on you

as in the past and will make you prosper more than before. Then you will know that I am the Lord. I will cause people, my people Israel, to live on you. They will possess you, and you will be their inheritance; you will never again deprive them of their children." (Ezekiel 36:6-12)

In these verses we find the answer to the question I asked at the outset of this essay: How might a mountain or valley flourish? In the Old Testament, when the Lord looks with favor on the land, it results in the land being plowed and sown and it supporting numerous humans and animals. To see a land that is blessed is to see a mountain or valley that exists in a vigorous sustaining relationship with the plants, animals, and people that live upon it. This is no excuse for ravaging the land, for stripping it of its wealth, or smothering it under cement or asphalt. But neither may we consider the land "untouchable." **The notion that wilderness is somehow the ideal for the land is just as wrongheaded as the idea that the land is there for humans to do whatever in their self-centered greed they want with it.**

The mountains and valleys prosper, according to the Word of the Lord, when the people that inhabit them become numerous and prosper. Perhaps in another essay we should ask what it means for people to truly prosper.

CHAPTER 38

"IT'S NOT YOUR FATHER'S ULCER"

FEBRUARY 3, 2003

Not too long ago I had the opportunity to join President Carl Zylstra on his Friday morning "Conversations" radio program on KDCR. I was one of three guests who were invited to talk about biotechnology. Northwest Iowa has been the birthplace of a number of biotech companies during the past two decades and now that the industry in general is getting a lot of press, the focus of many questions is right here in our own back yard – a community that has, in some important ways, retained its unique identity even while the rest of North America has been radically transformed by the many changes of the last century. But now it seems that modern technology has not only come to Sioux County, it is actually gestating here. Thus many questions are being raised. Technophiles envision all sorts of good things for Northwest Iowa, suggesting that Siouxland soon will give birth to an industry that will usher in a salvific cornucopia of riches and fame. Technophobes, on the other hand, ponder this gestation period and think of the film *Alien*, where demonic alien creatures use humans as incubators for their offspring. At the birth of the offspring – a violent bursting forth from the chest of the victim – the human is destroyed. Likewise Sioux County is envisioned losing its distinctiveness and becoming infected with the multitude of social evils that plague the great cities of the western world.

But technology is neither a savior nor a demon. Neither is it just a neutral tool, however. Like any human activity, it is a way that we respond to our Creator, and therefore it can always be characterized as either obedient – that is, in accordance with the Word of the Lord – or disobedient – in violation of God's Word. So as I prepared for that "Conversations" program, I was particularly concerned to walk a bit of a tightrope between those two contrary perceptions. Biotechnology in particular has become quite controversial in the last decade. Issues such as cloning and genetic engineering have raised alarm bells in the minds of many Christians. On the other hand, the benefits of biotechnology are already very significant and the future potential for good is enormous . . . or is it? That's the question I was asked a few months ago by a colleague of mine and its answer is at the center of this essay.

Has technology really been as beneficial as many technophiles suggest? What about advances in transportation technology, in communication technology, or in medical technology? It's possible to argue that the advances have brought at least as many problems as benefits. For example, the automobile has radically increased the mobility of people over the days when the horse and the bicycle were the only non-pedestrian modes of transportation. But with that benefit has come injury and death due to accidents, pollution of the air over large cities, the wasteful use of nonrenewable petrochemicals, and myriad social changes, many of which entail disruptions in societal units ranging from the family to the state. Advances in communication technology have brought about similar problems. While we may be able to instantly communicate both orally and visually over long distances, consider the impact of the telephone, the television, and the Internet on our young people. It seems to me that one of the greatest challenges for any young couple as they contemplate raising children is how to resist turning those children over to our modern Molek, the television, which seductively offers help in babysitting but which potentially enslaves our children with the addictive and hallucinatory powers that we associate with illicit drugs.

So it was that when my colleague challenged me regarding the pros and cons of modern technology – suggesting that the cons greatly outweighed the pros – it gave me pause to think. I'm an engineer and have always been fascinated by technical systems, both theoretical and practical. Could it be that I've been too dazzled to see that the balance was far and away on the other side; that is, with modern technology being much more of a deficit than an advantage in living an obedient Christian life? But recently an incident occurred that re-established for me what is, I hope, a properly balanced view of modern technology. Over the Thanksgiving holiday it became obvious that something was not right with this old body of mine. I couldn't climb a set of stairs without getting out of breath. I couldn't even get out of bed and into the shower in the morning without feeling like I was going to faint. A visit to the doctor indicated that my blood supply was deficient by almost 50 percent of its primary component: hemoglobin. Before I knew it I was in the stainless steel environment of a hospital, surrounded by a myriad of fascinating instruments for measuring and viewing things that I never realized could be measured and viewed. By the end of a single day my problem had been diagnosed, some donated blood had brought my hemoglobin to a safe level, and I was on my way home with a couple of prescriptions that would quickly bring me back to full health. All told I lost two days of work, no worse than if I had had a bad bout with the flu.

Now the reason why this incident impressed me so much was because I was diagnosed with the same ailment that had afflicted my father a generation ago: I had an ulcer in exactly the same part of the stomach that he did. The difference, however, was that when my father experienced the symptoms I experienced, it took him out of action for six weeks – two of those six weeks being spent in the hospital. Worse than that, for the last 40 years of his life he was placed on a very restrictive diet – a diet that was both unpalatable and unhealthy. My only dietary restriction is that I shouldn't take aspirin anymore.

So does this experience justify my enthusiasm for technology? Well it certainly confirms to me that there is much good in technology and that our role as stewards and caretakers of creation involves the further development of *biotechnology* in particular. But I'm still cautious. All of creation, including us, was brought into being in order to serve others. And by "others" I mean to include the nonhuman as well as the human creation. I'm cautiously enthused about the prospect of Northwest Iowa becoming a light to the world by bringing healing, peace, and shalom where there is sickness, suffering, and strife. There is no other justification for developing biotechnology.

CHAPTER 39

HOW WE COME TO KNOW OUR MISERY

SEPTEMBER 1, 2003

This past summer I drove over 4000 miles during a two-week trip to Florida and New Jersey. In doing so I deepened my understanding of Question and Answer #3 of the Heidelberg Catechism. Some may remember that question. It asks, "How do you come to know your misery?" And the answer: "The Law of God tells me."

Now when the authors of the Heidelberg Catechism wrote that particular question and answer, they had in mind Paul's discussion in the book of Romans (3:20) where, referring to the Old Testament law – particularly the Ten Commandments – he writes, "through the law we become conscious of our sin." My deepened understanding of the Heidelberg Catechism, in connection with my summer driving experience, is also related to God's law. But it is not so much the Ten Commandments as the law of God that the psalmist describes in Psalm 19:7–11, where we read:

> The law of the LORD is perfect, refreshing the soul. The statutes of the LORD are trustworthy, making wise the simple. The precepts of the LORD are right, giving joy to the heart. The commands of the LORD are radiant, giving light to the eyes. The fear of the LORD is pure, enduring forever. The decrees of the LORD are firm, and all of them are righteous. They are more precious than gold, than much pure gold; they are sweeter than honey, than honey from the honeycomb. By them is your servant warned; in keeping them there is great reward.

What the psalmist is describing is the law of God that is evidenced in the structure and behavior of the world around us. That's why he begins the Psalm by writing "The heavens declare the glory of God; the skies proclaim the work of his hands." Today it is not only the sky that evidences clearly God's law, it's everything on earth as well – particularly those technological products that he has called us, his image bearers, to help shape. You see, I'm an engineer. As such, I'm particularly interested in the way technological products work, behave, affect the environment, and influence us. I've come to understand that if you want to design an automobile engine to have 200 horsepower, you need to understand chemical kinetics, the physics of motion, Newtonian dynamics, and the behavior of deformable bodies under conditions of applied stress. All of these things are topics that undergraduate engineers study. And all of these things are expressions of God's law for his creation. If I were the one assigned to write Psalm 19 today, I might begin it by saying "Newtonian dynamics declares the glory of God, chemical kinetics proclaims the work of his hands." It may not have the same poetic ring to it, but it's just as accurate as those original words of the psalmist.

What interests me particularly is the way in which the properties of technological products influence us. In many instances, as one author has put it,* they "bite back." In other words, after we design some technological device to do something constructive, we discover, to our dismay, that it does other not so constructive things as well – it has unintended consequences. But we also realize that those unintended consequences might have been foreseen if we had spent enough time studying the law of God for his creation and trying to anticipate the behavior of our technological creations in response to that law.

But I'm digressing a bit. Let me tell you about my 4000-mile trip. Early on a Saturday morning in July my wife and I set out in our car for Nashville, Tennessee. We arrived in about 12 hours. After visiting Andrew Jackson's home, we continued on our way to Orlando, Florida, where the Christian Schools International Convention was held this year. Following the convention we drove up the east coast to Williamsburg, Virginia, where we spent a day touring its historic colonial village. Next day we continued north to New Jersey, to visit my wife's parents. After a few days there we headed back toward Iowa, stopping briefly in South Bend, Indiana, for a night's rest and to visit a niece of ours who had recently moved there. Finally, on a Saturday, two weeks after we began our trek, we arrived back home in Sioux Center – the odometer of the car registering over 4000 miles more than it did when we left.

While the trip was productive – and, at times, interesting – it was one of the worst traveling experiences that I can remember. And it was so because of the way the technological properties of automobiles, roads, and clocks interacted with my fallen human nature. In a nutshell, the problem was traffic. But "traffic" is simply a word that we have invented to describe a complex concept by which we understand a set of peculiar, interactive properties of automobiles, roads, and clocks. The clock, because of the way it very precisely measures and communicates time,

---

* Tenner, Edward, *Why Things Bite Back: Technology and the revenge of unintended consequences* (New York: Knopf, 1996).

amplifies my sensitivity to the efficiency of my actions. When on a trip, a clock is a constant reminder of the proportion of time I spend just sitting in the car driving compared to our time visiting Andrew Jackson's home, or simply relaxing in a hotel room. And that awareness is not something I need to have amplified. You see, I'm not a very greedy person when it comes to money or material things. But when it comes to time, it seems I can never have enough.

The automobile is a technological product that inhibits us from seeing our neighbor as we should: as an image-bearer of God whom we are called to serve. The geometric, motion, and physical properties of the automobile are such that they isolate us from our neighbor. When driving, my neighbor is transformed into "that car over there that is in my way," or "that crazy driver who doesn't know how to signal," or worse. When it comes to the struggle to "love one's neighbor as oneself," I am, at best, only an average Christian. I try. Sometimes I succeed. More often I don't. The properties of the automobile create an enormous barrier to my success by inhibiting me from seeing my neighbor – those occupying the other cars in traffic – as whole image-bearers of God.

Finally, the nature of the highway system is such that only a fixed number of automobiles can travel on it at the same time. Of course, far more can travel on four-lane or six-lane highways than on two-lane highways. But regardless of the number of lanes, once the automobile density reaches a certain level, you have "traffic." Put that density together with a desire to minimize time on the road and you have competition with your fellow image-bearers. Isolate those image-bearers from yourself by having all of you situated in different automobiles, and you have the perfect conditions for the cultivation of – that's right – "road rage." Experience road-rage, and – if you have any Christian sensitivity whatsoever – you will come to know your own sin and misery very quickly. So how do you come to know your sin and misery? The law of God, by which he structures the heavens and the earth and all things natural and artificial, tells you of it.

Next trip, maybe, I'll take a train.

CHAPTER 40

HIGH-TECH WORSHIP?

MAY 6, 2004

How much does modern communication technology drive worship style? Should it? Would a better understanding of God's Word for technology and his Word for aesthetics serve to enlighten our path toward biblically obedient worship practice? A few weeks ago I read a new book by Quentin Schultze (Grand Rapids: Baker, 2004) titled *High-Tech Worship? Using presentational technologies wisely.* In that book Schultze tries to weigh the pros and cons of using technology – particularly presentation technologies such as Microsoft's PowerPoint – in developing liturgies for worship services. The book is short, barely a hundred pages, and rather on the light side. It offers no profound analysis of either technology or worship. But it does suggest some good reasons why using such things as screens and computer projectors might enhance worship, and some valid warnings regarding how the same technology might seriously detract from worship.

Schultze says that "worship has its own, God-ordained purpose," and that is "gratefully expressing gratitude to the Creator in the most fitting means possible and inviting God's grace to move us to sacrificial lives of service" (23). In other words, when groups of Christians together give thanks to God for his goodness to us and ask him to help us become better servants, then we are – by definition – engaged in worship. That sounds reasonable to me.

Technology, however, is *also* something we do. It's the way that we respond to God's command to rule over and serve the rest of creation by shaping it, usually with tools and procedures, for practical ends and purposes. On the other hand, the word "technology" is also used to describe the products of that human activity, and that's what is meant by the phrase "presentation technologies" – artifacts that result from using tools and procedures with the objective of producing something practical, something that will help us communicate better.

One very important point that Schultze makes in his book is that "worship is necessarily liturgical *and* technological" (17). A liturgy is a prescribed form, a ritual, or an order for worship. To enable that liturgy, some kind of technology is necessary: a building, seats, a pulpit, lighting, musical instruments like an organ, and books like Bibles and hymnals are typical and familiar to most of us. And they are all the products of technological activity. So it is not unreasonable to assume that modern communication technology – like sound systems and video projectors – might enable our worship just as well as pulpits and hymnals.

However, it's very important to realize that technological artifacts are not neutral. They always predispose us to behave in particular ways. Back in the 1960s it was the automobile that greatly shaped my world and the activities of my adolescence. Consider how the cell phone shapes the behavior of teenagers today.

So the question I want to ask in this essay is: how do modern communication technologies like sound systems and video projectors affect our worship? In what behavioral directions do they predispose us to move as we seek to express our gratitude to God and seek his grace to move us to sacrificial lives of service? Do they enhance or inhibit our efforts? And how do they do so?

That's a question that will take more than an essay to answer. But let me suggest just one answer, just one way among many that modern technology influences us as we worship. Consider that all artifacts – technological and otherwise – have an aesthetic character, some more significant than others. And, according to Calvin Seerveld, a Christian philosopher of aesthetics, the central core of aesthetics is *allusivity*. In more straightforward language, *things subtly suggest to us other things that we have previously encountered*. People have differing tastes, differing aesthetic sensitivities, because they have different life experiences. We develop those sensitivities as we grow and mature. Now the aesthetic quality of a worship service is very important – not all-important, but very important. Therefore, the artifacts used in a worship service will subtly suggest things to the participants. Clearly the pulpit, the baptismal fount, and the character of the music all subtly (or not so subtly) suggest things to the worshippers.

What do the elements of presentation technology suggest to the worshipers. When you walk into the sanctuary and see a large, central projection screen, what does it suggest to you. All right, that was a leading question. But think about it. What does a microphone and amplified sound suggest? If these things can be arranged so as to suggest an attitude of thankful service to God, well then, good. But if these things suggest to us something else – dare I say, "profane things" – then we ought to think twice regarding how we use them.

I don't want to suggest that we ought not use modern technology in our worship services. What I *do* want to suggest is that we take great care regarding *how* we use such technology in worship. Otherwise I fear we will bring down on ourselves the judgment that the Lord spoke to his people Israel through the prophet Amos, where he said:

> "I hate, I despise your religious festivals; your assemblies are a stench to me. Even though you bring me burnt offerings and grain offerings, I will not accept them. Though you bring choice fellowship offerings, I will have no regard for them. Away with the noise of your songs! I will not listen to the music of your harps. But let justice roll on like a river, righteousness like a never-failing stream!" (Amos 5:21–24)

CHAPTER 41

BIOTECHNOLOGY, CREATION CARE, AND SUSTAINABLE AGRICULTURE

DECEMBER 1, 2004

"The earth is the LORD's, and everything in it, the world, and all who live in it. . . ." Those are the words of the psalmist at the beginning of Psalm 24. In Psalm 8:3–8, we read,

> When I consider your heavens, the work of your fingers, the moon and the stars, which you have set in place, what is mankind that you are mindful of them, human beings that you care for them? You made them a little lower than the angels and crowned them with glory and honor. You made them rulers over the works of your hands; you put everything under their feet: all flocks and herds, and the animals of the wild, the birds in the sky, and the fish in the sea, all that swim the paths of the seas.

There are numerous other places in Scripture where these basic themes are repeated. They are the foundation of agriculture. And in this essay I want to argue that they are the bases on which to build a proper understanding of biotechnology as well. God cares for and delights in the whole of his creation and he has charged his image-bearers to do likewise. Humans are called to appreciate, care for, maintain, unfold, and bring healing to the nonhuman creation. A stress within agriculture on stewardship, creation care, and sustainability is thus a biblically defensible stress. The perception of many, however, is that biotechnology runs counter to his stress. Is that a correct perception? Is biotechnology in opposition to stewardship, creation care, and sustainability within agriculture?

One reason for the perception that biotechnology runs counter to the biblical mandate to care for creation might be called "guilt by association." There is no question that post-Enlightenment humanity has, in general, looked upon the nonhuman creation as "neutral matter" that is here for the sake of humankind and with which humans can do whatever they please. Modern technological systems – including agricultural systems – do seem to treat the nonhuman creation with impunity. Biotechnology has emerged and is being developed from within that post-Enlightenment, human-centered tradition. But the callous and avaricious treatment of what God has called us to care for is sin, is disobedience to the Word of God, and thus runs counter to the order of creation, the way God created and expressly desires all things to be. While biotechnology certainly can be practiced sinfully, it can also be practiced obediently – unless there is something inherent in the practice of transforming the biotic creation that is, at root, evil.

Some Christians do believe that transforming the biotic creation is interfering with nature and, if not evil, ought to be always a path of last resort. That belief, however, rests on two assumptions – regardless of whether those assumptions are stated or even consciously held. The first is that the nonhuman creation is unaffected by humankind's fall into sin and the curse upon that sin, except through human action. The second assumption is that human action is independent of the natural order of things; it is autonomous, free, and arbitrary relative to the law structure for the biotic and physical worlds. But these assumptions owe more to naturalism than to a biblical, Christian view of creation. The "thorns and thistles" of Genesis 3:18 and the "bondage to decay" that we read of in Romans 8:21 may be metaphors, but they are metaphors that point unambiguously to a creation that suffers apparent structural brokenness. The word "apparent" is appropriate here because the creation is also faithfully upheld by its Creator – the *undergirding* structure *is* secure. But human perception is incapable of perfectly distinguishing the undergirding structure from the apparent structure. Thus science and technology must work with a creation that is both orderly and broken. The second assumption, that human action is independent of the natural order of things, is a distortion of the truth that humankind is unique among created beings in having been given freedom and responsibility. Human beings are subject to God's normative Word as well as to his determinative, structuring Word. And there is a significant difference between subjectivity to God's normative Word, which confronts us with what we *ought* to do, and his determinative Word, which structures our lives without giving us a choice in the matter. That difference is one that distinguishes clearly humans from animals, plants, and physical things and the passage I read earlier from Psalm 8 alludes to it. However, that difference does not transcend creation, and thus a biblically consistent view of creation is one that is integral or holistic. Human activity is different from nonhuman activity in that much of it is normed rather than determined. But human and nonhuman activities, taken together, constitute creational activity. To separate human activity from nonhuman activity and to see it as somehow "unnatural" is to fall prey to a dualism much like the

Cartesian mind-matter or the medieval soul-body dualisms.

Despite the reality that biotechnology has emerged from a human-centered cultural milieu wherein stewardship, creation care, and sustainable agriculture have often been trumped by an emphasis on profitability, it need not be so. Biotechnology, just like agriculture in general, *is* a form of stewardship, and as such, may be practiced either obediently or disobediently. It is our task as Christians living at the beginning of the twenty-first century to discover and articulate God's norms for biotechnology − his Word for how we ought to do biotechnology − and then to live in harmony with those norms.

CHAPTER 42

ICE HOUSE TECHNOLOGY

OCTOBER 7, 2005

Over the summer I had the opportunity to visit Sunnyside Manor House, the home of Washington Irving in Tarrytown, New York, overlooking the Hudson River. Washington Irving, you may know, is the author of such Colonial American stories as "Rip Van Winkle" and "The Legend of Sleepy Hollow." He was born in 1783 and became America's first full-time author; that is, he was the first person on this side of the Atlantic Ocean to be able to support himself solely by his writing. In 1835 he purchased a small, stone cottage on ten acres of land. He remodeled it and added to it until it became an architectural model of the Romantic Movement in nineteenth century America. It's not a castle or a mansion. In many ways it's quite modest. On the other hand, Washington Irving combined some of the best elements of romanticism with an enthusiasm for modern technological advancement. In addition to beautiful landscaping, walking paths, pastures, and gardens, the house had running water and a closed-fire cooking stove in the kitchen – advanced technology for the early nineteenth century.

But the example of innovative technology that caught my eye and remained in my thoughts for weeks afterward was found outside the house and just beyond the kitchen yard. At first glance it looked like a tool shed of sorts, except that it was fairly large and built very sturdily. When I opened the heavy wooden door and looked inside I noticed that the floor was well below grade level and filled with straw. The walls of the structure seemed thicker than the walls in the house. It became apparent to me that this small structure was designed to resist heat transfer. Looking at the small map of the Sunnyside Manor property that had been provided, I discovered that I was puzzling over an *ice house*, a piece of technology that the modern refrigerator displaced before I was born. "And we're glad it did," you might say. "After all, who would want to go through all the trouble of gathering and storing ice in the winter just to refrigerate one's food, or have a little ice for one's drink on a warm summer day?"

Ah, but wait. There's a lesson here. It may very well be that by relegating the ice house to the forgotten and neglected world of pre-modern technology, we have lost some valuable insights into how our world works; insights that might serve us well in these days of soaring gasoline prices and petrochemical depletion. You see, the basic principle by which the ice house functions is called seasonal thermal storage: making use of the earth's orderly rhythms for the purpose of ordering our lives – in the case of the ice house, gathering low thermal energy material in the winter, when it is plentiful, so as to meet our need for it in the summer, when it does not naturally exist. We don't consider this principle much today because we are impatient; we want instant gratification of our desires. And, at least for the past century, we've had the technological resources to provide that instant gratification when it comes to thermal energy storage. The electric refrigerator is a prime example.

But our impatience is catching up with us. It just so happened that the week I was visiting Tarrytown, New York, was the same week this past summer that the price of gasoline took an inordinate jump. I had bought gasoline in Sioux Center for $2.09 per gallon before I left, and then only four days later, found that the price in New York was $2.69 per gallon. This certainly made me reflect on the finitude of our petrochemical resources and, combined with Washington Irving's ice house, resurrected my enthusiasm for the idea of seasonal storage of thermal energy.

Let me make a prediction. Unlike the new houses that are being built now in Sioux Center, each with its individual heating system, homes in the future will be arranged in clusters, and each cluster will have a centralized heating system that will serve all the homes in that cluster – and that heating system will combine seasonal storage with solar energy collection, using petrochemical combustion only as an emergency backup. Consider a typical new housing development in Sioux Center where you have houses on parallel streets, each with about a 100 foot frontage. Each house has one house on either side and another house behind it. A cluster of six houses would have 300 feet of back yard in common. What I imagine is that where those back yards meet there will be, submerged beneath the ground, a well-insulated reservoir, filled with water. There will also be a system of solar collectors on the roofs of those six houses, and possibly above the reservoir area where the back yards intersect. All year round, but particularly during the summer, solar energy will be collected and stored in that reservoir as hot water. Because the reservoir is large, submerged well beneath grade level, and extremely well insulated, it will not lose much heat, even during the winter. Rather, heat from that reservoir will be pumped into the six homes. At first it will be used to increase greatly the efficiency of heat pumps, devices that use a modest amount of electrical energy to move much greater amounts of thermal energy from one place to another. But as electrical energy gets more and more expensive, the thermal energy from the reservoirs will be used – supplemented by solar energy – to directly heat the six homes in the cluster.

The first homes and buildings to use solar assisted seasonal storage will not only be demonstrating good stewardship of resources and energy independence, they will be harmonizing their functioning with the natural cycles we find in the creation around us. I think that is the kind of technology that the Lord intended us to develop all along. Isn't it interesting that such a technology was available to, and used by, Washington Irving back in the early 1800s. And isn't it sad that our desire for instant gratification has allowed that technology, like Rip Van Winkle, to fall asleep. Well, it's time to wake up!

CHAPTER 43

THE SANCTITY OF HUMAN LIFE

APRIL 1, 2005

What is meant by the phrase, "the sanctity of human life"? Traditionally it has been used as a kind of argument against abortion and euthanasia. More recently it has played a role in discussions surrounding issues such as in vitro fertilization, germline genetic manipulation, stem cell research, and cloning. Most recently it has been raised during discussions of the Terri Schiavo controversy, the case of a woman judged by medical experts to be in a persistent vegetative state, who was kept alive by means of a feeding tube.

The word "sanctity" is defined as the quality or condition of being considered sacred and therefore inviolable. We use the word to draw a line; a boundary that separates ordinary things from those that we believe ought not to be treated in the same way that we treat ordinary things. For example, Christians believe that there is a radical distinction – a "boundary," if you will – between God and his creation. Most of us who are Christians working in the natural sciences believe that it is legitimate to scientifically investigate any part of the creation but that it would be wrong to think that we can use the scientific method to "study" God. What kind of boundary might exist between "human life" and the rest of creation that would make human life sacred and inviolable? Before we can answer that question we need to more carefully come to understand what we mean by human life.

Life is the central meaning of the biotic aspect of creation. Just as motion is the central meaning of the kinematic aspect, mass/energy is the central meaning of the physical aspect, and perception is the central meaning of the sensory aspect, so "life" describes what is central to the biotic dimension of the creation. As such, life is no more sacred than sensation (which depends on life but is not reducible to it) or matter/energy (on which life depends but to which life is not reducible). So it is that biologists have no qualms probing and questioning the nature of biotic creatures from the simplest plants to the most complex animals. Animals are subject to the laws for the sensory aspect of creation – in other words, they have feelings – and by this might be differentiated from plants. And this, no doubt, plays a role in the special kind of care with which we treat animals. There is a society for the prevention of cruelty to animals but, as far as I know, no society for the prevention of cruelty to plants. But life – in and of itself – is not a quality that renders the living creature "sacred" or inviolable. We have no misgivings about ending the life of the mold that grows in our shower stalls or the mosquitoes and flies that invade our houses. We have only a little anxiety about ending the lives of deer that overpopulate the countryside and become a threat to automobiles on the highways. We do concern ourselves with the death of fish when we see them mysteriously washed up on the shore and with the extinction of animal and plant species. But these latter concerns are rightfully rooted in the norm of stewardship, our desiring to care for God's good creation. They have nothing to do with life per se.

Human life, on the other hand, is obviously different. When that phrase is used it means more than simply the biotic aspect of human beings. It means human existence in its totality, physical and biotic, as well as aesthetic, social, ethical, and faith (just to name six). And, I would argue, it means more than that. The phrase "human life" suggests a relationship of wholeness and totality that transcends the nonhuman creation by relating itself covenantally, as image bearer, to its Creator. The unique covenantal relationship that humans and no other creatures have means that human life is normed, that human life entails responsibility; in other words, that all of (human) life is religion.

I believe that it is this covenantal relationship of image-bearing responsibility to the Creator that gives meaning to the phrase "the sanctity of human life." Thus it is not the biotic aspect of any being that is sacred, it is rather human existence in its wholeness. The inviolability of "human existence in its wholeness," I suggest, is thus "an image" of the inviolability of the Creator. It alludes to the radical Creator-creature distinction and suggests a boundary, not equivalent to, but not unlike the boundary that prevents us from making God the subject of scientific investigation or technological manipulation. This boundary does not prevent us from scientifically investigating the physical aspect of humans, the biotic aspect of humans, the aesthetic aspect of humans, or even the faith aspect of humans. On the other hand, attempting to *reduce* any human being to a biotic object, an economic object, a political object, or even a faith object, crosses that boundary.

Abortion and euthanasia are wrong because they represent a community of image bearers of God treating other image bearers of God as if they were mere matter, or mere biotic things – like the mold that grows in your bathroom shower stall. That violates the covenantal relationship between God and his human creatures. But too often Christians are using simplistic arguments when they debate with those who are in favor of abortion or euthanasia. Too often the arguments are based merely on the biotic dimension of what it means to be human. And then we are just as guilty of violating that covenantal relationship because it is not life that is sacred, but that multidimensional covenantal relationship.

Finally, consider some of the new developments in biotechnology that are creating anxiety among many Christians and non-Christians alike. Too many arguments against germline genetic manipulation in humans, against stem cell research, and against cloning are based simplistically upon the notion that the biotic aspect of human life is sacred. Well, it's not. Otherwise we would have to argue against all kinds of medical treatments. If we are to show that these new biotechnological procedures ought to be banned, we will have to do so by showing that they violate the covenantal relationship that humans have with their Creator. I've heard a few good arguments of this sort; but far too few. It's time that we Christians begin studying biotechnology and giving direction to an area of life that has the potential for great good, but also for great evil if left in the hands of the secularists.

Chapter 44

Mediating Toys

September 8, 2006

One of the more fascinating aspects of grandchildren is that they give you the opportunity to imagine a very different perspective on the world – a perspective only allusively similar to one that is all but lost in the bowels of your own memory. Last week Mason and Malia – the one set of twins among our twelve grandchildren – turned four years old. Four is a delightful age. At four, one is old enough to talk, to display a wide range of emotions, and to interact with one's environment in varied and complex ways. And that's what set my mind to pondering last week: observing Mason and Malia as they scurried about their house, as they ate parts of the meal prepared for those of us who had gathered in celebration, as they opened their birthday presents, and as they interacted with people who they see only every so often. Observations like these cause you to reflect on the past. I thought about my three sons and tried to imagine any commonality of experience between their birthdays and those of Mason and Malia. I found that strangely hard to do. For a fleeting moment I tried to imagine the experiences of my younger brothers, which, as *first-born* son, I had the privilege to observe. But memories of fifty years ago are, at best, very hazy. My own memories of being four-years-old are wisps of uncertainty; confused with retrospections of photographs that I've viewed since that time.

After a brief and futile effort to recall past experiences, I began focusing on *how* Mason and Malia were interacting with their four-year-old world. I realized that something very important was going on before my eyes. The worldviews of these two little grandchildren of ours were being formed by that interaction. But – and here is the thought that fascinated me most – the interaction they were having with their world, the world of Sioux Falls, South Dakota, in August of 2006, was *mediated* by the objects in that world; objects that were for the most part technological artifacts.

No one interacts directly with the world, of course. Our experience of the world outside ourselves is mediated by the sensory apparatus the Lord has given us: eyes, ears, nose, taste-buds, and nerve endings in our fingertips. But beyond that our senses interact with a plethora of technological artifacts. The various nooks and crannies on the two levels of the house in which Mason and Malia spend a good portion of their time provide their worldviews with a rudimentary understanding of the concept of environment. The minivan with its car seats, and their perceptions – through the windows of that van – of the streets and highways of Sioux Falls, provide them with their first, primitive understandings of transportation. And the electronic world into which they've been born, replete with audio and video experiences unimagined fifty years ago, furnish their worldviews with models and ideas about the nature of communication. And all of these concepts, models, understandings, and ideas become the constituent elements of their worldviews. For Mason and Malia, growing and learning will be a process of refining and revising those elements. But these earliest, most primitive elemental formulations will define how they think and imagine the world for years to come.

Now if my belaboring the modern technological habitat of our twin, four-year-old grandchildren makes you pine for a simpler time, when our interaction with the world was mediated by things natural rather than things technological, then stop right now. There never was a time when our interactions with the world were not mediated by things technological. When I think of my baby brother, back in 1957, I remember a child in a crib, his three month old eyes seemingly fascinated by the mobile attached to the side of the crib. The mobile, the crib, and the walls and ceiling of the room he shared with my parents were artifacts that mediated my little brother's experience of the world, at least for a portion of time back in those early months of his existence. If that's not convincing, then think of Abraham and Isaac and the world of almost 4000 years ago. Little Isaac's worldview was shaped by experiences mediated by such technological artifacts as a tent, the fabrics used for clothing and other purposes, and the bowls, utensils, and tools used to provide and serve food. Even the animals, already domesticated by earlier agricultural technology, helped shape the elements of Isaac's worldview.

The point of all this reflection on the mediating function of technological artifacts is to emphasize how important – in ways we don't often consider – the technological artifacts are that surround us. They shape our view of the world and who we are – from a very early age. It's important then for us to reflect on that shaping function. We take artifacts like televisions and automobiles for granted. In fact, we would be lost without them. But, I would argue, our relationship with and reflections upon those artifacts are far too uncritical. And if you look carefully, you will find that those two artifacts are powerful tools in the reshaping of our environment and of our collective worldviews – a reshaping of which we are, for the most part, oblivious. Scariest of all, those artifacts are shaping the worldviews of four-year-old children; the Lord's children; of whom he said to us, "Let the little children come to me, and do not hinder them. . ." (Matthew 19:14). That worldviews are shaped by artifacts is inevitable. That's the Lord's responsibility because that's the way he created us. But the shaping quality of those particular artifacts, like televisions and automobiles, well, that's our responsibility. And as I think of Mason and Malia I can't help but remember another statement that the Lord made about four-year-olds when he walked this earth. He said, "If anyone causes one of these little ones – those who believe in me – to stumble, it would be better for them to have a large millstone hung around their neck and to be drowned in the depths of the sea" (Matthew 18:6).

CHAPTER 45

BEING HUMAN IN THE YEAR 2029

OCTOBER 6, 2006

It was Sunday morning and Ben sat and listened intently as his pastor went on describing the unique status of humankind in the world. He had heard Psalm 8 read many times before, even preached on once or twice. But somehow those verses about being made "a little lower than the angels" and being made "rulers over" everything that God created never had the impact that they had on this particular Sunday morning. Of course, in his twenty-one years on the planet, Ben had never had a week quite like the last one.

Ben was an engineering student at a big university, finishing up his fourth and final year and eagerly anticipating graduation. But in this his final semester the university had thrown him a curve ball. After three and a half years of primarily technical courses, he was now sitting through a course called "Issues in Technology." At first he found it to be quite interesting – even entertaining. The first assignment was to read through most of Ray Kurzweil's *The Age of Spiritual Machines: When computers exceed human intelligence* (New York: Viking, 1999). Its subtitle should have provoked Ben a bit more than it did. But then again, who reads subtitles? What caught Ben's interest was Kurzweil's discussion of Moore's Law: the observation that the number of transistors and the speed of computer chips were doubling every two years. Kurzweil was predicting that by the year 2009 "personal computers with high-resolution visual displays [would] come in a range of sizes, from those small enough to be embedded in clothing and jewelry up to the size of a thin book" (277). He was also predicting that a computer could be bought for $1000 in the year 2019 that would have the computational ability of the human brain (278). None of this seemed too far-fetched to Ben. After all, he could remember when, as a little kid, he played with computer games that were incredibly crude compared to the sophisticated computer gaming that now occupied much of his spare time. Ben reasoned that if technology advanced as much in the next ten years as it did in the last ten, well, then of course computers are going to be incredibly powerful – powerful in ways that maybe even Ray Kurzweil could not imagine. Of course, there were those other kinds of weird predictions that Kurzweil was making for the world of 2019. He predicted that virtual artists, with their own reputations, would be emerging in all of the arts (279). And weirder still was his prediction that people would begin to have relationships with automated personalities – in other words, robots – and use them as companions, teachers, caretakers, and lovers (279). Although that sounded like tacky science fiction to Ben, he accepted it with just a bit of a smirk. After all, his professor seemed to take the book and all of Kurzweil's predictions quite seriously.

But the next book Ben had to read for the course was *Flesh and Machines: How robots will change us*, by Rodney Brooks, the Director of the Artificial Intelligence Lab at MIT and the Chair of iRobot Corporation. Brooks sounded very much like Kurzweil, except he began to raise some puzzling questions, particularly for students like Ben, who wanted to remain faithful to their Christian beliefs. At one point in the book Brooks wrote the following confession. He said,

> When I was younger, I was perplexed by people who were both religious and scientists. I simply could not see how it was possible to keep both sets of beliefs intact. They were inconsistent, and so it seemed to me that scientific objectivity demanded a rejection of religious beliefs. It was only later in life, after I had children, that I realized that I too operated in a dual nature as I went about my business in the world.
>
> On the one hand, I believe myself and my children all to be mere machines. Automatons at large in the universe. Every person I meet is also a machine – a big bag of skin full of biomolecules interacting according to describable and knowable rules. When I look at my children, I can, when I force myself, understand them in this way. I can see that they are machines interacting with the world.
>
> But that is not how I treat them. I treat them in a very special way, and interact with them on an entirely different level. They have my unconditional love, the furthest one might be able to get from rational analysis. Like a religious scientist, I maintain two sets of inconsistent beliefs and act on each of them in different circumstances.

Ben considered himself to be a "religious scientist," well, at least "a religious engineering student." And he didn't like the idea of maintaining two sets of inconsistent beliefs.

But those two sets of inconsistent beliefs were only the beginning of the confusion for Ben. Shortly after reading the Brooks and Kurzweil books, Ben's professor assigned the class to read a paper by Bill Joy, the co-founder of Sun Microsystems and the designer of three important microprocessor architectures. Joy had grown up with a vision of the future based on Star Trek, the well-known science fiction television show featuring Captain Kirk, Mr. Spock, and the starship Enterprise. The world of Star Trek was a comforting one for Joy, a world of ethically inclined humans who dominated the future and governed themselves by norms such as "the prime directive," which prohibited them from interfering in the development of less technically advanced civilizations. But in Bill Joy's paper, published in *Wired Magazine** under the title, "Why the Future Doesn't Need Us," he raises the question, "Will our robot descendants be ethical?" Bill Joy's real concern about the future, as robots develop greater and greater intelligence and as humans begin to download their minds into robot hardware, is that we might lose our essential humanity as we evolve into our robotic creations. But that's when Ben took a step back and began to re-think all of the ideas he had accepted about science, technology, and the technological development that would occur in the next decade or so. He had to ask the more basic question, "What does it mean to be human?"

And that brings us back to the sermon on Psalm 8:5–6, 9 that he had listened intently to on that Sunday morning.

> You have made them a little lower than the angels and crowned them with glory and honor. You made them rulers over the works of your hands; you put everything under their feet. . . . LORD, our Lord, how majestic is your name in all the earth!

As Ben reflected on the writings of Ray Kurzweil, Rodney Brooks, and Bill Joy in the light of Psalm 8, he realized that his technical engineering education left much to be desired. Perhaps he should have gone to a Christian college after all. But it was too late for that. It began to look like this one course in "Issues in Technology" was about to require of him research and study that even his professor didn't plan on.

---

* August 4, 2000; see http://www.wired.com/wired/archive/8.04/joy_pr.html.

CHAPTER 46

THE HEARTH VS. THE CENTRAL HEATING SYSTEM
DECEMBER 8, 2006

In 1949 my parents bought a new house – their first house – and, with my baby brother and I, began life as homeowners in the Northern New Jersey suburbs, less than ten miles from the George Washington Bridge. It was a small "Cape Cod" style house, and definitely bare-bones. No finished attic, no finished basement, no garage, it had just four small rooms and the smallest bathroom you can imagine. Those four rooms – five, if you call the closet-sized bathroom a room – all opened into a tiny hallway that connected them. In the middle of the 4-foot by 5-foot floor of that hallway was a grate; an opening in the floor that you could manually adjust to allow air from the basement to rise into the main part of the house. And if you haven't already figured it out, the purpose of that grate was to enable heat from the coal-fired, pot-belly stove to rise into and warm the four small rooms of the house. After a few years my parents replaced the pot-belly stove with a gas-fired central heating system. But I still remember those early years in the house. The Kopper's Coke truck would come a couple times a year and deliver coal, down a chute, through the basement window, and into the one corner of our basement that served as a coal bin. Every night during the winter my father would carefully tend the fire in the stove, adjusting the damper so that the coal would burn slowly and last the night. Just recently my mother told me of one cold winter night when my father was away and she had to tend the fire. She shoveled in the coal but forgot to adjust the damper. Within a couple of hours the fire in the pot-belly stove was burning so hot that it turned part of the grate in the floor red hot. That was the closest we ever came to having the house burn down.

My reason for describing the heating system in my parent's first house has to do with a course I am preparing to teach next semester that deals with the philosophy of technology. In an article that I will assign my students to read, the author describes Albert Borgmann's philosophy of technology, particularly the distinction he makes between what he calls "technological things" and "technological devices." Technological things are simply the tools and artifacts that we design and use to do particular kinds of tasks and achieve particular ends. Technological devices, on the other hand, "disburden" people from a particular task and from the context that surrounds that task. And Borgmann uses the central heating system as an example of a technological device.

Without a central heating system, one has to concern oneself with the temperature in various areas of a building, with how heat is distributed from one or more sources – like from a coal-burning, pot-belly stove – to those various areas, and with maintaining one or more sources of heat. Using a coal-burning, pot-belly stove to heat your house takes focused attention, careful planning, and regular maintenance of the stove. A central heating system, on the other hand, requires almost no attention, no planning (other than setting the thermostat), and only annual or semi-annual maintenance. Borgmann argues that "technological devices" not only disburden us of time-consuming and dangerous tasks, they also disengage us from the world in a way that, paradoxically, reduces our freedom by alienating us from the environment in which we were created to live and flourish. The result, he says, is boredom.

Now, I can't say that I have ever really been bored with my central heating system. But a few years ago I designed and built a house that I wanted to conform to what I believe is the Lord's will for obedient living at this time in history and at this place on earth. In particular, I was concerned with designing a heating system that made stewardly use of energy resources while providing maximum comfort for the inhabitants of the house. And that turned out to be quite a challenge. To make a long story short, I found that I had to specify a highly efficient central heating system that was augmented by two or three efficient local heating systems. In other words, I chose a very efficient, but otherwise rather traditional gas-hot-air central heating system, coupled with gas fireplaces in two different rooms. And to be honest, my motivation for specifying the gas fireplaces was as much aesthetics as it was stewardship or comfort.

This winter is the sixth that my wife and I are spending in our new house; and so far we are quite pleased with both the physical comfort and aesthetic pleasure that the heating system has provided us. But just a few weeks ago I read in the newspaper how average homeowners spend $850 per year to heat their homes in the winter. That helped confirm for me that our heating system has also been quite stewardly, for my annual heating bill has always been far less than that. But what really pleased me was reading that article about Albert Borgmann's philosophy of technology, about the difference between "things" and "devices," and about how we need to design technological artifacts that do not disengage us from the world around us. You see, much of the aesthetic pleasure and physical comfort of my heating system comes from the gas fireplaces that augment my central heating system. They create a kind of zone heating in the house that contributes to efficiency (because only one area is heated at a time), comfort (because of the direct radiant heat that they provide), and aesthetics (because of the direct view of the fire). But now I realize that they have also created for me a certain level of "engagement." They need to be turned on and off manually, and the fire level and fan speed need to be adjusted. It's not the kind of burden my father had in shoveling coal into his pot-belly stove, stoking the fire, and being vigilant regarding the possibility of a house fire. But it does require my direct interaction with this part of the house.

So, while I don't have anywhere near the ideal heating system in my house, I have one that is relatively stewardly, aesthetically pleasing, comfortable, safe, and technologically engaging. And as my students will learn this coming term, those are norms for obedient technology.

CHAPTER 47

WHAT IS A "CREATURE?"

MARCH 9, 2007

A little over a week ago I had the privilege of giving a set of lectures at Calvin College in Grand Rapids, Michigan. The lectures dealt primarily with a Reformed, Christian approach to science and technology education. At one point I was describing how I try to enable my students in a course on fluid mechanics to understand the subject matter in the context of creation-fall-redemption. That is, I try to get them to see that fluid mechanics – its principles, theories, and artifacts – is an integral part of God's good creation; that fluid mechanics is a response to God's call that we care for creation, unfolding it and developing the potential that God has put within it. The implication here is that the products of fluid mechanics technology – products such as pumps and compressors – are likewise "creatures"; i.e., that they are a part of God's good creation that have arisen over time as a result of technological activity.

Well, the idea that a pump is a creature seemed a bit odd to some in my class, and so we took some time to discuss the creatureliness of pumps and other things that come about through the developmental efforts of humankind. So, in this essay I would like to consider the question, "What is a creature?" And I hope you will come to understand why I think it is important for us to see the products of technology, as well as naturally occurring animals, plants, and minerals, as "creatures."

When it comes to understanding important concepts, I don't always rely on the authority of the dictionary. Dictionary definitions always arise from within a given culture and carry with them the worldview baggage that is peculiar to that culture. In this case, however, I found the primary definition of the word "creature" to be quite helpful. *Webster's New World Dictionary* defines the word creature first as "anything created, animate or inanimate."[*] I also have a version of Microsoft Bookshelf dictionary on my computer. The first definition it gives for

---

[*] *Webster's New World Dictionary of the American Language*, 2nd College Edition (New York: World, 1972).

the word "creature" is "Something created."**

The real authority, of course, for understanding important concepts – or anything else in creation, for that matter – is the Word of God as we find it in the Scriptures of the Old and New Testaments. Now you're not going to go to the Bible to find out the nature of a pump – at least not in terms of its technical functioning. However, it is in the light of God's Word that I have come to understand that a pump is not simply some religiously neutral device dreamed up and constructed by humans. Rather, it has a relationship to God similar to that of the stars in the sky, the fish of the sea, and the birds of the air. Consider the words of the psalmist in Psalm 119:89–91:

> Your word, LORD, is eternal; it stands firm in the heavens. Your faithfulness continues through all generations; you established the earth, and it endures. Your laws endure to this day, for all things serve you.

In other versions of the Bible that last line reads, "for all things are your servants." Now, to be sure, the psalmist had no idea of what a pump might be. And there were not as many technological artifacts in his time as there are in ours. But that last line is rather all-inclusive. *All things* **are** God's servants. In the psalmist's day that would include the sun, moon, and stars, human beings, the wind and the rain, as well as the tools crafted by the people of that time to do the work that they were called to do.

But if that is not quite convincing enough, consider these words of the Apostle Paul from his letter to the Colossians as he describes the Lord Jesus Christ:

> For in him all things were created: things in heaven and on earth, visible and invisible, whether thrones or powers or rulers or authorities; all things have been created through him and for him. He is before all things, and in him all things hold together. . . . For God was pleased to have all his fullness dwell in him, and through him to reconcile to himself all things, whether things on earth or things in heaven, by making peace through his blood, shed on the cross. (Colossians 1:16, 17, 19, 20)

Notice how Paul stresses the comprehensive character of Christ's lordship and of his redeeming work. The phrase "all things" is repeated, and then, just in case we are a bit dense, Paul emphasizes that he *really means* "all things" by using the words "in heaven and on earth, visible and invisible." Clearly the very nature of existence is servanthood. Everything that exists – good things, bad things, natural things, technological things – all things exist in order to serve God.

** *Microsoft Bookshelf 98*, Microsoft Corporation, 1987.

So why does it sound strange to refer to a pump as a "creature"? A pump is part of creation, part of "all things." And "all things" have come into existence by the Word of God. All things exist with the mandate to serve God. Now, to be sure, pumps were not created by God "in the beginning." But neither was that tree in the park that gives shade in the summer. The tree came about through the natural process of biotic reproduction. And the pump came about through the natural process of technological development.

What's that you say? Technological development is not natural? Well, maybe not in exactly the same sense as biotic reproduction. But when we read the book of Genesis we learn that humankind was created to work the creation. That implies opening up the possibilities that God has put in creation: what we call "technological development."

But I think there is more to the concern about calling a pump a creature than the distinction between biotic reproduction and technological development. I think the concern has to do with the realization that in technology, we too often mess things up rather badly. We create monsters, creatures that we can't imagine as faithfully and obediently serving the Lord. Ah, all the more reason why we need to see our technological artifacts as creatures. For in doing so we become convicted of our responsibility for designing these artifacts in ways such that they render him obedient service and not disservice.

We need to explore and develop our understanding of technology much more in the light of God's Word. Then one day we will be perfectly comfortable in paraphrasing the psalmist in Psalm 19 and saying: "The heavens declare the glory of God; the pump proclaims the work of his hands."

CHAPTER 48

DEHUMANIZATION

MAY 4, 2007

In 1932 Aldous Huxley wrote *Brave New World*, now a classic. It's a novel that imagines a future society where the masses of citizens are provided with all their basic needs and with a series of diversions and entertainments ranging from an advanced form of video called "the feelies" to an allegedly "safe" recreational drug called "soma." Those in charge of the society are an elite group of technologists and technocrats. The theme of the novel is "dehumanization": the transformation of free human beings into a class of passive, pleasure-seekers who ask no questions, cause no incidents, and yield to the authority of the technocrats. For a number of years we have asked Dordt students to read *Brave New World* in a course that they take just before they graduate. It's one of the ways in which we attempt to sensitize them to the dangers of technicism and authoritarianism, and to have them reconsider their responsibility for being the image-bearing servants that the Lord calls them to be. In a course that I teach to engineering and computer science seniors, we also deal with the issue of dehumanization. But there I am teaching students who actually have the potential for becoming the technocrats of tomorrow. So we look at the issue not so much from the point of view of the average citizen as from the point of view of the engineer and technologist who has a responsibility for respecting – and, in fact, enabling – the image-bearing qualities of fellow citizens. For example, we probe the characteristics of the technological workplace, with particular emphasis on such artifacts as the assembly line.

But recently the term "dehumanization" has developed a new context – that of biotechnology, genetic engineering, and cloning. Last week I had my class read an article entitled "Preventing a Brave New World" by Leon Kass, the former Chairman of the President's Commission on Bioethics. In that article Kass suggests that "Human nature itself lies on the operating table, ready for alteration, for eugenic and psychic 'enhancement,' for wholesale re-design."* This is a different view of "de-

---

* Kass, Leon, 2001, "Preventing a Brave New World," in Winston and Edelbach,

humanization" than Huxley had in mind, and it forces us to ask some probing questions about what it means to be human, and therefore, what "*de*humanization" might possibly mean.

In Huxley's dystopian vision, and throughout recorded history, "dehumanization" has always started with the assumption that a given human being is being prevented somehow from exercising his or her essential humanity. In the secularized West, that "essential humanity" has been defined by the notion of autonomous, individual freedom. In contrast, a biblical view of what it means to be human is rooted in our creation as image-bearing servants of God and our responsibility for exercising that image-bearing character in obedient response to God's Word. The secular view is thus a distortion of the biblical view, retaining the notion of responsibility, but directing it toward self rather than toward God.

But today there is a new view of what it means to be human: what we might call a "naturalist" view. It assumes that the essence of humanity is found in our genetic make-up. What distinguishes us from other living creatures and what gives us the qualities that we associate with being human, is the particular chemical sequences in our genetic structure. On the basis of this kind of genetic reductionism, it is no wonder that concerns are raised about our essential humanity. With this view, "dehumanization" implies not the loss of human characteristics, but the transformation of the human into something other than human.

I think it's important to reject not only genetic reductionism, but also the view of dehumanization that is based on genetic reductionism. It is possible to dehumanize persons by treating them as economic objects, as sex objects, as political objects. These are the common forms of dehumanization that we find in slave camps and factories, in brothels and Miss America pageants, and in concentration camps. As Huxley describes so well in his book, it's also possible to seduce persons into dehumanizing themselves by using amusements such as "the feelies" and drugs such as "soma" – or our contemporary forms of the same: video, computer games, and the plethora of inebriants ranging from hard drugs to gourmet coffee. But as mere creatures, as servants of the Creator-Sustainer-Redeemer of the universe, we do not have the power to become something other than human. Our humanity may be shaped differently in the future. Just consider how dumbfounded a first century Christian would be to see us with our eyeglasses, artificial hearts, cell phones, and personal automobiles. But no matter how we change our outward appearance, or

---

Society, Ethics, and Technology, 3ʳᵈ Edition (Belmont, CA: Thomson-Wadsworth, 2006).

even the shape of our genetic makeup, we will remain God's image bearing servants, standing before him in either obedience or disobedience.

It's important, in these days when genetic engineering and cloning threaten dehumanization beyond even that of concentration camps, assembly lines, and Barbie Dolls, to remember that we will always remain God's children. And because of that, dehumanization will remain dehumanization, the treating of human beings as if they were not God's image bearers. We ought to fear the wrath of God if we attempt to reduce people to their genetic make-up. But we need not fear losing our essential humanity. In this regard it is interesting to consider Jesus words in Matthew 10:28 when he sent his disciples out into the cruel world of first century Palestine. He said, "Do not be afraid of those who kill the body but cannot kill the soul. Rather, be afraid of the One who can destroy both soul and body in hell."

CHAPTER 49

THE VATICAN'S "10 COMMANDMENTS FOR DRIVERS"

JULY 13, 2007

Earlier this summer the Vatican released a 36-page document titled "Guidelines for the Pastoral Care of the Road." Dubbed "the 10 Commandments for drivers" by the media, it was reported by many of the news services and became the subject of not a few editorial pieces. I found it interesting for a number of reasons: not the least was the traveling that my wife and I did this summer and the opportunity we had to compare flying with driving.

The most intriguing aspect of the document, however, is the way in which the media has reacted to it. As might be predicted, it was described as "unusual" by some news services and presented with just a bit of condescension and humor by most. In the modern world, with its wall of separation between church and state, between faith and reason, and between the private, sacred sphere and the public, secular sphere, it is taken as unusual for a sacred institution – the Roman Catholic Church – to be making pronouncements on secular topics like highways and the public transportation system. Moreover, one might have good reason to suspect that beneath that surface of condescension and humor lies a wee bit of animosity: resistance to the notion that anyone – least of all a religious institution – should tell us how we ought to drive our cars. After all, the driving of one's automobile perhaps epitomizes what it means to be a free individual in American Society.

But the puzzlement of the media is just one reason why I was delighted to learn of these "Guidelines for the Pastoral Care of the Road." Another is the fact that we people of the Reformation have often blamed Roman Catholics – particularly St. Thomas Aquinas – for the theological codification of the sacred-secular dualism. Yet here we have the Vatican presenting us with an example of how all of life is religion! God expects righteousness and obedience from us even as we drive our cars: perhaps especially as we drive our cars. Isn't it ironic that the particular sector of Christ's body that has most visibly embodied the sacred-sacred dualism has recently been wonderfully active in proclaiming Christ's lordship over the everyday areas of our lives? Previous proclamations by the Pope or by the Catholic bishops dealing with issues of war, poverty, materialism, and economic injustice have won my admiration. This latest proclamation dealing with obedient automobile driving has added to that.

But let me describe very specifically why I appreciate the Vatican's ten commandments for drivers. As an engineering instructor I teach a course every year called "Technology and Society" to our senior engineering students. In that course we deal with what we call "the bi-directional character of technological artifacts." What that means is that the products of technology, on the one hand, embody the values of the society that produced them and, on the other hand, reinforce those values as they are put to use in that society. The automobile and the supporting road system is perhaps the clearest example of this. Just think of how the SUV or the luxury car embodies particular North American values. What's important for us to realize is that they not only embody those values, they reinforce them as well.

But here's where my traveling experience this summer comes in. Early in the summer I was frustrated by a number of airline trips that I took. Delays, cancellations, long waits in airports, and cramped airplane seating all contributed to making the idea of highway travel seem much more palatable than it normally is. Then just two weeks ago my wife and I drove from Sioux Center to New Jersey and back, stopping in Sheboygan Wisconsin along the way out in order to visit family. When the time for the trip came, I dreaded the long drive and what I anticipated would be hopeless traffic going around Chicago on Route 80. On the whole, however, this turned out to be a relatively pleasant trip. A few minor delays due to road construction and a total of 48 hours in the car, but compared to many highway trips I've taken, this was among the least problematic. Yet, I was still very much aware of the properties of the automobile that I was driving and how those properties – dispositional properties, as we call them in my engineering class – inclined me to particular kinds of behavior. And here's also where the Vatican's ten commandments for drivers helped. The first of those commandments, interestingly, is "You shall not kill." Well, nobody gets into their car and heads for the highway with the intent of killing other people, so obviously what the Vatican has in mind with this first commandment is what Jesus taught us in Matthew 5:21–22:

> You have heard that it was said to the people long ago, "You shall not murder, and anyone who murders will be subject to judgment." But I tell

you that anyone who is angry with his brother or sister will be subject to judgment. Again, anyone who says to his brother, "Raca," is answerable to the court. But anyone who says, "You fool!" will be in danger of the fire of hell.

Well, "Raca" and "you fool" are not precisely the words that I use when encountering a fellow driver who is doing something inordinately stupid on the road – usually something that causes me to have to alter my driving pattern. But anger is certainly something that I find myself having to control or guard against. And I have found that the properties of the automobile and highway system – putting people in competitive situations where they perceive each other as impersonal beings encased in glass and steel – exacerbate the proclivity toward anger that results from our impatience and our desire to be in complete control when we are driving. Thus the Vatican's first commandment is most appropriate.

And speaking of control, the fifth of the Vatican's ten commandments for drivers states that "Cars shall not be for you an expression of power and domination, and an occasion of sin." In an editorial piece in the *Sioux City Journal* Dave Yoder quipped that "Except for pretty much knocking out the philosophical basis behind the American car industry, you can't argue with that." While intended as an off-hand bit of humor, there is an awful lot of truth in that remark. You see the automobile is designed in such a way that it necessarily provides a sense of power, control, and, yes, even domination to the operator. It's a property – I dare say, a value – that is embedded in the very technology. And it influences us. We can, of course, resist. But that takes concentration and effort. The Vatican's ten commandments for drivers alert us to the need for that concentration and that effort.

TECHNOLOGY AND THE TRANSFORMATION OF READING HABITS

DECEMBER 7, 2007

The food we eat is not neutral to our bodies. The effect it has on us is not dependent solely on how much of it we eat. Everyone pretty much understands that. And we also understand that some people are more susceptible to the effects of certain foods than others. Some people are allergic to peanuts, some to eggs. Other people suffer from indigestion with just a bite of onion or pepperoni. There are the biological properties of foods that cause biological effects in those who consume them. That relationship, between the properties and the effects, is fairly well understood and accepted by most people. Less well understood are the economic and aesthetic properties of food and the economic and aesthetic effects of consuming particular foods. But if we think about it, most of us will agree that there is an enormous aesthetic difference between eating filet mignon served on fine china over a tablecloth in a classy restaurant and eating a hot dog with sauerkraut served in red-striped wax paper at a baseball game. With a little thought we can also quickly be convinced that different foods have different economic properties. Consuming a meal of macaroni and cheese, cooked at home and served with a glass of orange drink, will make far less of a dent in your budget than enjoying a meal of filet mignon, jumbo shrimp, and Cabernet Sauvignon, even if these are prepared and served at home.

Well, OK, food has more than just biological properties and the relationship we have with our food is more than just nutritional. It's also aesthetic and economic. And I could go on to argue that we have a moral and even a faith relationship with our food as well – just think of the sin of gluttony on the one hand and the celebration of the Lord's Supper on the other. But the main point I want to make in this essay is not about food, it's about technology. And it's about a particular kind of technology that has been undergoing rapid transformation in the past hundred years: communication technology.

In 1985 Neil Postman wrote a book titled *Amusing Ourselves to*

Death in which he argued that the media of mass communication in our culture was moving from print to images, and that the change was transforming our thinking habits. He claimed, for example, that "Television has conditioned us to tolerate visually entertaining material measured out in spoonfuls of time, to the detriment of rational public discourse and reasoned public affairs."* I think Postman was right on target when he wrote that. But that was twenty-three years ago. Communication technology has not stopped evolving. Today, while the heart of Postman's message remains valid, the context has altered so drastically that his arguments might seem archaic. Postman's concern rang true for the twentieth century. In those one hundred years mass communication moved from print to images. Books, magazines, and newspapers, gave way to movies, television, and the World Wide Web. Print didn't disappear, but it yielded prominence to the moving images on the television screen and the multiplicity of fixed and moving images on the exponentially growing number of websites. The digital camera, inkjet printer, and digital camcorder are among the primary artifacts that symbolize the ascendancy of image-based communication in the 1990s.

But the first decade of the twenty-first century has witnessed another shift in mass communication. And curiously enough, that shift has, in a superficial way, returned to print. You see, images, by their very nature, are severely limited in the scope of what they can communicate. Images cultivate feelings in the viewer that *suggest* ideas. Despite the old saying that a picture is worth a thousand words images are incapable of conveying ideas directly. You need words to do that. In TV and film, media that incorporate both images and sound, you have both images and words. But by the end of the twentieth century, even the TV was yielding to the personal computer and the World Wide Web. And the nature of the computer and the Web are such – at least today – that words are best expressed visually rather than sonically. Hence websites will contain both images and text. But the nature of computer-based communication is to maximize the amount of data conveyed and to minimize the time needed to convey it. Let's call that the norm of efficiency. Thus we can conclude that communication during the first decade of the twenty-first century is heavily influenced by the norm of efficiency. The most obvious example of this is how the computer has transformed the telephone. Back in the 1960s – which to some must seem like the technological dark ages – teenagers could be found curled up on couches with telephone

---

* Postman, Neil, *Amusing Ourselves to Death: Public discourse in the age of show business* (New York: Viking Penguin, 1985), back cover.

receivers glued to their ears. It was sort of a standing joke that if you had a teenager in your household, phoning home would more often than not get you a busy signal. Today the phone once glued to the ear is yielding to fingers that do text-messaging. And that is due to computer technology's embodying the norm of efficiency.

So here's my main point. One of Neil Postman's fears was that the movement of text-based communication to image-based communication would severely inhibit reading, and that a reduction in reading would mean a decline in rational thought. Postman was only half-right. Today reading is back because text is back. But text is back in a way very different from when it existed in the nineteenth century. Writing an e-mail, a blog, or text-messaging is radically different from composing and writing a letter, an essay, or a book. And likewise, reading an e-mail, reading a blog, or reading a text-message is radically different from reading a novel or reading a fluid mechanics textbook. I mention fluid mechanics textbooks because this semester I learned – to my chagrin – that many of my engineering students are not reading the textbook assignments I have given them. More shocking to me than that: a few of my students have not even purchased the assigned text. I can only assume that this behavior evidences a transformation of reading habits that has occurred in response to recent changes in communication technology.

An emphasis on the norm of efficiency has given birth to the microwave oven and the fast-food restaurant chain. But despite our consumption of frozen meals and McDonald's hamburgers, we still appreciate a leisurely meal prepared at home or served in a quality restaurant. Sadly, the analogy does not hold with our reading habits. The god of efficiency and the technology that has given rise to e-mailing, blogging, and text-messaging, is transforming our reading habits so that we – as a culture – are becoming unable to intellectually digest a Dostoevsky novel or a fluid mechanics textbook. That's a serious cultural problem: a problem that Christians – particularly Christian educators and Christian parents – need to be dealing with.

PART V
WISDOM AND THE
SPIRIT OF ECCLESIASTES

It was a marvelous night, the sort of night one only experiences when one is young. The sky was so bright, and there were so many stars that, gazing upward, one couldn't help wondering how so many whimsical, wicked people could live under such a sky. This too is a question that would only occur to the young, to the very young; but may God make you wonder like that as often as possible!

– Dostoyevsky, *White Nights*[*]

---

[*] Dostoyevsky, Fyodor M., "White Nights," in *Notes from the Underground, White Nights, The Dream of a Ridiculous Man, and Selections from The House of the Dead* (New York: New American Library), 7.

CHAPTER 1

THE CREATION AS ART

AUGUST 3, 1990

For those who can see it, reality often appears as a majestic piece of artwork, a tapestry woven together with the finest of threads – "quantum threads," my physicist friend John might say – and containing an unimaginable assortment of jewels. These jewels sparkle, like the stars in the night sky, radiating a portion of the unfathomable creative imagination of the Master Artist to those who, as he said, "have eyes to see and ears to hear." In fact, some of those jewels in the fabric of reality *are* the stars – those distant nuclear furnaces whose light takes thousands of years to travel before it reaches and delights our upward gazing eyes. But the stars are only one kind of an infinite assortment of jewels that decorate this carefully woven tapestry we call creation. No less important are the moments and events that, when taken together, comprise the life of each one of us human creatures; whether it be an extended event, like that shared by two people who have pledged troth to each other and kept it for over fifty years, or a brief pleasure, as in the smell of honey-suckle, when a gentle summer breeze wafts it across your face. Each of these jewels is held fast in the fabric of the tapestry. Tied together by "quantum" threads, each has its place and significance; but again, to appreciate that significance you must have "eyes to see and ears to hear."

Without that special vision, reality will be perceived very differently from what I've just described. There may be, now and then, a twinkling glimpse of the light radiated from each of the jewels, but the fabric as a whole is invisible. And when the threads that tie the fabric of reality together go unseen, then the jewels cannot be fully appreciated, and the whole beautiful tapestry disintegrates into a chaotic array of unrelated creatures and events. The stars become merely distant nuclear reactions whose only significance to one's life is related to the chance that you happen to be looking at them at a given point in time. Each event in one's life is disconnected from other events, and from the events in the lives of other persons; none of them having any lasting significance. Shakespeare expressed it well when he had his tragic hero Macbeth say:

. . . all our yesterdays have lighted fools the way to dusty death. . . .

Life's but a walking shadow, a poor player that struts and frets his hour upon the stage and then is heard no more; it is a tale told by an idiot, full of sound and fury, signifying nothing. (Act V, Scene V)

This past week I couldn't help being dazzled by the brilliance of a number of jewels in the fabric of the creation, and I was particularly struck by the way they were tied to each other. For three days, invited by relatives, my wife and I stayed at a rather plush resort in the Ozarks. We were involved in a number of activities while there, but two in particular gleamed brighter than the rest. The first was when four of us left the resort and traveled to a little, out of the way, state preserve, where we and perhaps a dozen other people, guided by a state conservation officer, explored a cave. The simple beauty of a relatively unspoiled cave, half a mile into a mountain side and a quarter mile underground, contrasted sharply with the noise of the busy lake resort. The intelligent and genuinely enthusiastic commentary of the conservation officer, as she discussed the geology, biology, and history of the cave was totally different from the superficial vocabulary and "party" attitude of the resort workers, who, after teaching you to water ski, talked among themselves of the money to be made in the next month as the influx of tourists hit its yearly maximum.

The other activity occurred on the night before we left for home when my wife's brother-in-law took us out for an hour's spin around Lake-of-the-Ozarks in his boat. The night was softly illumined by moon and starlight filtering through a haze of clouds. The air was cool, and as the boat sped at 30 mph across the waves, I was immersed in a sea of sensations: the pounding of the waves just under my seat, the smell of fresh, lake-water air as the breeze blew it across my face, the sight of lights on one shore of the lake contrasting with the darkness of the woods on the opposite shore; and then there was this exceptionally pretty girl sitting across from me in the bow, her brown hair blowing backward as she gazed ahead into the 30 mph breeze. For a fleeting moment I thought how nice it would be to own a boat on a lake like the one I was riding in. But then I began to see those threads in the fabric of reality, and other thoughts filled my mind. The boat was just a kind of lens that enabled me to see, with a little different perspective, things that the Lord had given to me to enjoy long ago. The boat's owner could take it home, but he also would be responsible for keeping it up, for paying the fees associated with using it on the lake, and for justifying his ownership of it when for weeks it went unused, just sitting in the dock. I, on the other hand, would still have the moonlight and stars to appreciate. While the breeze in Sioux

Center on a summer's night would rarely compare to that experience while crossing the lake, it is still something to enjoy and be thankful for. And the boat ride itself would not be just one pleasurable but isolated past event. Rather, it is one of those creational jewels, tied neatly into the fabric of reality. In fact if you close your eyes you just might see that fine thread tying it, this essay, and you the reader together. And best of all, while my friend may have had the satisfaction of knowing that his boat was safely docked a hundred feet from his house after the ride was over, that pretty girl I mentioned went home with me – as she's done now for the past twenty-three years.

What enabled me to perceive even better the threads that tied these events into the fabric of reality was something that occurred shortly after we returned home to Sioux Center. After a long day's drive we unpacked, settled in, and began to go through the mail of the past few days. Under the usual mountain of junk mail, I found the latest issue of *The Reformed Journal*. In that issue was a gem of an article by Philip Yancey entitled *Ecclesiastes: The high counterpoint of boredom*. Reading the article reinforced my vision of the connecting threads that tied together our recent time spent at the resort, the cave exploration, the boat ride, the *Journal* article itself, and this essay (which I began anticipating).

To many people Ecclesiastes is a strange book and seems to not belong in the Bible. Over and over the writer of Ecclesiastes is driven to give us the same discouraging advice that Shakespeare's Macbeth gave us:

> "Meaningless! Meaningless!" says the Teacher. "Utterly meaningless! Everything is meaningless." . . . All things are wearisome, more than one can say. The eye never has enough of seeing, nor the ear its fill of hearing. (1:1, 8)

Yancey, in his *Reformed Journal* article, suggests that Ecclesiastes was written during a time of great material success for Israel. Whether or not it was Solomon who wrote it, it was written to portray the kind of despair that comes to anyone who honestly looks at reality with eyes blinded by the success of power, wealth, fame, or intellectual prowess – eyes blind to the threads in the tapestry of the Lord's creation. Contemplating the events of the past few days in the light of Ecclesiastes, I was reminded of those words of James 1:17:

> Every good and perfect gift is from above, coming down from the Father of the heavenly lights, who does not change like shifting shadows.

If one cannot see that life's pleasures are an integral part of God's good creation, that they have lasting significance because they are gifts, referring back to the Creator who gives them, then even the best of those pleasures will fade quickly with time, and be seen in retrospect as mere "chasing after wind," as the writer of Ecclesiastes says.

I want to end this essay by trying to make visible one other thread that the events at the lake and the essay on Ecclesiastes enabled me to see. Chapter 12 of Ecclesiastes begins with these words:

> Remember your Creator in the days of your youth, before the days of trouble come and the years approach when you will say, "I find no pleasure in them" –

Twenty-five years ago, when I first seriously studied the book of Ecclesiastes, I was troubled by those words. They almost seemed to reinforce that pagan notion that to be 18 is better than to be 80. Well experience has taught me to disbelieve such nonsense. But reading that *Reformed Journal* article enabled me to better understand what the Lord is telling us in this strange Old Testament book. Oddly enough, that thread tied together the last chapter in Ecclesiastes with the words of a song written by Bob Dylan about thirty years ago. It's true that reading Ecclesiastes chapter 12 troubled me many years ago: "Ah, but I was so much older then, I'm younger than that now."*

---

* Bob Dylan, "My Back Pages," Track #8 on *Another Side of Bob Dylan*, Columbia CD #CK8993.

Chapter 2

Undertaking Great Projects and Chasing After Wind

May 6, 1991

Ecclesiastes is one of the strangest books in the Bible. There are times however, when its strangeness turns to familiarity and it begins to sound like it was specifically written for those of us in twentieth century North America caught up in making "great plans" and undertaking "great projects." Graduation can be one of those times. In a few days, here at Dordt College, graduating seniors and their professors will don long robes and funny tasseled hats and enact a ceremony that represents, for the seniors, successful completion of a program of studies that has hopefully prepared them to undertake a variety of tasks. As each graduate receives their diploma, there will be a handshake, a smile, and, no doubt, the beaming faces of proud parents, some of whom will try to capture the event with their flash camera or camcorder. In the minds of some of the graduates there will be thoughts of great projects yet to come. An engineering graduate may contemplate his fellowship to graduate school, and the possibility of being part of a design team creating vehicles for interplanetary flight. A pre-med graduate may contemplate her days ahead in medical school, the opportunities to heal thousands of sick people by her medical breakthroughs, and the abundant income that such successes will bring. There may be a business graduate who looks forward to beginning corporate work in less than a week's time, and who hopes to make his first million in less than a decade. Perhaps a pre-sem graduate anticipates going to Calvin Seminary in the fall; and is already laying plans for being one of the first women pastors in the Christian Reformed Church. There will likely be an education major, only last week having been offered contracts to teach second grade by three different schools, with visions of eventually becoming the Secretary of Education, having the power and ability to institute great reforming projects in the schools of America.

These are the times when Ecclesiastes is not only relevant, but necessary. And wouldn't it be a brave but wise graduation speaker who opened his address by reading from Chapter 2 of that remarkable book. Listen to the words of the Teacher, the king over Israel in Jerusalem during the time when Israel was at its cultural zenith. And keep in mind, this is the Word of God:

> I undertook great projects: I built houses for myself and planted vineyards. I made gardens and parks and planted all kinds of fruit trees in them. I made reservoirs to water groves of flourishing trees. I bought male and female slaves and had other slaves who were born in my house. I also owned more herds and flocks than anyone in Jerusalem before me. I amassed silver and gold for myself, and the treasure of kings and provinces. I acquired male and female singers, and a harem as well – the delights of a man's heart. I became greater by far than anyone in Jerusalem before me. In all this my wisdom stayed with me.
>
> I denied myself nothing my eyes desired;
> I refused my heart no pleasure.
> My heart took delight in all my work,
>   and this was the reward for all my labor.
> Yet when I surveyed all that my hands had done
>   and what I had toiled to achieve,
> everything was meaningless, a chasing after the wind;
>   nothing was gained under the sun.

What we have here is a crushing critique of "ambition." The teacher (who we are to understand is Solomon himself) is telling us that our greatest plans and expectations will turn out to be nothing but *chasing after the wind*. He tells us that a day will come when our sight will be dimmed, our teeth will all have fallen out, and we won't hear without the help of an electronic device in your ear – and even then it won't sound very good. He wants us to think ahead to that day when our breath comes in short and quick puffs, when the IV in your arm dulls the pain but also dulls your ability to communicate with those who have come to say one final good-bye.

What's going on here? Is the writer of Ecclesiastes an existentialist, who, like the French thinkers Jean-Paul Sartre and Albert Camus, tries to convince us that life is absurd? Is God some kind of sadist who derives pleasure by putting this book in the Bible only to crush us with the knowledge that all our ambitious plans are vanity, futility, meaningless? Many Christians have answered those questions by pointing to the twelfth chapter where we read,

> Remember your Creator in the days of your youth, before the days of trouble come and the years approach when you will say, "I find no pleasure in them" –

Or the very last verses of that last chapter where we read,

. . . here is the conclusion of the matter: Fear God and keep his commandments, for this is the duty of all mankind. For God will bring every deed into judgment, including every hidden thing, whether it is good or evil.

The suggestion is then that Ecclesiastes is written in a sort of tongue-in-cheek style. In other words, what the teacher tells us in the first eleven chapters is *as if* it were written by a godless existentialist, until in the last chapter we have the man of God telling us, "Now remember, this is what life will be like unless you are faithful to the Lord." That's one way of understanding the book, I suppose. But I think there is a danger in seeing it that way, because it can lead us to discredit everything in the first eleven chapters as somehow non-authoritative, non-normative because it is expressed by a near atheist.

I think Ecclesiastes needs to be read as one piece. It is all the words of a man of God who, like the rest of us, has his ups and downs. The Lord uses his writing to teach us some very important things about ourselves, about our ambitions, and about our relationship to him. Listen to some other words from this wise man, taken from the third, fifth, seventh, eighth, and ninth chapters.

There is a time for everything, and a season for every activity under heaven: a time to be born and a time to die, a time to plant and a time to uproot. . . . (3:1–2)

[God] has made everything beautiful in its time. He has also set eternity in the human heart; yet no one can fathom what God has done from beginning to end. (3:11)

. . . when God gives someone wealth and possessions, and the ability to enjoy them, to accept their lot and be happy in their toil – this is a gift of God. They seldom reflect on the days of their life, because God keeps them occupied with gladness of heart. (5:19–20)

Do not say, "Why were the old days better than these?" For it is not wise to ask such questions. (7:10)

No one can comprehend what goes on under the sun. Despite all their efforts to search it out, no one can discover its meaning. Even if the wise claim they know, they cannot really comprehend it. (8:17)

Go, eat your food with gladness, and drink you wine with a joyful heart, for God has already approved what you do. Always be clothed in white, and always anoint your head with oil. (9:7–8)

What the writer of Ecclesiastes is telling us here is exactly what Jesus told us when he said, ". . . do not worry about tomorrow, for tomorrow will worry about itself. Each day has enough trouble of its own" (Matthew

6:34). And that's important to remember, especially at this time of year. It's wrong for a student to seek after a high grade and lose out on the enjoyment of simply learning. It's wrong for a sprinter to seek after a gold medal and lose the simple joy of running today's race. It's wrong for an educational institution to seek after recognition, honors, and prizes while losing the satisfaction to be had in the day-by-day interaction of teachers and students. Summa cum laude's, accreditations, Nobel Prizes, and various kinds of scientific, technological, or medical breakthroughs are things to be enjoyed "in their time." But when they become the object of our craving or the reason for our pride, they become empty, meaningless, and vain.

So let's drink our graduation wine with joyful hearts, remembering that it is God who favors and judges what we do. And let's not take our plans or ourselves too seriously.

CHAPTER 3

A SUMMER MORNING SUNRISE . . .
AND CHARLES DICKENS

JULY 1, 1988

The sun rises early on midsummer days like today. You have to get up
before 6 a.m. if you want to catch a glimpse of that reddish oblong sphere
as it slowly lifts itself above the cornfield horizon in the northeast. You are
rewarded richly for your effort, however. For in addition to the sunrise
you are greeted with the cool and calm morning air, the sound of birds,
and perhaps the smiling face of a neighbor jogging past your house whose
expression seems to say "What are you doing up so early?" This is that
time of day – before car engines roar, the wind picks up, and the heat
becomes oppressive – when you get just an inkling of what the world was
intended to be like when it was first created, and what it will be like when
the One who designed it and laid its foundations returns to complete its
redemption.

There are evidences of Eden and the New Jerusalem all around us if
we have the eyes to see and the ears to hear them. Unfortunately our eyes
and ears, as well as the rest of us, are usually so busy with the concerns of
day-to-day life that we have no time for such exquisite perceptions. And
what's worse, we so often fall prey to the sinful passions of our fallen state
that our sensitivities to things of eternal significance are dulled to the
point of ineffectiveness. But for many of us the summer provides an opportunity to retreat from the oppressive busyness of the workweek/week-
end routine. Whether it's a two-week vacation or simply the slackening
of the pressure and pace that characterizes the rest of the year, summer is
a time when we can engage in some form of relaxing reflection. And that
means increased opportunities for appreciating God's creation the way it
was intended to be appreciated.

I've already mentioned that one marvelous creature, the sun, which
when perceived especially at its rising or setting is capable of opening our
eyes to the glories of the new heavens and the new earth. I trust that most
of you have experienced similar perceptions. I'd like to call your attention to a couple of other creatures whom I met earlier this summer. They

stirred me to a greater awareness of God's creative wisdom and what we can look forward to when he welcomes us into his golden city.

Unlike the sun, the summer breeze, or singing birds, these were human creatures, image bearers of their Creator. I didn't actually meet them in the flesh. In fact one of them never had flesh. But I feel as if I've known them as well as I've known any person other than my wife.

Near the end of the school year I began reading *David Copperfield*. I was about a third of the way through its 830 pages when I finished grading my last exam. So leaving engineering, Dordt College, and Sioux Center behind, I spent my first week of the summer in nineteenth century England. There I met Wilkins Micawber, Betsey Trotworth, Ham Peggotty, Uriah Heep, and Agnes Wickfield. I was continually amazed at the feelings the book was capable of cultivating within me. Most amazing of all was the sense I had, every so often, that in reading *David Copperfield* I was somehow catching a glimmer of the new heavens and the new earth, similar to watching the sun rise on a cool summer morning. I wondered how that could be. Certainly the New Jerusalem will not be like nineteenth century England!

In *David Copperfield*, Charles Dickens presents to us a human life – an idealistic life to be sure – where evildoers get punished, virtue is rewarded, sorrow quickly gives way to joy, and even tragedy is viewed through a holistic perspective whereby it is rendered temporal and ultimately gives way to the eternal. It is semi-autobiographical in that many of the experiences of Copperfield parallel those of Dickens as he grew up. Many of the characters bear a close resemblance to real life people who Dickens knew well. Surely Charles Dickens was a literary genius with the ability to communicate real life experiences in written words.

What struck me most of all, however, was the sense in which these snatches of human experience seemed more real, in an eternal sense, than much of everyday experience. It's kind of like comparing the actual sky at 11 a.m. with a beautiful painting of sunrise. The painting, though in one sense less real, seems to capture God's intentions for reality better than the real thing. But how could Dickens do that unless he was a strong Christian with a deep-seated understanding of God's Word?

That question, and the great appreciation I had for *David Copperfield* after finishing it, led me next to pick up a biography of Dickens.* His biographer, Edgar Johnson, didn't have quite the literary talent of Dickens. Nonetheless, the book was extremely well-written and I de-

---

*   Johnson, Edgar, *Charles Dickens: His tragedy and triumph* (New York: Little, Brown and Company, 1952).

voured all 600 pages of it in a week's time. To my surprise I found that while Dickens was a Christian and was certainly influenced by the Bible (especially the Sermon on the Mount), he was what we would call today an extreme liberal. When visiting the United States he was most comfortable in the company of Unitarians. So for a while I was puzzled, and perhaps a little disappointed. It would have been nice to discover that Dickens was a reformational writer who battled with the secularists and pietists of nineteenth century England. But such was not the case. And now on a cool summer morning, with the birds singing and the sun peeking up over the horizon, I think I understand.

You see, the sun is not a Christian. It's a sustained nuclear reaction taking place 93 million miles from here. But it's God's creature, and even in a fallen world it can't help but declare his glory. It's the same with my two friends: the real one, Charles Dickens, and the imaginary one, David Copperfield. They too are God's creatures, made in his image. And they too, in spite of themselves, cannot help but declare his glory.

Would you like to read a really good Christian book this summer, one that declares God's glory and allows you to catch glimpses of both Eden and the New Jerusalem? Read *David Copperfield*. It's long. Its vocabulary makes demands on you. You may have to work pretty hard for the first 200 pages. But it will stay with you the rest of your life, for you will have a sense of what God wanted to do when he created humankind. You may even increase your appreciation of the summer morning sunrise.

CHAPTER 4

EXPLORING THE CREATION

JUNE 4, 1986

Well, summer is finally here, and many of us, because of increased vacation opportunities, will find a week or two in which we are able to
leave our usual daily activities and do something different. For some this
will mean fishing at a quiet lake or visiting areas of the country that we
haven't visited before. For others it will mean staying at home and engaging in a hobby, like woodworking or photography, that requires sustained
attention and a fair bit of time, and for that reason is impossible to do
during non-vacation times. These activities have one thing in common –
they involve the exploration of God's good creation.

One of the attributes of being human, of being created in the image
of God, is that we are called to appreciate the creation. We can do that in
many ways, of course. One way is in our work. But we also can do that
just as much in our play.

The equivalence of work and play for appreciating the creation is no
more clearly stated than in Proverbs 8:27–31 where we hear "wisdom"
declaring:

> I was there when he set the heavens in place,
>> when he marked out the horizon on the face of the deep,
> when he established the clouds above
>> and fixed securely the fountains of the deep,
> when he gave the sea its boundary
>> so the waters would not overstep his command,
> and when he marked out the foundations of the earth.
> Then I was constantly at his side.
> I was filled with delight day after day,
> rejoicing always in his presence,
> rejoicing in his whole world
> and delighting in mankind.

When we think of vacationing, of appreciating the creation, we usually think about the part of the creation that we refer to as "nature":
mountains, lakes, trees, gentle summer breezes, ocean waves, flowers, and
star-filled evenings. And this, no doubt, has been the case ever since the

Garden of Eden. Even in this post-fall, sin-filled, and often violent world, the natural harmony between us and the rest of God's good creation, and the glory of God as it is manifested in the works of his hands, is clearly seen by anyone who cares to look.

But the creation is more expansive than what we, perhaps somewhat ambiguously, refer to as "nature." There exists a world that God created only in terms of its possibilities during those six days of creation − a world that was dependent on God's image bearers, humans, to unfold. It includes such things as imaginative novels, computers, poems, bridges, skyscrapers, symphonies, paintings, and automobiles. And it's important that we appreciate this part of God's handiwork as much as those that have not been dependent on humans for their coming into full being.

Unfortunately there is a problem with these things, which I will call "cultural nature" to distinguish them from the rest of creation and indicate that their origin is in humankind's cultivating the creation. Since cultural nature comes into being as a result of a human activity, and since human activities are normed (meaning they can be done obediently or in disobedience to will of the Lord), there will necessarily be many examples of cultural disobedience. These are usually very ugly, very hard to appreciate, and sometimes completely unable to be used as a vehicle for praising the Lord. Instead of displaying the harmony in the creation, they are evidence of disharmony, of humankind's estrangement from the creation, from one's neighbor, and from God. As a result of such things as congested highways, X-rated movies, cruise missiles, and the like, we have become very suspicious of many parts of cultural nature and have retreated to the safety of the nonhuman parts of creation. But in doing so we miss many opportunities that the Lord makes available to us for our growth, enjoyment, and praise to him.

Thoughts about how we might make better use of cultural nature in our summer vacationing came to me a week ago, shortly after I had finished reading a science fiction novel. The novel had a lot to do with computers and left me with the feeling that I ought to explore the computer a bit more this summer. But very shortly after finishing that novel I picked up the June issue of *National Geographic* and read an excellent article on "The World of Tolstoy." It beautifully described the Russian cities and countryside of the late 1800's that were home to Leo Tolstoy, perhaps the greatest novelist of all time. It occurred to me that not too many people fully appreciate both Tolstoy and the computer, that is, classic literature and modern technology. And worse, almost no one would think of taking a two-week vacation to read *War and Peace* or teach themselves PASCAL.

So if you're planning a vacation this summer, let me suggest that you consider reading a piece of good literature as part of that vacation, let's say *War and Peace, Anna Karenina,* or *Resurrection* by Tolstoy; or that you consider teaching yourself a computer language, let's say either BASIC or PASCAL. Maybe you can even do both! You will be surprised to find how the richness and beauty of the creation extends beyond the beauty of mountains, lakes, and flowers.

The psalmist knew how to praise God for his majesty in the nonhuman creation. But because he lived in a culturally primitive society, he didn't have many examples of cultural nature with which to praise God. I've always been impressed with the first four verses of Psalm 19. But living in the twentieth century, I think we can read those verses and understand the majesty of the sun, moon, and stars as metaphors for the glory of our Lord throughout all creation, including cultural nature:

> The heavens declare the glory of God;
>> the skies proclaim the work of his hands.
> Day after day they pour forth speech;
>> night after night they display knowledge.
> They have no speech, they use no words;
>> no sound is heard from them.
> Yet their voice goes out into all the earth,
>> their words to the ends of the world.

That voice can also be heard in the great cultural achievements of humankind, both artistic and technological. I urge you this summer to take some time to listen to it.

CHAPTER 5

SPACE THE UNCONQUERABLE

JANUARY 3, 1996

A few weeks ago I saw the film *Apollo 13*. Like many, I was impressed with the way it was able to convey some of the unique fear that the three astronauts must have felt as they careened around the moon in their disabled spacecraft. The film also made me realize how our perceptions of what is going on in the world are dependent on the media; for, I must admit that, when the Apollo 13 misadventure was actually occurring, I was hardly aware of it. I certainly had no idea of how close the mission had come to a calamitous and tragic end. More importantly, however, the film reminded me of how much faith people often put in science and technology, and how that faith can be shaken when an event like Apollo 13 or the Challenger disaster occurs.

I spent most of the seventies teaching physics to high school students in a suburban, New Jersey community, just outside New York City. It was a time when technological expectations were high. The anxiety that accompanied the Apollo 13 mission in 1970 was soon forgotten, having never really shaken our belief that those earlier Apollo missions were, indeed, major leaps for humankind. We thought that writer Arthur C. Clarke was pretty much on the money when he predicted in his *2001 A Space Odyssey* (New York: New American Library, 1968) that humans would colonize the moon and send manned spacecraft to the outer planets by the end of the century. In fact, I recall that, to cultivate enthusiasm for my courses, I would tell my students that one day they would very likely travel regularly in space, perhaps even work at mining the asteroid belt in order to provide raw materials for real estate development on the moon. And I believed that. After all, the end of the century was then 25 years away, and considering the technological advances that had occurred from 1950 to 1975, space travel seemed a sure thing.

Around 1975, however, I discovered a wonderful little nonfiction essay by that same author, Arthur C. Clarke: "Space the Unconquerable."*

---

*   Clarke, Arthur C., "Space the Unconquerable," in *Profiles of the Future* (New York: Popular Library, 1977).

I assigned it to my physics students every year thereafter, along with the 40ᵗʰ chapter of Isaiah, when we studied astronomy and space exploration. Written in 1958, in many respects Clarke's essay bore the same technological optimism as his fictional *2001*. But that optimism was tempered by knowledge of basic physics and a kind of scientific humility that is rare among modern scientists and engineers.

Clarke makes the point that although we have in some ways conquered the earth – for example, global distances are now only minor inconveniences that are easily overcome by the telephone and jet airplane – we can never do the same thing in space. He writes,

> Our age is in many ways unique, full of events and phenomena which never occurred before and can never happen again. They distort our thinking, making us believe that what is true now will be true forever, though perhaps on a larger scale. Because we have annihilated distance on this planet, we imagine that we can do it once again. The facts are far otherwise. . . .
>
> The marvelous telephone and television network that will soon enmesh the whole world, making all men neighbors, cannot be extended into space. *It will never be possible to converse with anyone on another planet.* (130, 132)

The reason for this, he explains, is the finite speed of light and the fact that nothing, neither material nor signal, can travel faster than the speed of light. Thus, if you have a friend on Mars with whom you would like to communicate, there would be about a six minute lag between the time you uttered your "Hello John, how are you?" and the time you heard your friend John respond. Communicating with people on the moon wouldn't be quite as bad because there would only be an annoying 2.5 second delay. But speaking to old friends who left to work in the colonies around the star system Alpha Centauri, the stars nearest to our own sun, would be impossible, for there are almost 5 light years separating that system from earth, and that would require you to wait ten years after sending your message before you could hope to get a response.

Clarke also considers the problems of governing from earth, colonies in some of the nearer star systems just beyond Alpha Centauri. He writes

> Try to imagine how the War of Independence would have gone if news of Bunker Hill had not arrived in England until Disraeli was Victoria's prime minister, and his urgent instructions on how to deal with the situation had reached America during President Eisenhower's second term. Stated in this way, the whole concept of interstellar administration or culture is seen to be an absurdity. (136)

Of course there are those whose faith in science and technology will simply not be shaken and who will argue that the speed of light limit is just another hurdle to be overcome by human ingenuity. In the essay, Clarke picks up that challenge and, assuming the discovery of a means for instantaneous transport over any distance (which, by the way, is as baffling to a physicist as is the notion of being in two places at the same time), he uncovers the next hurdle that space puts in our path: its sheer complexity. Consider that there may be more star systems in the universe than there are grains of sand on the earth. How does one keep track of them all? Well, I won't go further down that road because I believe that the speed of light limit is sufficient to make the point.

I noted earlier that I used to assign this essay by Arthur C. Clarke along with Isaiah 40 to my high school physics students. I will close by trying to show what the two have in common.

At the end of his essay, Clarke makes this marvelous statement,

> Space can be mapped and crossed and occupied without definable limit; but it can never be conquered. When our race has reached its ultimate achievements, and the stars themselves are scattered no more widely than the seed of Adam, even then we shall still be like ants crawling on the face of the Earth. The ants have covered the world, but have they conquered it – for what do their countless colonies know of it, or of each other? (138)

Now consider the words of Isaiah 40:21–23, where the prophet is trying to remind prideful humanity that they are but finite creatures, and that the Lord is the creator of the heavens and the earth:

> Do you not know? Have you not heard? Has it not been told you from the beginning? Have you not understood since the earth was founded? He sits enthroned above the circle of the earth, and its people are like grasshoppers. He stretches out the heaven like a canopy, and spreads them out like a tent to live in. He brings princes to naught and reduces the rulers of this world to nothing.

The point of Isaiah 40 is not just to humble humankind in its scientific and technological hubris. It is also to comfort us, and not just *spiritually* either, but to comfort and encourage us as we seek to unfold and develop the Lord's creation, in obedience before his face. For consider the last three verses of Isaiah 40. They provide a context of comfort and encouragement for events such as the Apollo 13 mission and for the vision that today's young scientists and engineers may have for exploring the stars:

> He gives strength to the weary and increases the power of the weak. Even youths grow tired and weary, and young men stumble and fall; but those who hope in the LORD will renew their strength. They will soar on wings like eagles; they will run and not grow weary, they will walk and not be faint.

CHAPTER 6

"LORD, KEEP MY MEMORY GREEN"
REFLECTIONS ON MATTHEW 6:34,
INSPIRED BY A DICKENS CHRISTMAS NOVEL,
WITH APPLICATION TO THE NEW YEAR

JANUARY 1, 1997

At the beginning of a new year, among the less likely words that one may contemplate are those in Matthew 6:31–34, where, after warning against being anxious about food and clothing, the Lord tells us, "Therefore do not worry about tomorrow, for tomorrow will worry about itself. Each day has enough trouble of its own." Hardly words to build a New Year's worship service around. After all, the arrival of the new year is usually the occasion for anticipation, for putting behind the past and cultivating thoughts and plans about the future. But the message of Scripture – particularly those words of Jesus – seems to, if not rebuke, at least discourage any fixation on the future that neglects the past and present. Our anxious or zealous attempts to insure – or even contemplate – a successful, secure, or triumphant future, would appear to be at odds with the basic biblical theme that enjoins us to trust in the Lord, and not in our own strength. Likewise, any attempt we might make to forget the past, to distance ourselves from where we have been or where we are today, by taking refuge in fantasies about the future, would appear to warrant equal censure.

Charles Dickens harbored some insights into these basic biblical themes, and was blessed by his Creator with the skill to articulate them in a form that makes them accessible to anyone who can read the English language.

For the past few years I have taken the opportunity, during the winter recess, to read one of Dickens' so-called "Christmas novels." Starting seven years ago with his well-known *A Christmas Carol*, I have worked my way through two volumes of those novels. This year I read the last novel in those two volumes. In fact, it was the last one that Dickens wrote, and is titled *The Haunted Man*.

Bearing some similarity to *A Christmas Carol*, the main character of the novel is Mr. Redlaw, a chemistry teacher who lives a lonely, gloomy life, haunted by painful memories of suffering that occurred in his youth: memories of growing up orphaned and poverty stricken, of being wronged by a close friend, and of a beloved sister who died just at the time that his hard work extricated them from poverty. His gloom and despair are evidenced in his response to being reminded of the season:

> "Another Christmas come, another year gone!" murmured the Chemist, with a gloomy sigh. "More figures in the lengthening sum of recollections that we work and work at to our torment, till Death idly jumbles all together, and rubs all out."*

There are numerous other characters, all described in classic Dickens style. Two are of particular importance. One is Philip, an eighty-seven-year-old man whose joy in life is his ability to remember the past. In the following quotation, he instructs his grown son, William, to head back from Redlaw's chambers to their own apartment, while Dickens has him recite what becomes the punch line of the novel:

> ". . . William, you take the lantern and go on first, through them long dark passages, as you did last year and the year before. Ha, ha! *I* remember – though I'm eighty seven! 'Lord keep my memory green!' It's a very good prayer, Mr. Redlaw. . . ." (263)

The second character of note is a young boy – but a far cry from Tiny Tim. Dickens describes him as

> A bundle of tatters, held together by a hand, in size and form almost an infant's, but, in its greedy, desperate little clutch, a bad old man's. A face rounded and smoothed by some half-dozen years, but pinched and twisted by the experiences of life. Bright eyes, but not youthful. Naked feet, beautiful in their childish delicacy, – ugly in the blood and dirt that cracked upon them. A baby savage, a young monster, a child who had never been a child, a creature who might live to take the outward form of man, but who, within, would live and perish a mere beast. (272)

The basic plot of the novel is simple. Redlaw is visited by a ghost who grants the chemist his wish that he forget his past. The ghost, however, grants him more than he wishes for. He also makes that "gift" contagious in Redlaw. In other words, anyone who Redlaw associates with has his memory similarly "cleansed." At first Redlaw sees this as a bonus. In fact he asks the "baby savage" to lead him to one of the worst parts of the city, where people are suffering from all kinds of want. Redlaw intends to

---

* Dickens, Charles, 1848, *The Haunted Man*, in *The Christmas Books, Vol. 2* (Penguin English Library, 1971), 255.

alleviate some of that suffering by passing on to them the "gift" of forget-fulness. On the way to that center of suffering, however, he begins to get a glimpse of what is really going on. Dickens describes Redlaw and the child's short journey as follows:

> Three times in their progress, they were side by side. Three times they stopped, being side by side. Three times the Chemist glanced down at his face, and shuddered as it forced upon him one reflection.
>
> The first occasion was when they were crossing an old churchyard, and Redlaw stopped among the graves, utterly at a loss how to connect them with any tender, softening, or consolatory thought.
>
> The second was, when the breaking forth of the moon induced him to look up at the Heavens, where he saw her in her glory, surrounded by a host of stars he still knew by the names and histories which human science has appended to them; but where he saw nothing else he had been wont to see, felt nothing he had been wont to feel, in looking up there, on a bright night.
>
> The third was when he stopped to listen to a plaintive strain of music, but could only hear a tune, made manifest to him by the dry mechanism of the instruments and his own ears, with no address to any mystery within him, without a whisper in it of the past, or of the future, powerless upon him as the sound of last year's running water, or the rushing of last year's wind.
>
> At each of these three times, he saw with horror that, in spite of the vast intellectual distance between them, and their being unlike each other in all physical respects, the expression on the boy's face was the expression on his own. (309–310)

Throughout the rest of the novel Dickens has Redlaw diffusing his gift of forgetfulness to most of the other characters. As he does so, and wit-nesses the change that comes over them, he grows to abhor the power that he has and to see how foolish he was to wish for forgetfulness. There is far more to the novel than I have described here, including another main character that plays the role of a Christ-figure, and is Dickens at his most idealistic, and, some might say, his best. Suffice it to say that, like all Dickens' Christmas novels, there is a happy ending and a moral. The moral, of course, is that we need to pay more heed to – indeed, even cherish – the events of our past. The most terrible events in our lives, says Dickens, are used by our Creator to bring forth good. Thus even our most painful memories, while yet painful, can be cherished for the good that proceeds from them. As you might imagine, the last line of the novel ends with that prayer, uttered earlier by old Philip, "Lord, keep my memory green."

Thus, at the beginning of a new year, I suggest that we take those words of Matthew 6 to heart. Let's not be so caught up with our future, but let's take some time to reflect on the past and present. And regardless of the pain or joy that such reflection entails, let *us* take joy in the fact that our heavenly Father works all things together for the good of his children. Martin Luther had it right when he wrote

> And though this world, with devils filled, should threaten to undo us, we will not fear, for God has willed his truth to triumph through us.**

It would appear that Charles Dickens had that same thought in mind when he wrote his last Christmas novel.

** Luther, Martin, 1529, *A Mighty Fortress is Our God*, vs. 3.

CHAPTER 7

HAPPY ENDINGS

DECEMBER 1, 1998

The month of December is distinguished by shortened days, often the first significant snowfall of the winter, and, for those of us in education, that interval between Thanksgiving and Christmas when another semester of school is brought to a close. It's an interval of frenzied anxiety as students and teachers alike attempt to tie up the loose ends of their academic work. But it's also an interval of anticipation. Christmas break will soon be here; a kind of annual Sabbath that presents us with opportunities for refreshment and recreation. It makes possible opportunities to rest from the ordinary work of the semester, to read, and to reflect on ideas at a pace that is prohibited by the semester's mad scramble of lectures, homework, and tests.

This Christmas break, as has been my custom in the past, I plan to read some fiction, which I have not had the time to read since the summer. First I intend to finish a novel by Philip Roth that I am currently reading – *American Pastoral* (New York, Vintage, 1997). Then I intend to re-visit my favorite English author: Charles Dickens. For a number of years Christmas break has given me the opportunity to read one of his Christmas novels, novels like *A Christmas Carol*, *The Cricket on the Hearth*, or, most recently for me, *The Haunted Man*. This year I will be making what I hope to be a sizeable dent in an anthology of Dickens' Christmas stories. As I contemplate doing so, however, I'm struck by the radical difference in writing style between Charles Dickens and modern fictional writers like Philip Roth. That difference says something about the novelist, to be sure. But it also says something about our culture and about the times in which we live.

In a nutshell, the difference between reading Charles Dickens and a modern author is the absence of a happy ending in the latter. There are many other differences, of course, but this one seems to epitomize the rest. Consider *American Pastoral* on the one hand, and *The Haunted Man* on the other.

American Pastoral is about a man who grows up near the city of

Newark, New Jersey, in the 1940s. He is the star athlete at his high school, who never allows his stardom to go to his head. Always in control of his life, he serves in the marines just after World War II has ended. He returns home to attend a small, local college, where he meets his wife, a former Miss America contestant. He takes over his father's leather glove business in the city of Newark and buys a 200-year-old house in rural northern New Jersey where he has 15 acres of land, acceptance into an historic upper-middle class community, and an hour commute to his work. He and his wife have one child, a daughter, who they adore, and for whom they have great expectations. But that is where the story turns from "pastoral" to existentially wretched. In the late sixties the daughter gets involved with the most radical of the counter-cultural groups, the Weathermen. Her defining act of rebellion is blowing up the local grocery store, killing the grocery storeowner, and disappearing into the counter-cultural underground to avoid prosecution. The whole novel consists of an unsuccessful attempt on the part of the main character to reconcile the pastoral character of the first half of his life with the angst and misery that typifies its second half. Although I have not quite finished the book, it's fairly obvious that its basic message is that the pastoral quality of the main character's life is an illusion – life is really meaningless and bleak. The book won the 1998 Pulitzer Prize and has an artistry about it that one has to admire. But its disposition is the polar opposite of that of a Dickens novel.

In Charles Dickens' *The Haunted Man*,[*] the main character is a bitter and lonely chemistry professor who is haunted by painful memories of suffering that occurred in his youth. He grew up orphaned and poverty stricken. A close friend wronged him. And his beloved sister died just at the time when his hard work extricated them from their poverty. The basic plot of the novel has the chemist visited by a ghost who not only grants the chemist's melancholy wish that he forget his past, but also gives him the power to bestow this "gift of forgetfulness" on others. But as the chemist goes about interacting with the other characters in the novel, infusing them with forgetfulness and observing the terrible changes that result, he soon comes to abhor the power that he has and begins to see how foolish he was to wish for forgetfulness. Thus the novel does not end in despair. Like Ebenezer Scrooge at the end of *A Christmas Carol*, the last chapter finds the chemist penitent and reformed, with his memory returned. Likewise, all those who he infected with forgetfulness

---

[*] Dickens, Charles, 1848, *The Haunted Man* in *The Christmas Books*, Volume II (New York: Penguin, 1984), 235–353.

regain their memories. There is a happy ending. And the moral of the story is that we ought to cherish the events of our past, even those that were the cause of our suffering. The most terrible events of our lives, says Dickens, are used by our Creator to bring forth good. Our most painful memories, while yet painful, can be cherished for the good that proceeds from them. To drive that point home Dickens puts into the mouth of one of his characters a line that appears early in the novel and as the closing punch line. It is a simple prayer: "Lord, keep my memory green."

Now the point of this essay is to contrast the style of Dickens with that of modern authors like Philip Roth. I have heard Dickens criticized for being unrealistic, for developing characters that are caricatures of moral ideals rather than real people. So I have had to ask myself, even though I appreciate the existential, angst-embodied work of authors like Philip Roth, why is it that I much prefer to read Charles Dickens? And why do I believe that my life is more enriched by reading a Dickens short story than by reading a Pulitzer Prize winning novel? Well, the answer may be that I am simply a sentimental, literary Neanderthal who has not yet matured from the fairy tale stage to that of accepting the realities of the human condition and their necessary embodiment in good literature. But I think not. I am no expert on literature. But I have been around for over half a century. And during that time I have been doing a lot of observing and a lot of reading. And the evidence I've gathered tells me that Dickens, not the modern authors, paints the more faithful picture of reality.

One case in point: It was just under two years ago when I first wrote an essay about Dickens' novel *The Haunted Man* – January 1, 1997 to be exact. I wrote the essay because I was moved by reading the novel. But I was moved by the novel, in part, because it spoke to the suffering that I was observing in a friend of mine who had just lost his wife to cancer. Both the novel and my essay could be construed as prayers for delivery from despair – and strength to trust in the Lord's deliverance. Well, this past weekend that friend of mine was married again. And the joy occasioned by that marriage – in my friend, his wife, his young daughters, and in his friends – is far better portrayed by Charles Dickens than by any modern author that I have read. You might say it's a "happy ending." Or better, you might say that it is evidence of God's grace to his image-bearing creatures, evidence that encourages us to see that his creation – which he once called "very good" – is, despite all the anguish and suffering, destined for a happy ending.

Listen as I close this essay by reading the last paragraph from another one of Charles Dickens' Christmas novels, titled *The Chimes*. Here Dickens seems to be reflecting on the very question that I am raising here. Trotty Veck is one of the main characters in the story.

> Had Trotty dreamed? Or are his joys and sorrows, and the actors in them, but a dream; himself a dream; the teller of this tale a dreamer, waking but now? If it be so, oh Listener, dear to him in all his visions, try to bear in mind the stern realities from which these shadows come; and in your sphere – none is too wide, and none too limited for such an end – endeavor to correct, improve, and soften them. So may the New Year be a Happy one to You, Happy to many more whose Happiness depends on You! So may each Year be happier than the last, and not the meanest of our brethren or sisterhood debarred their rightful share, in what our Great Creator formed them to enjoy.**

** Dickens, Charles, 1844, *The Chimes* in *The Christmas Books*, Volume I (New York: Penguin, 1982), 246.

CHAPTER 8

GOD'S PROVIDENCE:
JEWELS AND QUANTUM THREADS REVISITED
MARCH 4, 1998

Almost a decade ago, in an essay responding to the creational beauty that I was privileged to enjoy while on a three-day trip to the Ozarks, I compared the creation to "a majestic piece of artwork, a tapestry woven together with the finest of threads – 'quantum threads' . . . – and containing an unimaginable assortment of jewels."[*] Events occurring over the past two weeks have caused me to revisit that metaphor, although my motivation for the revisit is very different than it was earlier. Then I was dazzled by the brilliance of the jewels, awe-struck by the beauty in the event structure of the creation and the gracefulness with which those events relate to each other, tied together as it were, by the quantum threads of the creational tapestry. This time, however, it was not beauty, but disappointment; it was not delight, but sorrow that pointed me back to the tapestry metaphor.

Two weeks ago I assigned one of my classes to read a 1984 essay of mine having to do with energy efficiency and Christian witness. The last paragraph of that essay reads as follows:

> In the year 2000, only sixteen years from now, Lord willing, I hope to pull this essay out of my files and use it as a basis for a new one. Perhaps at that time, something called Sioux Solar Corporation, instead of being an idealistic vision in the head of a dreamer, will be a viable institution and a model of biblically normed energy service, environmental stewardship, and Kingdom-oriented business practice.[**]

Well, it's almost the year 2000 and no such model of Kingdom-oriented business practice is in sight. My first reaction to this was one of disappointment and almost a sense of embarrassment. Was this an indictment upon the ideals to which I was seeking to have my students aspire?

Then, a week and a half ago, I listened as a colleague of mine grieved and lamented over the death of a young church leader who, in the midst

---

[*] Supra, 401–403.
[**] Supra, 241–243.

of his considerable service to the Christian community, was taken from his family, church, and social justice ministry. The disappointment, grief, and bewilderment of my colleague was so great that he described the event as among the most profound threats to his faith in God's providence that he had ever experienced.

It was trying to make sense of these two events that drove me back to the "jewels and threads" metaphor that, a decade earlier, had helped me appreciate God's providential goodness. For I was struck by the realization that the two different disappointments had something in common. In both cases, a significant part of the disappointment resulted from *crushed expectations*. In the first case, the expectation was that a model Christian technological business enterprise would arise to shine its light for all North America to see. In the second case, the expectation was that a talented servant of the Lord would lead the organizations he served to greater accomplishments. But in both cases those expectations had, quite naturally, overlooked the fact that God's providence often works in such a way that it is the humble things, the seemingly insignificant and weak things, in this world that are used to accomplish his greater purposes. I say "quite naturally" because it seems to me that it is good and proper for us to strive for success in Kingdom endeavors, and success is often evidenced by its visible manifestations. Thus we ought to aspire to long lives of service as God's saints. And we ought to seek the brilliant light of effectiveness for our Christian service organizations. But then again, perhaps we ought not to be so surprised when, paradoxically, the Lord, in his providence decides to use the small things, the humble and seemingly insignificant, to accomplish his purposes. In the end, it is not a long and successful life or a headline-making Christian service organization that is of ultimate value. It is the contribution that any life or organizational structure might make to the cause of God's Kingdom. To see that contribution, however, is to have a glimpse of the fabric of creation – to perceive the jewels in the tapestry and the threads tying those jewels together into a unity. Our finite and fallen eyes cannot see these things at all without the spectacles of faith. And even with those spectacles, we only see, as the Apostle Paul says, "as through a glass, darkly."

So I return to my tapestry metaphor. Each event in the life of God's saints is potentially a jewel in the fabric of the tapestry. Those events may be extremely brief – just long enough for us to perceive them. On the other hand, the event may be the life of one of his saints. And it is neither the length of the event nor the extent of its visible manifestation that constitutes its jewel-like character. The quality of the jewels is determined by the Creator, not by the expectations of his myopic creatures.

Thus it is our task to put on the spectacles of faith and to trace those quantum threads that weave in and out of the creation tapestry, holding the various jewels in place. While disappointments and sorrows at crushed expectations, no matter how trivial or profound, are natural to us, those disappointments and sorrows need to be relativized by our perception of the creational tapestry as a whole, and our awareness that some of the brightest jewels of God's good creation may only be perceived once the clouds of disappointment and sorrow are dissipated by the sun's redemptive light.

This past weekend I was privileged to trace out a few of the threads in the creation tapestry and to experience some of the jewels embedded there. Along with about twenty Dordt professors and students, I attended the centennial celebration of the Abraham Kuyper Stone Lectures at Princeton University. As some of the threads in the creation tapestry became visible to me, I was once again impressed that the Lord works through his most humble servants and uses the most modest means to bring about his purposes. For the brightest jewels to which the threads led were not big names like Abraham Kuyper or Nicholas Wolterstorff, but rather those servants whose faithfulness and enthusiasm for God's Kingdom made the overall event a jewel. Some of those servants had eighty-year-old faces, faces that I had not seen for over thirty years. Others had the young faces of students, the next generation of Kingdom servants. In each of the faces the light of the Kingdom was beautifully reflected, for creational jewels are independent of age and time.

And so I return to Sioux Center to write this essay, encouraged to face the disappointments of the new week and hopeful that the eyeglasses of faith that I wear – which seem to have a strengthened prescription since the weekend – will allow me to perceive more clearly the jewels in the fabric of the creation and the quantum threads that hold them fast and tie them together.

CHAPTER 9

BASEBALL AND AESTHETIC LIFE

JULY 1, 1998

When I was a child, summer provided an abundance of days during which I was required to do nothing except "go outside and play." More often than not, that play involved a pink rubber ball and a broom handle – playing stickball in the street where I lived. Or – if my friends and I were more ambitious – we would gather together bats, balls, and gloves and walk a half-mile to our grade school playground to play "hardball." There were other team sports like football and basketball that would become important during high school. But in northern New Jersey, in the early 1950s, the only one that mattered to me was baseball. When I was not playing it, I was rooting for my favorite team, the Brooklyn Dodgers.

The philosopher Calvin Seerveld writes that *playfulness* is an integral element of aesthetic life* and that the aesthetic dimension is one of a few fundamental characteristics by which we exercise our humanness. It's therefore no wonder to me that I've experienced resonance with my early baseball activities at various times during the four subsequent decades.

So it was natural last winter, when the opportunity presented itself, that I purchased a copy of Doris Kearns Goodwin's book *Wait Till Next Year: A Memoir* (New York: Simon & Schuster, 1997), with the intent of reading it this summer. Its title has to do with the Brooklyn Dodgers, who, particularly in the late 1940s and early 1950s had excellent teams but, prior to 1955, were unable to win the World Series. Goodwin is a respected historian who has taught at Harvard and written a number of other books, one of which won the 1995 Pulitzer Prize in History. More important for motivating me to buy her book, however, was her role in the 1994 Public Television documentary *Baseball*, produced by Ken Burns. What made that documentary especially enjoyable was the vibrancy and authenticity of the interviews with three unforgettable characters. One was Buck O'Neil, a black player from the Jim Crow era in baseball. Another was the relatively young sportscaster for NBC television, Bob Costas. The third was Doris Kearns Goodwin. In the documentary, she describes growing up in the late

---

* Seerveld, Calvin, *Rainbows for the Fallen World* (Toronto: Tuppence Press, 1980/2005), 52.

1940s and early 1950s as a Brooklyn Dodger fan, and how her enthusiasm for the game and for the Dodgers was passed on to her by her father. The enthusiasm she displays in the interviews, as well as the fact that our childhoods seem to bear some interesting parallels, encouraged me to buy her book. It turns out we have more in common than I originally thought.

My father was born in 1915, in the Bay Ridge section of Brooklyn, into a relatively large but rather poor family. In 1929 the depression forced him to leave high school after his freshman year in order to help support the family. It seems that the two external institutions that lent greatest stability to his upbringing were the Catholic Church and the Brooklyn Dodgers. Some of the earliest summer memories of my own childhood include listening to Dodger games on the radio and, sometime after 1951, watching games on our little black and white TV. And every summer Saturday morning I would be given a nickel and sent around the corner to the "candy store" to buy a copy of the *New York Daily News*. Thus, when I was old enough to read, I developed the habit of reading the newspaper by starting at the back, at the sports section.

So it was that the year 1955 turned out to be a big one in my life. During that summer I was between the third and fourth grades. I enjoyed reading and had a Brooklyn Dodger stamp book. That was one of those large paperbacks with three pages of picture-stamps in the front. As you read the book you would tear off the big, colorful, stamps, moisten the back, and paste them on the particular page where they belonged. The really big event of the 1955 summer was when my father took me to Ebbets Field to see the Dodgers play. It was the first time I had ever been to a baseball game, and I still remember the awe I felt when we emerged from the tunnels and stairways of the stadium into the sunlight. Nothing in my experience has ever been as incredibly green as the grass of Ebbets Field was at that moment. During the game my father spent a whole dollar to buy me a Dodger yearbook. I studied that yearbook and my stamp book for the rest of the summer, and was well prepared when the World Series rolled around. The Dodgers, having won the National League Pennant, were pitted against their arch-nemeses, the all-powerful New York Yankees. 1955 was the year that Dodger fans didn't have to "wait till next year." They won the World Series for the first and only time. A vision of Roy Campenella, the Dodger catcher, running toward the pitchers' mound to embrace pitcher Johnny Podres after the last Yankee grounded out is indelibly fixed in my mind.

Doris Goodwin's book is a memoir of her childhood that centers on Brooklyn Dodger baseball from 1949 to 1955. But the book is far more than just another sports commentary. It's a book that describes how the influential social institutions in our lives shape our childhood years and, ultimately, the people we become. For Goodwin, those influential social institutions were: her family – her mother, two sisters, and especially the father that taught her to love baseball; her neighborhood community – a block of houses in Rockville Centre, on the South Shore of Long Island; her church – the St. Agnes Roman Catholic Church; and the Brooklyn Dodgers. And Goodwin's book is more than just an autobiography. It's good literature. It's so well-written that on completing the "Epilogue," I continued on, reading the "Acknowledgments." And I was not disappointed! I whole-heartedly and unreservedly recommend the book to anyone. It is at once delightful, compelling, and allusively insightful. By reading it you will not only learn about Doris Kearns Goodwin and the Brooklyn Dodgers, you will learn more about yourself.

But reading the book held an additional, unexpected pleasure for me. When I was nearing the end – just about the time that the Dodgers won the World Series – I received a package in the mail. It was a Father's Day gift from my oldest son. Unwrapping it, I discovered to my delight, a book – and not just any book. It was one that he picked up on a recent vacation out east, when his family visited the Baseball Hall of Fame in Cooperstown, New York. The book was Stewart Wolpin's *Bums No More! The championship season of the 1955 Brooklyn Dodgers* (New York: St. Martin's Press, 1995). Written inside, a note recalled our watching the New York Yankees and Los Angeles Dodgers play in the 1977 World Series when my son was nine years old. The note went on to remind me how I told him all about the 1955 World Series and how his love for baseball stemmed, in part, from those conversations.

So there you have it. Not coincidence, but rather an example of how an allusive playfulness is fundamental to the very fabric of creation. Here, on the plains of Northwest Iowa in the summer of 1998, the delight that my father passed on to me, that Doris Goodwin's father passed on to her, and that I apparently passed on to my son, resonates with the baseball play of the old Brooklyn Dodgers. Is it any wonder that in the book of Proverbs we hear Wisdom describing the creation of the universe and saying

> There I was enjoying myself day after day,
> Playing around all the time in front of God's face,
> Playing through the hemispheres of his earth,
> Having fun with all mankind. . . ."

** Proverbs 8:30–31 (translated by Calvin Seerveld).

CHAPTER 10

CHRISTMAS CAMOUFLAGE

DECEMBER 8, 1980

Once again the time of year is upon us when Christians must prepare for the onslaught of the "principalities and powers" who battle to diminish the presence and witness of the body of Christ in the world. It is at this time of the year that we are most vulnerable, because the Evil One has succeeded in transforming what originally was a simple and pious celebration of the incarnation into a highly sensual festival of self-centeredness and greed. The success of this transformation is owing to the paradoxical capability of this yearly exercise in hedonism to retain a flimsy but effective camouflage of genuine Christian piety.

The questions I wish to consider are ones of self-examination. In our celebration of what is generally called "the Christmas holiday," do we contribute to the success of Satan's subtle devices, which camouflage unrighteousness in a façade of piety? If so, what are we to do about it?

I think we must be very specific in answering these questions. We must examine the concrete activities and attitudes that have become a part of our Christmas activities, in whatever form they take. And we must do so with a ruthless disregard for the sentimentality that often accompanies thoughts of Christmas but that might dull our critical insight; for sentimentality that stands in the way of obedience is sin.

Two very important biblical principles should guide us as we subject our holiday activities to self-examination. The first is simply the recognition that as Christ's disciples we are called to be the "light of the world." Consider these oft-quoted words of Jesus recorded in Matthew 5:14–16:

> "You are the light of the world. A town built on a hill cannot be hidden. Neither do people light a lamp and put it under a bowl. Instead they put it on its stand, and it gives light to everyone in the house. In the same way, let your light shine before others, that they may see your good deeds and glorify your Father in heaven."

Clearly, Christians must be visible to the rest of the world. In order to be visible, our activities must be clearly distinguishable from the activities of the masses.

The second principle that we must consider, closely related to the first, is the simple assertion that "Christ Jesus came into the world to save sinners." Thus, Christians may never regard the celebration of Christmas as a purely parochial exercise, engaged in without concern for the lost masses. If anything, Christmas is a wonderful opportunity for the body of Christ to shine its light brightly in our modern world of darkness.

Now let us examine carefully the phenomenon we call Christmas. Certainly what we want it to be is a joyous celebration of the birth of Christ, an opportunity for us to humbly reflect on the fact that our Creator did not leave us in the darkness of our sin, but by humbling himself and taking on the form of his own creature, he brought hope and meaning to a world otherwise doomed to death. Surely, then, the central thrust of the celebration must be the re-proclamation of the words of the angels as they announced Christ's birth to the shepherds: "Glory to God in the highest heaven, and on earth peace to those on whom his favor rests" (Luke 2:14).

But as we look around us, what does Christmas really mean to the masses of people who celebrate it today? What is the concrete impact of this holiday on our lives? Underneath the pretty, pious camouflage of "a baby born in a manger," or "good will among men," what do we really find as the essential defining characteristic of this annual event?

To find out, one only needs eyes that can see and ears that can hear; just leaf through a December issue of *Time, Scientific American*, or *Good Housekeeping*, paying particular attention to the advertisements. You need only watch primetime commercial television any night during a December week; the commercials, the "holiday specials," even the news programs join in proclaiming the same message loud and clear. Or else visit any reasonably sized merchandising area. Whether it be the "downtown" area of a humble municipality or the spread-out shopping mall of a large metropolitan area – it really doesn't matter. The spirit is still the same.

The essential defining characteristic of modern American Christmas – the spirit that drives it and is in turn served by all the pretty paraphernalia and customs – is hedonism; the sensual self-centered seeking after pleasure and things. It is this time of the year when humankind unabashedly devotes all its energies to *getting* – the merchants get their money, the kids get their toys, and the adults get high on alcohol, nostalgia, and other self-indulgent stimulants.

Surely we Christians reject this spirit and deplore the hypocrisy that uses the birth of the One who came to serve as an excuse for *self*-service.

But what do we do about it? I fear that at best we are guilty of inwardly rejecting the spirit of hedonism, while outwardly surrendering to the activities and symbols that constitute its real power. Thus, while proclaiming to ourselves and each other our allegiance to the Spirit of Christ, we effectively give the activities of our lives over to the service of the Evil One.

Allow me to give a concrete example. Many of us continue to use a "Christmas tree" in our holiday celebration, having done so since we were small children. If the thought ever occurred to us that we were one with the pagan masses in doing so, we quickly rationalized it away by pretending that the tree had some Christian symbolism, or at least that the tree was "nothing in itself," so we could give it the pious symbolism we wanted it to have. Surely a tree, even one cut off from its roots and placed in your living room, is not in itself evil. (Although one may legitimately question the wisdom of such a practice in terms of Christian stewardship.)

But what we fail to see is that no activity takes place in a cultural vacuum. The problem is that you simply cannot bring that tree into your home without bringing in the symbolism that our hedonistic culture has attached to it as well. The cut-off tree, decorated with colored balls and tinsel, is by its very nature a symbol. In twentieth century North America it is a symbol of decadence. It symbolizes humankind's turning away from God and turning inward on itself. It symbolizes the very worst kind of prostitution that can exist, the prostituting of the incarnation itself; taking what is most holy and using it to satisfy the sensualistic cravings of self-centered humans. And it is a most subtle and effective instrument for diverting the focus of Christians (particularly young Christians) from the Gospel message to the message of Madison Avenue.

I've picked on only one symbol here. There are many others we ought to question: the ritual of shopping, the "giving" of "gifts," the mass mailings of "Christmas cards," the increased consumption of food and alcohol; even our "built to scale" "nativity scenes" may serve to counteract our distinctiveness, or worse, foster a humanistic sentimentality that dulls the cutting edge of the Gospel message.

How does the Lord respond to the pagan festivities engaged in by his people while they are camouflaged in Christian piety? That is not a difficult question answer, since the people of God have been guilty of this same sin of unfaithfulness from earliest times. In the writings of the prophets we read of God's response to the Israelites when they were guilty of reducing once holy observances to a camouflage for their deeds of injustice and unfaithfulness. In Amos 5:21–24 we read:

> "I hate, I despise your religious festivals;
> your assemblies are a stench to me.
> Even though you bring me burnt offerings and grain offerings,
> I will not accept them.
> Though you bring choice fellowship offerings,
> I will have no regard for them.
> Away with the noise of your songs!
> I will not listen to the music of your harps.
> But let justice roll on like a river,
> righteousness like a never-failing stream!"

How then shall we celebrate Christmas? Are we to simply ignore it, keeping ourselves untainted by refusing to participate in any of the season's activities? No, that would only replace darkness with darkness and would in no way increase our effectiveness as the light of the world.

We must do two things. First, we must regain a sense of our prophetic office. Being careful not to condemn our neighbors, we must not hesitate to judge the activities engaged in during this holiday season. And then we must be prepared during this season of "nice-guy-ism" to humbly but firmly pay the social price of violating one of our culture's most cherished precepts. I would have you consider that the preaching of "nice-guy-ism," recurrent phrases such as "if you can't say something nice, say nothing," and constant admonitions about how terrible it is to be so "negative" are simply modern methods used by our culture to stone and silence the prophets.

Second, we must develop new ways of celebrating the Lord's birth. These ways must clearly contradict the currently accepted "holiday activities," and must be "signposts of the Kingdom of God" that faithfully call the attention of ourselves and our children to the full covenant meaning of the incarnation, while bringing the light of the Gospel to the cultural world of darkness in which we find ourselves.

Since I spent so much time discrediting one element of the traditional Christmas celebration, namely the "tree," I will close by suggesting an alternative that two families I know have instituted in order to reform that particular activity.

They began by recognizing that symbols are valid, that the aesthetic dimension of life is not to be deprecated, and that celebration of any kind calls for the use of symbols. Next came the admission that the "Christmas tree" cannot be stripped of the symbolic content given to it by our self-centered society and should be viewed with the reservation with which we view other symbols: for example, swastikas. So the question was, what symbol or symbolic object, if any, can be faithfully used to represent the meaning of the incarnation? The answer of the two families I mentioned is a rock. The rock, as a metaphor, is used very often in the Scriptures to symbolize God's covenant faithfulness to his people. His being born as a man in order to atone for our sins and to bring us freedom is the ultimate expression of faithfulness. A rock, large enough to convey a feeling of firmness and stability, but small enough to be carried into a house by one or two persons, is the beginning of the efforts of some Christians to reform their Christmas celebrations.

Finally, as we face together the onslaught of the principalities and powers that currently direct Christmas celebration in our culture, we might heed carefully the words of Paul, in Ephesians 6:11–17:

> Put on the full armor of God, so that you can take your stand against the devil's schemes. For our struggle is not against flesh and blood, but against the rulers, against the authorities, against the powers of this dark world and against the spiritual forces of evil in the heavenly realms. Therefore put on the full armor of God, so that when the day of evil comes, you may be able to stand your ground, and after you have done everything, to stand. Stand firm then, with the belt of truth buckled around your waist, with the breastplate of righteousness in place, and with your feet fitted with the readiness that comes from the gospel of peace. In addition to all this, take up the shield of faith, with which you can extinguish all the flaming arrows of the evil one. Take the helmet of salvation and the sword of the Spirit, which is the word of God.

Chapter 11

A BABY'S SMILE

JUNE 6, 2002

The nature of Sunday worship seems to be in a state of continual change. Some of that change arises from a renewed understanding of who we are, as the Body of Christ. Some is driven by our conformity to the culture around us. But there is other change that correlates with the stage we are at in our pilgrimage through life. This is perceptual change, having as much to do with the changing sensitivities of the individual worshiper as with altering liturgical styles. For example, on a Sunday evening a few weeks back, I completed my ushering duties and slipped quietly into the seat next to my wife – reserved for me by the strategic placement of her pocketbook. What I noticed first were the two young families occupying seats in the rows ahead of me. Directly in front of me sat a charming young lady, approximately two years old, who had the uncanny ability to remain unobtrusively animated throughout the worship service. The source of her animation was not the happenings at the front of the church. Rather, it was the playful exchange of her three brothers with the four children in the row ahead of them. So while others freely joined in singing words and music (that for these eight held no allusion to teenage love songs of the early 1960s) and focused their eyes on the four worship leaders at the front of the church (who for them were free of parallels to Perry Como, The Shirelles, or "Pete and Arlo together in concert"), my not-so-free perceptual psyche found worshipful distraction in the nuanced variation of smiles on the face of this two-year-old image bearer of God. I began reflecting on how unique we human creatures are, how different we are from other biotic beings, even at very early stages in our development.

Think about it. Have you even seen an animal smile? Oh, I know that the facial muscles on a dog will give its mouth different form depending on whether it is yelping in pain or barking excitedly, and that certain primates use facial expression to send biologically determined signals from one to another. And there probably are chimpanzees that imitate their human trainers by contorting their faces into what appears to us as variations of human expressions. But a true smile? No, a smile is a very complex part of creation. On the surface it is a response to a particular stimulus. But beneath the surface that response is being suggestively connected – sometimes consciously and sometimes unconsciously – to a stimulating event in the past of the one who is smiling. If you will allow me just a wee bit of philosophical terminology, a smile is an aesthetically qualified response. Now certainly all creatures have an aesthetic *aspect* to their existence. I would never want to be accused of denying the aesthetic quality of a morning sunrise, a thunderstorm, a butterfly flitting from flower to flower, a wolf howling at the moon, or even a common sparrow as it pecks its way across my front lawn looking for something to eat. But only human creatures have been called by God to *respond* to his Word for the aesthetic dimension of his creation. So what provided particular depth to my reflections, during that worship service, was the realization that we humans have this capacity for aesthetic response *even at early stages of development.*

One example of this that delighted me recently was when Luke and Josh, my two oldest grandsons, phoned one evening to tell about Josh's having won a poetry reading contest among his fellow third-graders. Not to be outdone, eleven-year-old brother Luke took the phone near the end of the conversation to say, "Wait Grandpa, don't hang up! Listen to this!" And after a brief pause, he continued, "Once upon a midnight dreary, while I pondered weak and weary, over many a quaint and curious volume of forgotten lore. . . ." He went on to recite four stanzas of Edgar Allen Poe's *The Raven*, confirming to me that back six months ago, when we had visited a Barnes & Noble store together, my reading to him the complete poem was not in vain. Despite the obvious vocabulary mismatch, Luke had caught on to the poem's cadence and rhyme scheme and had responded appropriately – an immature but genuine aesthetic response.

But Luke is at an age where we expect to see aesthetic sensitivity developing. Perhaps more amazing are those babies, six-months or younger, who become the subject of so much photographic effort. And the photos that we like the best are those that have captured a baby's smile – in other words, where we have evidence of a person's earliest response to God's Word for the aesthetic dimension of life.

I confess to having snapped hundreds of those kinds of photographs after each of my three sons were born, in an attempt to record evidence of that early aesthetic response. But since I've become a grandfather I've developed a far more refined method for eliciting and marveling at such primordial aesthetic activity. It's a subtle but sophisticated form of playing "peek-a-boo," and it is effective even with many babies to whom one may only be a stranger. Here's how it works. Let's say you are standing around after a church service, greeting people, when a young couple that you know walks up holding a baby who you really haven't met yet. The first thing you do is stay in the background. As those more effusive friends and relatives surround the baby with their overbearing attempts at baby-talk greetings, the child will seek refuge in the arms of its parent while exploring in frightful curiosity the surrounding crowd. At this point you should be positioned close to the child, but with someone else – preferably your spouse – between you. As the baby's gaze passes your way you quickly peek out from behind your spouse, catch the baby's eyes, and then duck back out of sight again. After no more than a second or two you peek out from behind your spouse in the other direction. This time, catching the child's eyes with a timid half-smile flashed just before you duck back once more behind your spouse, you are sure to elicit heightened curiosity free from the usual fear of strangers. Continue this, increasing the "surprise" quality of your smile each time, and you will certainly obtain a responsive smile from the baby. Then you can join me in marveling at God's faithfulness to his creation and at the uniqueness of his image-bearers who are capable of responding to his Word for playfulness even at such early stages in their development.

Oh, and if you get good at this playful little technique, you may even be inclined to try it out – discretely, of course – during an appropriate time in a worship service.

CHAPTER 12

LISTENING TO THE HEAVENS

OCTOBER 1, 2002

It was a very busy day in a very busy week. The last meeting of the day ended at about 5:10 p.m. and I rushed home to bolt down a microwaved dinner so that I could be back at school by 6:20 p.m. At 6:30 I was scheduled to join a class called *Perspectives in Physical Science*, which I had decided to sit in on during the course of the semester. The class was starting earlier than usual and was being shortened somewhat to accommodate those who wanted to attend a special lecture given by a guest speaker from Japan, who was on campus for just a couple of days. As I entered the class, I realized that the guest speaker would be participating in our class as well.

Perspectives in Physical Science is a course that is team taught by a member of the Physics Department and a member of the Philosophy Department. That may be a little unusual at a typical university. But at Dordt College it is the kind of thing that one expects, since we view the whole of creation as belonging to God, as unified, and as being the legitimate subject of our academic investigations. At Dordt, at least, there ought to be no artificial divisions between science and technology on one side and the humanities and social sciences on the other.

The class began with the usual housekeeping announcements, clarifications, and reminders. Then one of the two professors introduced the guest speaker, Inagaki Hizakazu. Hizakazu is a theoretical physicist, but he was about to lecture us on the philosophy of science. In particular, he zeroed in on a point made by the author of one of our textbooks, *Religion and the Rise of Modern Science* by Professor Reijer Hooykaas. Hooykaas's point was that nature does not function on its own, but that "the Bible attributes all events, however insignificant, immediately to God." Hizakazu agreed with that point, but wanted to show us that the language used by Hooykaas – particularly when "natural things" are referred to a God's "instruments" – betrayed a tendency toward a mechanistic view of creation. Hizakazu thought that creation ought to be viewed more holistically, more integrally, more as an organism than a machine. To demonstrate and reinforce his point, he asked that we read Psalm 8 and Psalm 19.

At this point I should tell you something about the weather. The late afternoon was cloudy and overcast. By suppertime, there was the threat of rain in the air. I remember having looked up at the gathering clouds and deciding to bring my umbrella to class so I wouldn't be caught in a downpour when the class ended at 7:45. Little did I know then of the fascinating events those gathering clouds portended.

But back to the class: Hizakazu asked for volunteers to read the two Psalms, and the first student to volunteer read Psalm 8 from the King James Version of the Bible. Soon into the reading of the psalm I was hearing these words:

> When I consider thy heavens, the work of thy fingers, the moon and the stars, which thou hast ordained; What is man, that thou art mindful of him? and the son of man, that thou visitest him? For thou hast made him a little lower than the angels, and hast crowned him with glory and honour. Thou madest him to have dominion over the works of thy hands; thou hast put all things under his feet. (Psalm 8:3–6)

As I listened to these words, I realized that two of the classroom windows were open, and powerful crashes of thunder were pealing in through them. But somehow it did not disturb the reading, and I was suddenly struck by the harmony of it all. Then another student began to read from the New International Version of Psalm 19:1–4 –

> The heavens declare the glory of God; the skies proclaim the work of his hands. Day after day they pour forth speech; night after night they reveal knowledge. They have no speech, they use no words; no sound is heard from them. Yet their voice goes out into all the earth, their words to the ends of the world.

As the peals of thunder echoed 'round the classroom, accentuating the words of the psalmist, I almost wished I had been brought up in the Baptist tradition so that I could uninhibitedly exclaim "amen!" Instead, like an introverted sports enthusiast whose team has just scored the winning goal, I shouted "yes!" inside the confines of my academically proper and Reformed head.

What I found most striking about this whole incident was the co-operation of the surrounding creation with the point being made by the guest professor. Surely the creation *is* a whole. There is no "natural" part of creation that functions separately from some "spiritual" part. And the Creator – the faithful, covenant God of Abraham, Isaac, and Jacob – orchestrates it all in both freedom and sovereignty; that is, the creation – and certainly humankind – has a certain freedom about it, yet God is behind it all, upholding it all, governing it all. There is lawfulness to creation, rooted in God's faithfulness, so that even that which looks to us to be randomness is not chaos, but is evidence of God's fatherly hand. And yet it's not just a machine and we are not puppets. There is indeed an organic unity to it all, and the Apostle Paul has it just right when he says, "He is before all things, and in him all things hold together" (Colossians 1:17). Lawfulness and freedom, uncertainty and sovereignty: these are concepts not easily reconciled in our finite, creaturely brains. Yet the contemplation of such concepts points one back to the faithful Creator, Sustainer, Redeemer of all things – as do his servants the thunderclouds.

After the reading of the Psalms Hizakazu returned to his lecture and I noticed that one student got up to close the windows so that we could hear him. Not too much later, during the class discussion, one of the two course instructors asked that the windows be opened again so that "we could hear the Lord speak." I wondered how many in the class really understood that remark and how many took it as simply a comment on an interesting coincidence. It's true that there is always a certain spontaneity to the events that surround us – but there is no such thing as *mere coincidence*. And as that thought passed through my mind on that interesting evening, I remembered the words of the Lord as recorded in Mark 4:23 – "If anyone has ears to hear, let them hear."

CHAPTER 13

A CHARLES DICKENS CHRISTMAS

JANUARY 5, 2007

Christmas Eve to New Year's Day: A time to celebrate with family and friends that historic event without which family and friends would have no meaning. In addition, for those educators among us, it is a time without classes to teach and without meetings to attend: a time when you can prepare for the coming semester at your own pace and still have time left over for "dessert."

"Christmas dessert" means for me reading a novel or at least a couple of stories by Charles Dickens. Dickens has fascinated me ever since, as a sophomore in high school, I met Pip, Magwitch, and Joe Gargery, in the early chapters of *Great Expectations*. Having now read most of his novels, many of his stories, and two Dickens biographies, my fascination has grown and, to a certain extent, has been joined by puzzlement. What puzzles me is how this author, who was known as a rather liberal Christian, could have conveyed in his writings what seems to me to be the very essence of biblical Christianity far better than any other writer of fiction that I have read. Let me probe that paradox a bit by telling you about Christmas 2006, which for me was very much a Charles Dickens Christmas.

Last year I had re-read *A Christmas Carol,* and so thought it would be appropriate to start Christmas break this year by viewing one of the two film versions of that short novel that I have on videotape. One version, made in 1951 and filmed in black & white, was shown on TV every Christmas eve when I was a child. But this year my wife and I decided to watch the 1984 version, starring George C. Scott as Ebenezer Scrooge. Among its greatest qualities is its faithfulness to the text when Scrooge is confronted by the ghost of Jacob Marley. Those familiar with the story may recall that when the ghost goes on lamenting about lost opportunities for doing good, Scrooge replies,

"But you were always a good man of business, Jacob."

"Business!" cried the Ghost, wringing its hands again. "Mankind was my business. The common welfare was my business; charity, mercy, forbearance, and benevolence, were, all, my business. The dealings of my trade were but a drop of water in the comprehensive ocean of my business!"

A Christmas Carol was first published in London around Christmas, 1843. Its success encouraged Dickens to make a practice of writing a short novel or a long story to be ready for publication each Christmas. In part, he – and certainly his publisher – was motivated by the lure of success in the form of fame and monetary profit. But those years in England were what have been called "the hungry forties": a time of cruel injustice, when the Darwinian notion of survival of the fittest seemed to justify the enormous chasm between wealthy men of business and the poverty stricken masses. Dickens was sensitive to that injustice and felt called upon to point it out. He used the Christmas novels and stories to do just that.

In *Mrs. Lirriper's Lodgings*, *Mrs. Lirriper's Legacy*, and *Dr. Marigold*, the three stories I read this year, the social justice theme remains present. But these three stories were written twenty years after *A Christmas Carol*, for the 1863, 1864, and 1865 Christmas seasons respectively. In these stories the themes of kindness, gentleness, self-sacrifice, and familial love are more prominent. In *Mrs. Lirriper's Legacy*, perhaps the most memorable of the three, the themes of repentance and forgiveness join kindness and familial love to evidence Dickens' mature Christian sensibilities. In a "twist" of the plot from *Oliver Twist*, an orphan – whose father deserts the family and whose mother subsequently dies – is raised lovingly and self-sacrificially by a widow and her lodger. A pensioner, the lodger pretends comically but most sincerely to be a gentleman; much like that lovable character from *David Copperfield*, Wilkins Micawber, or like Dickens' own father, John Dickens.

And here we have two of the three qualities that I believe are most central to the character and teachings of Jesus, and which are so prominent in Dickens' writings: compassion and self-sacrifice. The other quality is sensitivity to hypocrisy. These appear as themes in *A Christmas Carol* and, to a greater or lesser extent, in all the other Christmas stories and novels. But they are developed most fully in Dickens' full-length novels. Two of those novels – which I had read more than 20 years ago, and of which, thanks to Public Television's *Masterpiece Theatre*, I also have dramatizations recorded on videotape – my wife and I viewed during a couple of evenings this past Christmas break.

The first was *Great Expectations*, written in 1861. In *Great Expectations*, Pip, the main character is not the most likeable or noble person in the story. That person is Joe Gargery, a blacksmith, and the adoptive father of Pip. Joe is humble, compassionate, and self-sacrificing – in every way a Christ figure. There are others who evidence compassion, humility, and self-sacrifice as well, but not as consistently as Joe. There is also selfishness and hypocrisy embodied in major characters of the novel. Perhaps the best known are Miss Havisham and her protégée, Estella. In a way, Joe and Pip on the one hand, and Miss Havisham and Estella on the other, represent parallel but opposite paths. It is Joe's Christ-like influence that ultimately brings about the redemption of Pip. It is Miss Havisham's devilish influence that brings about the corruption of Estella. In the beginning of the novel both Pip and Estella have "great expectations." But in a "first shall be last and the last first" turn of the plot, we come to see that truly great expectations are not of a monetary, vengeance-seeking, or self-aggrandizing variety, but are rather led by love, faithfulness, and self-sacrifice.

The other video dramatization was also one I recorded from *Masterpiece Theatre* many years ago: *A Tale of Two Cities*. This is Dickens' ultimate tale of redemption and self-sacrifice, with one of the main characters, the originally dissolute Sydney Carton, going to the guillotine to save his look-alike, Charles Darnay. You have in *A Tale of Two Cities* portraits of great evil, hypocrisy, and self-delusion in the French aristocrat brothers Evrémonde as well as in the French revolutionary leader, Madame Defarge. And you have two tales of redemption: the redemption of the life of Charles Darnay by the sacrifice of Sydney Carton, and the redemption of Sydney Carton's eternal life – and thereby his brief, temporal life, by his losing the latter. Everyone knows that famous last line of *A Tale of Two Cities*, which the novel's narrator conjectures might have been the last thoughts of Sydney Carton: "It is a far, far better thing I do, than I have ever done; it is a far, far better rest that I go to than I have ever known." But few remember that the actual last words of Sydney Carton, just before he stepped up to be guillotined, were to quote the Lord's words from the eleventh chapter of the Gospel of John: "I am the Resurrection and the Life, saith the Lord: he that believeth in me, though he were dead, yet shall he live: and whosoever liveth and believeth in me shall never die."

I noted above that I was puzzled at how a writer who was known as a rather liberal Christian could have embodied in his writings the very essence of biblical Christianity. But perhaps in writing this *brief essay* I've answered my own puzzlement, at least to some extent. For, in a way, the transformation of the dissolute Sydney Carton into a Christ-figure is simply a shadow of the twist in the great cosmic plot whereby the first become last and the last first. Thus it makes some kind of biblical sense that a theologically weak Christian would be the greatest in expressing the essence of biblical Christianity. In any case, if you want a glimpse of Jesus, I suggest you read the fictional writings of his servant Charles Dickens. In *A Christmas Carol, A Tale of Two Cities, David Copperfield, Great Expectations*, and many others of Dickens' novels and stories, you will find righteous indignation at pharisaical hypocrisy, compassion for the poor, and self-sacrificing love. There are other writings of other authors – like Herman Melville's *Moby Dick* – where the biblical imagery is more intense, more frequent, and aesthetically more refined. But in Dickens' writings you will see the face of Jesus.

CHAPTER 14

MUGBY JUNCTION

JANUARY 4, 2008

For Christmas in 1866 Charles Dickens wrote a delightful story about a man who is running away from his birthday. Unlike Ebenezer Scrooge, of *A Christmas Carol* fame, Mr. Young Jackson was a self-conscious and reflective type from his youth. In school he had been motivated to become "a great healer" and he describes himself as "almost happy," despite "the horrible mask" that he wore and the "silence and constraint" that characterized all his actions. In Mr. Jackson, Dickens has fashioned for us a man with the potential for goodness, but whose heart is buried deep beneath layers of façade, pretense, and duplicity. In modern terminology we would say that he spent much of his time crafting an image – and that his life had become one without substance.

Mr. Jackson had early in his career joined the firm of Barbox Brothers, an "irregular branch of the Public Notary and bill-broking tree" and a firm that "had gained for itself a griping reputation before the days of Young Jackson, and the reputation had stuck to it and to him." Eventually Jackson "found himself a personage held in chronic distrust, whom it was essential to screw tight to every transaction in which he engaged, whose word was never to be taken without his attested bond. . . ."

But the story begins just after Mr. Jackson has taken a step in the direction of extricating himself from his miserable condition. Dickens describes it this way:

> He broke the oar he had plied so long, and he scuttled and sank the galley. He prevented the gradual retirement of an old conventional business from him, by taking the initiative and retiring from it. With enough to live on (though after all with not too much), he obliterated the firm of Barbox Brothers from the pages of the Post-office Directory and the face of the earth, leaving nothing of it but its name on two portmanteaus.[*]

In Dickens' story, Mr. Jackson winds up in Mugby Junction – which also happens to be the name of the story – a small, railroad station town

---

[*] Charles Dickens, 1866, "Mugby Junction," *Christmas Stories* (Oxford University Press, 1991), 482–483.

with seven sets of railroad tracks leading from the town to other places. In the railroad station Jackson meets a man called "Lamps." That's a nickname, of course, given to him because of the nature of his work: he is responsible for refilling oil in the lamps of trains that stop briefly at the station. But that name "Lamps" has a deeper meaning for Jackson and for the story, for it's through the kindness and Christ-likeness of Lamps and his invalid daughter Phoebe that Mr. Jackson comes to "see the light." I won't tell you the whole story because I want to encourage you *Plumbline* listeners to read "Mugby Junction" for yourselves. But I will say that it is a typical Dickens Christmas story where sadness is dispelled and brokenness is healed by goodness, mercy, kindness, and patience – and where a number of remarkably memorable characters have their lives intertwined in a plot that has an even more remarkable outcome.

But this *Plumbline* is not meant merely as a book review. For like the character of the same name in "Mugby Junction," a Dickens' Christmas book can serve as a "lamp" to focus the light of God's Word, enabling us to see ourselves – to our shame – in stark contrast to the One who came to redeem us and to show us how to live obediently.

For those of us in the Reformed community, it would be a rare Christmas that would pass without hearing the words of the prophet Isaiah (9:2 and 6) telling us that: "The people walking in darkness have seen a great light; on those living in the land of deep darkness a light has dawned. . . . For to us a child is born, to us a son is given. . . . And he will be called . . . Prince of Peace." Reading "Mugby Junction" helps those words to resonate for us with Jesus' words where he says,

> "Blessed are the meek, for they will inherit the earth. Blessed are those who hunger and thirst for righteousness, for they will be filled. Blessed are the merciful, for they will be shown mercy. . . . Blessed are the peacemakers, for they will be called children of God." (Matthew 5:5–9)

Looking up from the pages of "Mugby Junction" I see in myself and in those around me a discouraging excess of impatience, of grudge-holding, of presuming the worst, of failing to overcome evil with good. I look up and am amazed how a writer from the nineteenth century could have fashioned characters that, in essence, open God's Word for us, pointing us to passages like the prophecy of Isaiah or the Beatitudes by their words and actions – and thereby chastise us twenty-first century Christians for our unloving words and actions. There are characters like Lamps and his daughter Phoebe scattered throughout the writings of Charles Dickens. They are Christ-figures who, in the context of the story, make possible the redemption of the main character and who serve as models of righteousness and love.

By the end of "Mugby Junction," the main character – Young Jackson – is no longer running away from his birthday. He has found himself because he has found Christ in the words and actions of Lamps and his daughter Phoebe. And the reader of the story cannot help but examine himself and his colleagues with those words of Paul ringing in his ears:

> Love is patient, love is kind. It does not envy, it does not boast, it is not proud. It does not dishonor others, it is not self-seeking, it is not easily angered, it keeps no record of wrongs. Love does not delight in evil but rejoices with the truth. It always protects, always trusts, always hopes, always perseveres. (1 Corinthians 13:4–7)

For Plumbline, I'm Charles Adams.

Scripture Index

Subject Index

About the Author

Dr. Charles Adams is professor of engineering emeritus at Dordt College in Sioux Center, Iowa. As he explains in one of his essays: "I grew up in the 1950s as a kind of rootless, middle class American, in the suburbs outside of New York City. My mother was a nominal Protestant, my father a Catholic, and the identities of my grandparents, great-grandparents, or other ancestors were essentially unknown and irrelevant to me. My worldview was shaped more by the Brooklyn Dodgers, Davy Crockett, and Hopalong Cassidy than by the ethnic or cultic origins of my relatives. But by God's grace I came under the influence of Reformed preaching as a child. Then, at just the right time in my adolescent development, I was befriended by a radical, counter-cultural Dutchman, who made me aware of the Kuyperian, Dutch neo-Calvinist worldview and community. . . . Since that time I've come to realize that the Lord works his covenant faithfulness through generations of people in community with each other, people bound together by a common desire to live obediently before his face" (70).

After marriage and graduating from Newark College of Engineering, Charles, along with his wife, Pam, moved to Connecticut where Charlie worked for Pratt and Whitney Aircraft. There, two sons were born to them and they adopted a third. Charles and Pam looked for a church in Connecticut similar to the Orthodox Presbyterian church they had attended in New Jersey and found the Avery Street Christian Reformed Church in South Windsor. Charles taught junior high Sunday School there and came to realize that teaching was what he wanted to do.

The Adamses decided to move back to New Jersey where Charles began his teaching career at Eastern Christian High School. He often said that he was amazed that Eastern Christian would pay him for doing something he found so enjoyable. In 1979, after eight years of teaching high school, he moved to the college level and took on the challenge of starting Dordt College's engineering program. Charles designed the courses, helped hire professors, and was instrumental in designing the original building. Today, several of his former students are professors of engineering at the college.

His technical expertise and philosophical bent notwithstanding, Charles celebrated the "tapestry" of God's good creation with his family and with his students. At home, he became a photographer and built furniture from wood. In addition to his artistic abilities, Charles was a reader – novels, theology, philosophy, you name it, including authors like John Calvin, Søren Kierkegaard, and Charles Dickens. While not a performer, Charles' love of music is evident in the broad range of classical to folk music recordings he collected over the years (the largest collection of which is by his favorite musician Bob Dylan).

Throughout his life, Charles has been thankful to the Lord for the ways the spectacles of Christian faith allowed him to glimpse the fabric of God's good creation – as he says in another essay, "to perceive the jewels in the tapestry and the threads tying those jewels together into a unity." At the same time, he often reminded himself and others that "even with those spectacles, we only see, as the Apostle Paul says, 'as through a glass, darkly'" (427).

CPSIA information can be obtained at www.ICGtesting.com
Printed in the USA
LVOW12s1914051214

417396LV00006B/18/P